Proceedings
of the Steklov Institute of Mathematics

edited by
S. M. Nikol′skiĭ

Number 126 (1973)

Boundary value problems for differential equations. III

edited by

V. P. Mihaĭlov

AMERICAN MATHEMATICAL SOCIETY
Providence, Rhode Island
1975

Академия Наук

Союза Советских Социалистических Республик

ТРУДЫ

ордена Ленина

МАТЕМАТИЧЕСКОГО ИНСТИТУТА

имени В. А. СТЕКЛОВА

CXXVI

Ответственный редактор

член-корреспондент АН СССР *С. М. НИКОЛЬСКИЙ*

КРАЕВЫЕ ЗАДАЧИ

ДЛЯ ДИФФЕРЕНЦИАЛЬНЫХ УРАВНЕНИЙ. III

Сборник работ

под редакцией В. П. МИХАЙЛОВА

Издательство „Наука"

Москва 1973

Translated from the Russian by S. Smith

AMS (MOS) subject classifications (1970). Primary 35B40, 35K20, 35K35, 35L05, 35L20, 35M05, 35R05, 35Q10, 76D05; Secondary 35A05, 35B30, 35B40.

Library of Congress Cataloging in Publication Data (Revised) CIP

Main entry under title:

Boundary problems for differential equations.

(Proceedings of the Steklov Institute of Mathematics, no. 91, 103, 126

Originally published as v. 91, 103, 126 of Trudy of Matematicheskiĭ institut of Akademii͡a nauk SSSR.

Vol. 3 has title: Boundary value problems for differential equations.

Includes bibliographies.

1. Differential equations, Partial—Addresses, essays, lectures. 2. Boundary value problems—Addresses, essays, lectures. I. Mikhaĭlov, V. P., writer on geometry, ed. II. Title: Boundary value problems for differential equations. III. Series: Akademii͡a nauk SSSR. Matematicheskiĭ institut. Proceedings, no. 91, 103, 126.

QA1.A413 no. 91, etc. 510′8s [515′353] 70–9052

ISBN 0-8218-3026-0

PROCEEDINGS OF THE STEKLOV INSTITUTE OF MATHEMATICS
IN THE ACADEMY OF SCIENCES OF THE USSR

(Труды математического института им. В. А. Стеклова, т. CXXVI, 1973)

TABLE OF CONTENTS

FOREWORD

The present volume of "Proceedings of the Steklov Institute of Mathematics" contains papers in which the behavior of the solutions of boundary value problems in unbounded regions and the Cauchy problem for nonstationary equations are investigated. The majority of them are devoted to a study of the stabilization of the solutions for $t \to \infty$.

In the paper by V. N. Maslennikova the asymptotic behavior for large t of the solution of the Cauchy problem for a linearized Navier-Stokes system describing the motion of a rotating viscous fluid is investigated. Exact estimates characterizing the decrease of the solution as $t \to \infty$ that are uniform in the spatial variables are obtained. Owing to the presence of a Coriolis term in the system, the estimates obtained in the paper characterize the rate of decay of a vortex in a viscous fluid. A corresponding theorem is proved by solving the problem with the use of the Fourier transform and uniformly estimating some definite integrals of strongly oscillating functions over unbounded regions.

The papers by A. K. Guščin, L. A. Muraveĭ and D. V. Proka are devoted to a study of the asymptotic behavior for large t of the solutions of nonstationary exterior boundary value problems.

For a parabolic equation of second order, A. K. Guščin establishes estimates of the solution of boundary value problems (i) in domains with boundaries that are noncompact in the spatial variables, and (ii) with an initial function in L_1. He shows that his estimates are sharp in regard to the order of decrease as $t \to \infty$ of the solution of a second boundary value problem for a selfadjoint equation without minor terms. In the paper by L. A. Muraveĭ asymptotic expansions for $t \to \infty$ of the solution of the second and third exterior boundary value problems for the two-dimensional wave equation are obtained in the case of a convex boundary contour and initial data with compact support. The method of investigating the asymptotic properties of the solutions makes use of (i) the possibility, established in the paper, that the Green function correspond to a boundary value problem for the operator $\Delta + k^2$ can be analytically continued into a part of the halfplane $\operatorname{Im} k < 0$, and (ii) an investigation of the behavior of the Green function as $k \to \infty$ and $k \to 0$.

In the paper by D. V. Proka the stabilization for $t \to \infty$, $x > 0$ of the solutions of the equation $\partial u/\partial t = (-1)^p \partial^{2p} u/\partial x^{2p}$ with a fixed linear differential boundary operator is studied. The author establishes a necessary and sufficient condition on the boundary operator under which a solution satisfying the homogeneous initial condition is stabilized for any stabilizing boundary functions.

In the paper by V. I. Čehlov a mixed problem for the wave equation with a discontinuous boundary operator is studied. Along with existence and uniqueness theorems, he investigates the dependence of the smoothness of a solution on the manifold of discontinuity of the boundary operator. He indicates cases when the discontinuity of the boundary operator does not affect the smoothness of the solution in the interior of the region and cases when the discontinuity propagates into the interior of the region.

V. P. Mihaĭlov

Trudy Mat. Inst. Steklov.
126 (1973)

Proc. Steklov Inst. Math.
126 (1973)

ON ESTIMATES OF THE SOLUTIONS OF BOUNDARY VALUE PROBLEMS FOR A PARABOLIC EQUATION OF SECOND ORDER

UDC 517.947.4

A. K. GUŠČIN

ABSTRACT. In this paper sharp estimates are found characterizing the decrease for large time of solutions of the second and third boundary value problems for a second-order parabolic equation in the case of a boundary noncompact in the space variables and an initial function belonging to L_1.

Bibliography: 7 items.

Let Ω be an unbounded domain of the n $(\geqslant 2)$ dimensional space E_n ($x = (x_1, \cdots, x_n)$ is a point of this space) whose boundary Γ consists of a finite number of nonintersecting, closed or unbounded $(n-1)$-dimensional Ljapunov surfaces and satisfies the following condition: There exist constants $\delta_0 > 0$ and $a_0 > 0$ such that for any point $x_0 \in \Gamma$ the intersection of the ball $|x - x_0| < \delta_0$ with the surface Γ is connected and $\Sigma_1^n \nu_i(x_1)\nu_i(x_2) \geqslant a_0$ for $x_1 \in \Gamma, x_2 \in \Gamma$, $|x_1 - x_2| \leqslant \delta_0$, where $\nu(x) = (\nu_1(x), \cdots, \nu_n(x))$ is the outward (relative to Ω) unit normal to Γ.

We will consider in the cylinder $D = \Omega \times (t > 0)$ the solution of the following boundary value problem:

$$\frac{\partial u(t, x)}{\partial t} = L[u] = \sum_{i, j=1}^{n} \frac{\partial}{\partial x_i}\left(a_{ij}(t, x)\frac{\partial u(t, x)}{\partial x_j}\right) + \sum_{i=1}^{n} b_i(t, x)\frac{\partial u(t, x)}{\partial x_i} + c(t, x)u(t, x), \tag{1}$$

$$\frac{\partial u(t, x)}{\partial N} + q(t, x)u(t, x)|_{x\in\Gamma} = 0, \tag{2}$$

$$u(0, x) = \varphi(x). \tag{3}$$

The symmetric matrix $a_{ij}(t, x)$ satisfies the condition

$$0 < \gamma \leqslant \sum_{i, j=1}^{n} a_{ij}(t, x)y_i y_j \leqslant \gamma^{-1} \tag{4}$$

for any unit vector $y = (y_1, \cdots, y_n)$.

AMS (MOS) subject classifications (1970). Primary 35K20, 35B40.

The coefficients $b_i(t, x)$ and $c(t, x)$ are bounded and satisfy the inequalities

$$c(t, x) \leqslant 0, \qquad M \geqslant \sum_{i=1}^{n} \frac{\partial b_i(t, x)}{\partial x_i} - c(t, x) \geqslant 0,$$

$$-q(t, x) + \sum_{i=1}^{n} b_i(t, x)\nu_i(x)|_{x\in\Gamma} \leqslant 0, \tag{5}$$

where M is a positive constant, N is the outward conormal to Γ of the elliptic operator L and $q(t, x)$ is a bounded continuous nonnegative function. The initial function $\phi(x)$ is assumed to be a member of $L_1(\Omega)$.

Let $g(v)$, $v > 0$, be a monotone nondecreasing continuous positive function satisfying one of the following conditions:

a) There exist positive constants ϵ_0 $(\leqslant 1/n)$, δ_1 and C such that the estimate $g(v) \geqslant Cv^{1-\epsilon_0}$ holds for $v \leqslant \delta_1$.

b) There exists a positive constant ϵ_0 $(\leqslant 1/n)$ such that the function $v^{1-\epsilon_0}/g(v)$ is monotone nondecreasing.

We note that condition b) is stronger than condition a). For suppose $g(v)$ satisfies condition b). Then $g(v) \geqslant g(1)v^{1-\epsilon_0}$ for $v \leqslant 1$, i.e. condition a) is satisfied with $\delta_1 = 1$ and $C = g(1)$.

Let Q be an arbitrary open subset of Ω and consider the function

$$l(v) = \inf_{\operatorname{meas}_n Q = v} \operatorname{meas}_{n-1}[\partial Q \cap \Omega],$$

where ∂Q is the boundary of Q and meas_k is k-dimensional measure.

We will say that *a domain* Ω *belongs to the class* $\mathfrak{A}(g)$ ($g(v)$ satisfies condition a)) if the inequality

$$l(v) \geqslant g(v) \tag{6}$$

is valid for all $v > 0$.

The following assertions are valid if the coefficients of equation (1) are sufficiently smooth functions (condition A) of §2 is satisfied).

THEOREM 1. *Suppose* $\Omega \in \mathfrak{A}(g)$. *Then the Green function* $G(t, x; 0, \xi)$ *of problem* (1), (2) *satisfies the inequality*

$$G(t, x; 0, \xi) \leqslant \frac{4}{\mathcal{J}_{-1}(\gamma_1 t)} \tag{7}$$

for all $t > 0$ *and* $\xi \in \Omega$, *where* $\mathcal{J}_{-1}(t)$ *is the inverse of the function*

$$\mathcal{J}(v) = \int_0^v \theta^{-1} \int_0^\theta \eta^{-1} \int_0^\eta \frac{\zeta}{g^2(\zeta)}\, d\zeta\, d\eta\, d\theta \tag{8}$$

and $\gamma_1 = \gamma \ln^2 2/32$.

If the function $g(v)$ *satisfies condition* b), *then, in addition to* (7), *the estimate*

$$G(t,\ x;\ 0,\ \xi) \leqslant \frac{C_0}{\mathscr{F}_{-1}(\sqrt{t})} \tag{9}$$

is valid, where $\mathfrak{F}_{-1}(t)$ *is the inverse of the function*

$$\mathscr{F}(v) = v/g(v) \tag{10}$$

and $C_0 = 4(8\epsilon_0^3\gamma_1)^{-1/2\epsilon_0}$.

THEOREM 2. *Suppose* $\Omega \in \mathfrak{A}(g)$. *Then the solution of problem* (1)–(3) *satisfies the inequality*

$$|u(t,\ x)| \leqslant \frac{4\,\|\varphi\|_{L_1(\Omega)}}{\mathscr{F}_{-1}(\gamma_1 t)} \tag{11}$$

for all $t > 0$ *and, if* $g(v)$ *satisfies condition* b), *the inequality*

$$|u(t,\ x)| \leqslant \frac{C_0\,\|\varphi\|_{L_1(\Omega)}}{\mathscr{F}_{-1}(\sqrt{t})}, \tag{12}$$

where $\gamma_1 = \gamma \ln^2 2/32$ *and* $C_0 = 4(8\epsilon_0^3\gamma_1)^{-1/2\epsilon_0}$.

Let ξ be an arbitrary point of Ω and let $s_\xi(R)$ denote the set

$$\operatorname{meas}_{n-1}\{(|x-\xi| = R) \cap \Omega\}$$

We will say that *a domain* Ω *belongs to* $\mathfrak{B}(g)$ ($g(v)$ satisfies condition b)) if $\Omega \in \mathfrak{A}(g)$ and there exists a positive constant d (depending on ξ) such that

$$s_\xi(R) \leqslant d\mathscr{F}_{-1}(R)/R \tag{13}$$

(the function $\mathfrak{F}(v)$ is defined by equality (10)). We assume that $d(\xi)$ is continuous in Ω and bounded for bounded ξ.

Suppose the domain Ω contains the origin of coordinates. Then the following assertion is valid.

COROLLARY. *If* $\Omega \in \mathfrak{B}(g)$, *equalities* (9) *of Theorem* 1 *and* (12) *of Theorem* 2 *can be rewritten in the form*

$$G(t,\ x;\ 0,\ \xi) \leqslant \frac{C_0 d(0)}{v(\sqrt{t})}, \qquad |u(t,\ x)| \leqslant \frac{C_0 d(0)\,\|\varphi\|_{L_1(\Omega)}}{v(\sqrt{t})}, \tag{9$'$}$$

where $v(R) = \operatorname{meas}_n\{(|x| < R) \cap \Omega\}$.

Suppose $b_i(t, x) \equiv c(t, x) \equiv 0$ in equation (1) and $q(t, x) \equiv 0$ in the boundary condition (2); i.e. consider the boundary value problem

$$\frac{\partial u(t,\ x)}{\partial t} = \sum_{i,\ j=1}^{n} \frac{\partial}{\partial x_i}\left(a_{ij}(t,\ x)\frac{\partial u(t,\ x)}{\partial x_j}\right). \tag{1$'$}$$

$$\left.\frac{\partial u(t,\ x)}{\partial N}\right|_{x\in\Gamma} = 0. \tag{2$'$}$$

In this case the following assertions, which establish that estimates (9) of Theorem 1 and (12) of Theorem 2 are sharp for $\Omega \in \mathfrak{B}(g)$, are valid.

THEOREM 3. *Suppose* $\Omega \in \mathfrak{B}(g)$. *Then there exists a constant* k *depending only on* γ, ϵ_0 *and* $d = d(\xi)$ *(from conditions* (4) *and* b) *and inequality* (13)) *such that the Green function* $G(t, x; 0, \xi)$ *of problem* (1′), (2′) *satisfies the estimate*

$$\sup_{x \in \Omega \cap (|x-\xi| < 2k\sqrt{t})} G(t, x; 0, \xi) \geqslant \frac{1}{2v_\xi(2k\sqrt{t})}$$

$$\geqslant c_0/\mathcal{F}_{-1}(\sqrt{t}) \geqslant \epsilon_0^{1/\epsilon_0} c_0/v(\sqrt{t}) \qquad (14)$$

for all $t > 0$ *and* $\xi \in \Omega$, *where*

$$c_0 = 2^{-(1+1/\epsilon_0)} k^{-1/\epsilon_0} d^{-1}, \quad v_\xi(R) = \operatorname{meas}_n \{(|x - \xi| < R) \cap \Omega\},$$

$$= \operatorname{meas}_n \{(|x| < R) \cap \Omega\}. \qquad v(R) = v_0(R)$$

Thus, under the conditions of Theorem 3, estimates (9), (9′) and (14) imply

$$\frac{c_0 \epsilon^{1/\epsilon_0}}{v(\sqrt{t})} \leqslant \frac{c_0}{\mathcal{F}_{-1}(\sqrt{t})} \leqslant \sup_{x \in \Omega} G(t, x; 0, \xi) \leqslant \frac{C_0}{\mathcal{F}_{-1}(\sqrt{t})} \leqslant \frac{C_0 d(0)}{v(\sqrt{t})}. \qquad (15)$$

An analogous assertion is valid for the solution $u(t, x)$, it being sufficient to assume that the coefficients of the equation are measurable functions satisfying condition (4).

THEOREM 4. *Suppose* $\Omega \in \mathfrak{B}(g)$ *and the initial function* $\varphi(x)$ $(\varphi \in L_1(\Omega))$ *is such that*

$$\int_\Omega \varphi(x)\, dx > 0 \qquad (16)$$

and the integrals

$$\int_\Omega |\xi| |\varphi(\xi)|\, d\xi, \quad \int_\Omega |\varphi(\xi)| \ln(1 + d(\xi))\, d\xi \qquad (17)$$

are finite, where $d(\xi)$ *is from inequality* (13). *Then the solution* $u(t, x)$ *of problem* (1′), (2′), (3) *satisfies the estimates*

$$\frac{c_1}{d(0)\, \mathcal{F}_{-1}(\sqrt{t} + a_1)} \leqslant \frac{c_1}{v(\sqrt{t} + a_1)} \leqslant \sup_{x \in \Omega} u(t, x)$$

$$\leqslant \frac{C_0 \|\varphi\|_{L_1(\Omega)}}{\mathcal{F}_{-1}(\sqrt{t})} \leqslant \frac{C_0 d(0) \|\varphi\|_{L_1(\Omega)}}{v(\sqrt{t})} \qquad (18)$$

for all $t > 0$, *where the constant* C_0 *is the same as in Theorem* 1 *and the positive constants* c_1 *and* a_1 *depend on* ϵ_0, γ, $d(0)$ *and the integrals* (16) *and* (17).

Results analogous to the above theorems were established in the case of the less complete "power" scale of classes of domains in [1], [2]. Thus it was proved in [1] that if $\Omega \in \mathfrak{A}(\alpha, n)$, where

$$\mathfrak{A}(\alpha, n) = \bigcup_{d_0 > 0} \mathfrak{A}\left(d_0 \min\left\{v^{\frac{\alpha-1}{\alpha}}, v^{\frac{n-1}{n}}\right\}\right), \quad 1 \leqslant \alpha \leqslant n,$$

then estimate (12), in which the function $\mathcal{F}_{-1}(\sqrt{t})$ will be equal to const $\min(t^{\alpha/2}, t^{n/2})$, is valid. Estimates (14) and (18) have been established in the note

[2] for $\Omega \in \mathfrak{B}(\alpha, n)$, where

$$\mathfrak{B}(\alpha, n) = \bigcup_{d_0 > 0} \mathfrak{B}\left(d_0 \min\left\{v^{\frac{\alpha-1}{\alpha}}, v^{\frac{n-1}{n}}\right\}\right),$$

We note that $\mathfrak{A}(g_1) \subset \mathfrak{A}(g_2)$ if $g_1(v) \geqslant g_2(v)$. Analogously, $\mathfrak{A}(\alpha_1, n) \subset \mathfrak{A}(\alpha_2, n)$ for $\alpha_1 \geqslant \alpha_2$.

Consider a domain (we will denote it by Ω^β) of the form $x_2 > f(x_1)$, $f(0) = 0$, $n = 2$, where the smooth even function $f(x_1)$ has a monotone nondecreasing derivative and is equivalent to the function $x_1^\beta \ln x_1, \beta \geqslant 1$, for $x_1 \to \infty$. We have (see §6)

$$\Omega^\beta \in \mathfrak{B}\left(\min\left\{2\sqrt{\frac{v}{\pi+4}}, \tilde{f}_{-1}\left(4\frac{v}{\pi+4}\right)\right\}\right),$$

where $\tilde{f}_{-1}(v)$ is the inverse of the function $\tilde{f}(x_1) = x_1 f(x_1)$. Estimates (15) will have the form

$$\text{const} \max\{t^{-1}, (\sqrt{t}\tilde{f}_{-1}(\sqrt{t}))^{-1}\} \leqslant \sup_{x \in \Omega} G(t, x; 0, \xi)$$
$$\leqslant \text{const} \max\{t^{-1}, (\sqrt{t}\tilde{f}_{-1}(\sqrt{t}))^{-1}\},$$

where the function $(\sqrt{t}\tilde{f}_{-1}(\sqrt{t}))^{-1}$ behaves like $t^{-(\beta+1)/2\beta}\ln^{1/\beta} t$ for large t, i.e. for large t

$$\text{const}\, t^{-\frac{\beta+1}{2\beta}} \ln^{1/\beta} t \leqslant \sup_{x\in\Omega} G(t, x; 0, \xi) \leqslant \text{const}\, t^{-\frac{\beta+1}{2\beta}} \ln^{1/\beta} t.$$

Consequently an estimate of the Green function for the domain Ω^β cannot be sharp in the power scale, and Ω^β does not belong to a class $\mathfrak{B}(\alpha, n)$ (a sharp (from above and from below) power estimate holds in the case of domains from $\mathfrak{B}(\alpha, n)$).

We note that $\Omega^\beta \in \mathfrak{A}(\alpha, n)$ for any $\alpha < 1 + \beta^{-1}$, whereas $\Omega^\beta \notin \mathfrak{A}(1 + \beta^{-1}, n)$. Consequently the "power" scale of classes of domains is not complete in the sense that $\bigcap_{\alpha<\alpha_0} \mathfrak{A}(\alpha, n) \neq \mathfrak{A}(\alpha_0, n)$. The domains from $\bigcap_{\alpha<\alpha_0} \mathfrak{A}(\alpha, n) \setminus (\mathfrak{A}(\alpha_0, n))$ do not belong to a class $\mathfrak{B}(\alpha, n)$. The scale of classes $\mathfrak{A}(g)$, on the other hand, is complete in the indicated sense since $\bigcap \mathfrak{A}(g_\sigma) = \mathfrak{A}(g)$, where the intersection is taken over all of the functions $g_\sigma(v)$ satisfying the condition $g_\sigma(v) \leqslant g(v)$, i.e. the condition $\mathfrak{A}(g_\sigma) \supset \mathfrak{A}(g)$.

Suppose a domain Ω belongs to the class $\mathfrak{A}(g)$, where the function $g(v)$ satisfies condition b). We wish to determine the meaning of condition (13), i.e. when $\Omega \in \mathfrak{B}(g)$.

Let $v_\xi(R)$ denote the set $\text{meas}_n\{(|x - \xi| < R) \cap \Omega\}$.

From (6) we have

$$s_\xi(R) = \frac{dv_\xi(R)}{dR} \geqslant l(v) \geqslant g(v), \tag{19}$$

which implies

$$\int_0^v \frac{d\eta}{g(\eta)} \geqslant R. \tag{20}$$

Since $g(v)$ satisfies condition b),

$$\int_0^v \frac{d\eta}{g(\eta)} \leqslant \frac{v^{1-\varepsilon_0}}{g(v)} \int_0^v \frac{d\eta}{\eta^{1-\varepsilon_0}} = \frac{1}{\varepsilon_0}\frac{v}{g(v)} = \frac{1}{\varepsilon_0}\mathcal{F}(v). \tag{21}$$

Since $\mathcal{F}(v)$ is monotone increasing (condition b)), it is uniquely invertible and its inverse $\mathcal{F}_{-1}(R)$ is monotone increasing. Consequently from (20) and (21) we have

$$v_\xi(R) \geqslant \mathcal{F}_{-1}(\varepsilon_0 R) \geqslant \varepsilon_0^{1/\varepsilon_0}\mathcal{F}_{-1}(R), \tag{22}$$

and, substituting (22) into (19), we get

$$s_\xi(R) \geqslant g(\mathcal{F}_{-1}(\varepsilon_0 R)) = \frac{\mathcal{F}_{-1}(\varepsilon_0 R)}{\varepsilon_0 R} \geqslant \varepsilon_0^{\frac{1-\varepsilon_0}{\varepsilon_0}} \frac{\mathcal{F}_{-1}(R)}{R}, \tag{23}$$

since condition b) implies $\mathcal{F}_{-1}(\epsilon_0 R) \geqslant \epsilon_0^{1/\epsilon_0}\mathcal{F}_{-1}(R)$ (see §2).

Thus estimate (23) holds for a domain Ω from the class $\mathfrak{A}(g)$. In order for $\Omega \in \mathfrak{B}(g)$ it is necessary that $s_\xi(R)$ satisfy the analogous estimate from above (13). We note that (23) implies that $d(\xi) \geqslant \epsilon_0^{(1-\epsilon_0)/\epsilon_0}$ in (13).

In addition, from inequalities (16), (13) and (22) we get that

$$g(v) \leqslant l(v) \leqslant s_\xi(R(v)) \leqslant d(\xi)\frac{\mathcal{F}_{-1}(R(v))}{R(v)}$$

$$= d(\xi)\, g(\mathcal{F}_{-1}(R(v))) \leqslant d(\xi)\, g(\varepsilon_0^{-1/\varepsilon_0} v) \leqslant \varepsilon_0^{-\frac{1-\varepsilon_0}{\varepsilon_0}} d(\xi)\, g(v).$$

Consequently, in order for a domain Ω from $\mathfrak{A}(g)$ to be a member of the class $\mathfrak{B}(g)$ it is necessary that estimate (6) be sharp, i.e.

$$g(v) \leqslant l(v) \leqslant \text{const}\, g(v),$$

and that the function $s_\xi(R(v))$ be for each $\xi \in \Omega$ of the same order with respect to v as the infimum $l(v)$, i.e.

$$l(v) \leqslant s_\xi(R(v)) \leqslant \text{const}\, l(v),$$

where the constant depends on ξ. It is not difficult to see that these conditions are also sufficient.

§1. An inequality for functions from $W_2^1(\Omega) \cap L_1(\Omega)$ when $\Omega \in \mathfrak{A}(g)$

We will say that a continuous function $w(x)$ in Ω is *almost linear* if it is linear in each of a finite number of sets Q_i such that $\Omega = \bigcup_i Q_i$. Let

$$A(y) = \{x : w(x) > y\} \text{ and } B(y) = \{x : w(x) < y\}.$$

LEMMA 1. *Suppose the domain Ω belongs to the class $\mathfrak{A}(g)$. Then the following inequality is valid for any almost linear function $w(x)$ with compact support and any positive number ϵ:*

$$\int_{A(\varepsilon)} (w(x)-\varepsilon)^2 dx \leqslant c_2 \mathscr{P}(\text{meas}_n A(\varepsilon)) \int_{A(\varepsilon)} |\nabla w(x)|^2 dx, \tag{1.1}$$

where

$$\mathscr{P}(v) = \int_0^v \frac{1}{\zeta}\left(\int_0^\zeta \frac{\theta}{g^2(\theta)}\, d\theta\right) d\zeta \tag{1.2}$$

and c_2 *is an absolute constant.*

The proof of the lemma differs little from the proof of Lemma 1 in [1] and makes use of a method applied by De Giorgi in several other situations [3].

Since

$$\int_{A(\lambda_1)\setminus A(\lambda_2)} |\nabla w(x)|\, dx = \int_{\lambda_1}^{\lambda_2} \text{meas}_{n-1}(\partial A(y) \cap \Omega)\, dy$$

for $\lambda_2 > \lambda_1$ (see [3]), it follows by making use of estimate (6) and the monotonicity of $g(v)$ that

$$\int_{A(\lambda_1)\setminus A(\lambda_2)} |\nabla w(x)|\, dx \geqslant \int_{\lambda_1}^{\lambda_2} g(\text{meas}_n A(y))\, dy \geqslant g(\text{meas}_n A(\lambda_2))(\lambda_2 - \lambda_1). \tag{1.3}$$

Take any $\epsilon > 0$ and let $\lambda_k = \inf\{\lambda\colon \text{meas}_n A(\lambda) \leqslant 2^{-k}\, \text{meas}_n A(\epsilon)\}$. Then $\lambda_0 = \epsilon \leqslant \lambda_1 \leqslant \lambda_2 \leqslant \cdots$. Next take a sequence of sets D_k satisfying the conditions $\Omega\setminus B(\lambda_k) \supset D_k \supset A(\lambda_k)$ and $\text{meas}_n D_k = 2^{-k}\, \text{meas}_n A(\epsilon)$. Then

$$\lim_{\delta\to 0+} \text{meas}\, A(\lambda_k + \delta) = \text{meas}\, A(\lambda_k)$$

$$\leqslant \text{meas}\, D_k \leqslant \text{meas}[\Omega \setminus B(\lambda_k)] = \lim_{\delta\to 0+} \text{meas}\, A(\lambda_k - \delta). \tag{1.4}$$

From Hölder's inequality and formulas (1.3) and (1.4) we get

$$\int_{A(\varepsilon)} |\nabla w|^2 dx = \sum_{k=1}^{\infty} \int_{D_{k-1}\setminus D_k} |\nabla w|^2 dx$$

$$\geqslant \sum_{k=1}^{\infty} \frac{2^k}{\text{meas}\, A(\varepsilon)} \lim_{\delta\to 0+} \left(\int_{A(\lambda_{k-1}+\delta)\setminus A(\lambda_k-\delta)} |\nabla w|\, dx\right)^2$$

$$\geqslant \sum_{k=1}^{\infty} \frac{2^k}{\text{meas}\, A(\varepsilon)} \lim_{\delta\to 0+} [g(\text{meas}\, A(\lambda_k - \delta))(\lambda_k - \lambda_{k-1} - 2\delta)]^2$$

$$= \sum_{k=1}^{\infty} \frac{2^k}{\text{meas}\, A(\varepsilon)} (\lambda_k - \lambda_{k-1})^2 g^2(\lim_{\delta\to 0+} \text{meas}\, A(\lambda_k - \delta)).$$

Let $v_k = 2^{-k}\mathrm{meas}_n A(\epsilon)$. Then by virtue of (1.4) and the monotonicity of $g(v)$ the latter inequality can be rewritten in the form

$$\int_{A(\varepsilon)} |\nabla w|^2\, dx \geqslant \frac{1}{v_0}\sum_{k=1}^{\infty} 2^k(\lambda_k - \lambda_{k-1})^2 g^2(v_k),$$

which implies

$$\sup_k 2^k(\lambda_k - \lambda_{k-1})^2 g^2(v_k) \leqslant v_0 \int_{A(\varepsilon)} |\nabla w|^2\, dx. \tag{1.5}$$

Consider the integral

$$\int_{A(\varepsilon)} (w(x) - \varepsilon)^2\, dx = \sum_{k=1}^{\infty} \int_{D_{k-1}\setminus D_k} (w(x) - \varepsilon)^2\, dx \leqslant \sum_{k=1}^{\infty} (\lambda_k - \lambda_0)^2 v_0 2^{-k}. \tag{1.6}$$

From the Schwarz-Bunjakovskiĭ inequality we get

$$(\lambda_k - \lambda_0)^2 = \left(\sum_{m=1}^{k} \lambda_m - \lambda_{m-1}\right)^2 \leqslant k\sum_{m=1}^{k} (\lambda_m - \lambda_{m-1})^2.$$

Consequently (1.6) takes the form

$$\begin{aligned}\int_{A(\varepsilon)} (w(x) - \varepsilon)^2\, dx &\leqslant v_0 \sum_{k=1}^{\infty} 2^{-k} k \sum_{m=1}^{k} (\lambda_m - \lambda_{m-1})^2\\ &\leqslant v_0 \sum_{k=1}^{\infty} (\lambda_k - \lambda_{k-1})^2 \sum_{m=k}^{\infty} 2^{-m} m\\ &\leqslant v_0 \sum_{m=0}^{\infty} (m+1)2^{-m} \sum_{k=1}^{\infty} 2^{-k}k(\lambda_k - \lambda_{k-1})^2 = 4v_0 \sum_{k=1}^{\infty} k2^{-k}(\lambda_k - \lambda_{k-1})^2,\end{aligned}$$

from which, using (1.5), we get

$$\begin{aligned}\int_{A(\varepsilon)} (w(x) - \varepsilon)^2\, dx &\leqslant 4v_0 \sum_{k=1}^{\infty} k2^{-k} \frac{v_0}{2^k g^2(v_k)} \int_{A(\varepsilon)} |\nabla w|^2\, dx\\ &= 4v_0^2 \int_{A(\varepsilon)} |\nabla w|^2\, dx \sum_{k=1}^{\infty} \frac{k2^{-2k}}{g^2(v_k)}.\end{aligned} \tag{1.7}$$

Consider the function

$$\mathscr{P}_0(v) = 4\sum_{k=1}^{\infty} \frac{k2^{-2k}v^2}{g^2(2^{-k}v)},$$

Since $g(v)$ satisfies condition a), there exists a positive integer N $(2^{-N}v_0 < \delta_1)$ such that $g(v_k) \geqslant Cv_k^{1-\epsilon_0}$ for all $k \geqslant N$. Therefore

$$\frac{k2^{-2k}}{g^2(v_k)} \leqslant \frac{k2^{-2k}}{C^2 2^{-2(1-\varepsilon_0)k} v_0^{2(1-\varepsilon_0)}} = \frac{k2^{-2\varepsilon_0 k}}{C^2 v^{2(1-\varepsilon_0)}}$$

and consequently the series in the right side of (1.7) converges and the function $\mathcal{P}_0(v)$ is defined for all $v > 0$.

For $k-1 \leqslant \eta \leqslant k$ we have

$$\frac{k2^{-2k}}{g^2(v2^{-k})} \leqslant \frac{(\eta+1)\,2^{-2\eta}}{g^2(2^{-(\eta+1)}v)},$$

so that

$$\mathcal{P}_0(v) \leqslant 4\int_0^{\infty} \frac{(\eta+1)\,2^{-2\eta}v^2}{g^2(2^{-(\eta+1)}v)}\,d\eta.$$

Making the change of variable $v2^{-(\eta+1)} = \zeta$ and integrating by parts, we get

$$\mathcal{P}_0(v) \leqslant 4\int_0^{v/2} \frac{\left(\log_2 \frac{v}{\zeta}\right)4\zeta}{g^2(\zeta)\ln 2}\,d\zeta \leqslant \frac{16}{\ln^2 2}\int_0^{v} \frac{\zeta \ln \frac{v}{\zeta}}{g^2(\zeta)}\,d\zeta$$

$$= c_2\int_0^{v} \frac{1}{\zeta}\left[\int_0^{\zeta} \frac{\theta\,d\theta}{g^2(\theta)}\right]d\zeta = c_2\mathcal{P}(v),$$

where $c_2 = 16/\ln^2 2$. Here we have made use of the fact that condition a) implies the relation

$$\lim_{\zeta\to 0} \ln\zeta \int_0^{\zeta} \frac{\theta\,d\theta}{g^2(\theta)} = 0.$$

Thus, substituting the estimate $\mathcal{P}_0(v) \leqslant c_2\mathcal{P}(v)$ into inequality (1.7), we get (1.1) and (1.2). The lemma is proved.

LEMMA 2. *Suppose* $\Omega \in \mathfrak{A}(g)$. *Then the following inequality holds for any function* $f(x) \in W_2^1(\Omega) \cap L_1(\Omega)$:

$$\int_\Omega |\nabla f(x)|^2\,dx \geqslant \frac{\int_\Omega f^2(x)\,dx}{2c_2\mathcal{P}\left(\frac{\left(2\int_\Omega |f(x)|\,dx\right)^2}{\int_\Omega f^2(x)\,dx}\right)}, \tag{1.8}$$

where $\mathcal{P}(v)$ *is defined by formula* (1.2) *and* $c_2 = 16/\ln^2 2$.

PROOF. Since the function $\mathcal{P}(v)$ is continuous and any function $f(x) \in W_2^1(\Omega) \cap L_1(\Omega)$ can be approximated by almost linear functions with compact supports in the normed spaces $W_2^1(\Omega)$ and $L_1(\Omega)$ simultaneously, it suffices to prove inequality (1.8) for an almost linear function $f(x)$ with compact support.

Take any $\epsilon > 0$ and let $A(\epsilon) = \{x: |f(x)| > \epsilon\}$. Then

$$\int_\Omega f^2(x)\,dx \leqslant \varepsilon \int_{\Omega\setminus A(\varepsilon)} |f(x)|\,dx + \int_{A(\varepsilon)} f^2(x)\,dx. \tag{1.9}$$

Since

$$\int_{A(\varepsilon)} f^2(x)\,dx = \int_{A(\varepsilon)} (|f(x)| - \varepsilon)^2\,dx + 2\varepsilon \int_{A(\varepsilon)} |f(x)|\,dx$$

$$- \varepsilon^2 \operatorname{meas}_n A(\varepsilon) \leqslant \int_{A(\varepsilon)} (|f(x)| - \varepsilon)^2\,dx + 2\varepsilon \int_{A(\varepsilon)} |f(x)|\,dx,$$

it follows from (1.9) that

$$\int_\Omega f^2(x)\,dx \leqslant 2\varepsilon \int_\Omega |f(x)|\,dx + \int_{A(\varepsilon)} (|f(x)| - \varepsilon^2)\,dx. \tag{1.10}$$

The almost linearity of $f(x)$ implies that $|f(x)|$ is almost linear; consequently the second summand in the right side of (1.10) can be estimated by Lemma 1. Thus

$$\int_\Omega f^2(x)\,dx \leqslant 2\varepsilon \int_\Omega |f(x)|\,dx + c_2 \mathscr{P}(\operatorname{meas} A(\varepsilon)) \int_{A(\varepsilon)} |\nabla f(x)|^2\,dx.$$

And from the fact that $\operatorname{meas}_n A(\epsilon) \leqslant \epsilon^{-1} \int_\Omega |f(x)|\,dx$ we get

$$\int_\Omega f^2 dx \leqslant 2\varepsilon \int_\Omega |f|\,dx + c_2 \mathscr{P}\left(\frac{1}{\varepsilon} \int_\Omega |f|\,dx\right) \int_\Omega |\nabla f|^2\,dx. \tag{1.11}$$

Take as ϵ the root (it is unique) of the equation

$$2\varepsilon \int_\Omega |f|\,dx = c_2 \mathscr{P}\left(\frac{1}{\varepsilon} \int_\Omega |f|\,dx\right) \int_\Omega |\nabla f|^2\,dx.$$

Then, denoting $\epsilon^{-1} \int_\Omega |f|\,dx$ by z, we get

$$z \mathscr{P}(z) = \frac{2\left(\int_\Omega |f|\,dx\right)^2}{c^2 \int_\Omega |\nabla f|^2\,dx}. \tag{1.12}$$

Since $\mathscr{P}(z)$ is monotone increasing, the function $\mathfrak{U}(z) = z\mathscr{P}(z)$ and its inverse $\mathfrak{U}_{-1}(z)$ are also monotone increasing. From (1.12) we get

$$\frac{1}{\varepsilon} \int_\Omega |f(x)|\,dx = \mathscr{U}_{-1}\left(\frac{2\left(\int_\Omega |f(x)|\,dx\right)^2}{c_2 \int_\Omega |\nabla f|^2\,dx}\right).$$

And from inequality (1.11) we obtain the estimate

$$\int_\Omega f^2(x)\,dx \leqslant \frac{4\left(\int_\Omega |f(x)|\,dx\right)^2}{\mathscr{U}_{-1}\left(\frac{2\left(\int_\Omega |f(x)|\,dx\right)^2}{c_2 \int_\Omega |\nabla f(x)|^2\,dx}\right)},$$

which implies

$$\frac{2\left(\int_{\Omega}|f(x)|\,dx\right)^2}{c_2\int_{\Omega}|\nabla f(x)|^2\,dx}\leqslant\mathcal{U}\left(\frac{\left(2\int_{\Omega}|f(x)|\,dx\right)^2}{\int_{\Omega}f^2(x)\,dx}\right)=\frac{4\left(\int_{\Omega}|f(x)|\,dx\right)^2}{\int_{\Omega}f^2(x)\,dx}\mathcal{P}\left(\frac{\left(2\int_{\Omega}|f(x)|\,dx\right)^2}{\int_{\Omega}f^2(x)\,dx}\right),$$

which gives the required inequality (1.8).

§2. Estimates from above

In this section we will assume that the coefficients of equation (1) are sufficiently smooth. More precisely, we will assume that in addition to (4) and (5) the following condition is satisfied:

A) The $a_{ij}(t, x)$ satisfy a Hölder condition with respect to t and x; the derivatives $\partial a_{ij}(t, x)/\partial x_i$ and $\partial b_i(t, x)/\partial x_i$ exist, are bounded and together with the coefficients $b_i(t, x)$ and $c(t, x)$ are continuous and satisfy a Hölder condition with respect to x.

We recall that the $a_{ij}(t, x)$, $b_i(t, x)$, $c(t, x)$ and $q(t, x)$ are bounded and $q(t, x)$ is continuous.

These conditions ensure the existence of a Green function (we will denote it by $G(t, x; \tau, \xi)$) for problem (1), (2).

The following lemma, which is analogous to Lemma 3 of [1], characterizes the behavior of the Green function (and hence also the solutions) and its derivatives for large $|x|$ when $t \in (0, T]$.

LEMMA 3. *Let $G(t, x; \tau, \xi)$ be the Green function of problem* (1), (2) *in the cylinder D. Then for arbitrary numbers T and δ $(T > \delta > 0)$ the integrals*

$$\int_{\Omega} G(t, x; 0, \xi)\,dx, \quad \int_{\Gamma} G(t, x; 0, \xi)\,dS_x, \qquad\qquad (2.1)$$

$$\int_{\Omega}\sum_{i=1}^{n}\left(\frac{\partial G(t, x; 0, \xi)}{\partial x_i}\right)^2 dx$$

converge and the relation

$$\int_{\Omega\cap(R<|x|<R+1)}\sum_{i=1}^{n}\left|\frac{\partial G(t, x; 0, \xi)}{\partial x_i}\right| dx \to 0 \quad \text{as} \quad R\to\infty \qquad (2.2)$$

is satisfied uniformly with respect to $t \in [\delta, T]$ and ξ from any compactum $K \subset \Omega$. Further, for any continuous initial function $\varphi(x)$ with compact support the integral

$$\int_{\Omega}\left[\int_{\Omega} G(t, x; 0, \xi)\,\varphi(\xi)\,d\xi\right] dx$$

converges uniformly with respect to t from the segment $[0, T]$.

We note that since the convergence of the integrals (2.1) and the fulfilment of relation (2.2) are uniform with respect to ξ from any compactum, analogous properties also hold for the solution of problem (1)–(3) when the initial function $\varphi(x)$ is continuous and has compact support.

The proof of Lemma 3 will be carried out in §7; in the meanwhile, we will use it to prove the following assertion.

LEMMA 4. *Suppose $\Omega \in \mathfrak{A}(g)$ and the initial function $\varphi(x)$ is continuous and has compact support. Then the solution $u(t, x)$ of problem (1)–(3) satisfies the inequality*

$$\int_\Omega u^2(t, x)\, dx \leqslant \frac{4\left(\int_\Omega |\varphi(x)|\, dx\right)^2}{\mathcal{J}_{-1}(\gamma_2 t)}, \tag{2.3}$$

where $\mathcal{J}_{-1}(t)$ is the inverse of the monotone increasing function $\mathcal{J}(v)$ defined by formula (8) *and* $\gamma_2 = 2\gamma_1 = \gamma \ln 2/16$.

We note that

$$\mathcal{J}(v) = \int_0^v \frac{\mathcal{P}(\theta)}{\theta}\, d\theta, \tag{2.4}$$

where $\mathcal{P}(v)$ is defined by formula (1.2).

PROOF. Suppose that the initial function $\varphi(x) \geqslant 0$ and hence that $u(t, x) \geqslant 0$. Take a smooth function $\chi(z)$ on E_1 such that $0 \leqslant \chi(z) \leqslant 1$ for $z \in E_1$, $\chi(z) = 1$ for $z \leqslant 0$ and $\chi(z) = 0$ for $z \geqslant 1$.

Multiplying equation (1) by $\chi(|x| - R)$ and integrating over Ω for arbitrary $t > 0$, we get

$$\frac{d}{dt}\int_\Omega \chi(|x| - R)\, u(t, x)\, dx = \int_\Omega \chi u_t\, dx$$

$$= \int_\Gamma \chi u\left(-q + \sum_{i=1}^n b_i \nu_i\right) dS_x + \int_\Omega \chi\left(c - \sum_{i=1}^n \frac{\partial b_i}{\partial x_i}\right) u\, dx$$

$$- \int_{\Omega \cap (R < |x| < R+1)} \left(\sum_{i,j=1}^n \frac{d\chi}{dx_i} a_{ij} \frac{\partial u}{\partial x_j} + u \sum_{i=1}^n b_i \frac{\partial \chi}{\partial x_i}\right) dx.$$

Since the coefficients of equation (1) and the function $q(t, x)$ are bounded and satisfy conditions (4) and (5), Lemma 3 implies that as $R \to \infty$

$$\int_\Omega \chi(|x| - R)\, u(t, x)\, dx \rightrightarrows \int_\Omega u(t, x)\, dx \tag{2.5}$$

and

$$\int_\Gamma u\chi\left(-q+\sum_{i=1}^n b_i\nu_i\right)dS \rightrightarrows \int_\Gamma u\left(-q+\sum_{i=1}^n b_i\nu_i\right)dS,$$

$$\int_\Omega\left[\left(c-\sum_{i=1}^n\frac{\partial b_i}{\partial x_i}\right)u\cdot\chi-\sum_{i,j=1}^n\frac{\partial\chi}{\partial x_i}a_{ij}\frac{\partial u}{\partial x_j}-\sum_{i=1}^n b_i u\frac{\partial\chi}{\partial x_i}\right]dx$$

$$\rightrightarrows\int_\Omega\left[c(t,\ x)-\sum_{i=1}^n\frac{\partial b_i(t,\ x)}{\partial x_i}\right]u(t,\ x)\,dx$$

uniformly with respect to $t\in[\delta,\ T]$. Therefore, as $R\to\infty$,

$$\frac{d}{dt}\int_\Omega u(t,\ x)\chi(|x|-R)\,dx$$

$$\rightrightarrows\int_\Omega\left[c(t,\ x)-\sum_{i=1}^n\frac{\partial b_i(t,\ x)}{x\partial_i}\right]u(t,\ x)\,dx$$

$$+\int_\Gamma\left[-q(t,\ x)+\sum_{i=1}^n b_i(t,\ x)\nu_i(x)\right]u(t,\ x)\,dS \tag{2.6}$$

uniformly with respect to $t\in[\delta,\ T]$. But since δ and T are arbitrary, it follows from (2.5) and (2.6) that the function $\int_\Omega u(t,\ x)dx$ is differentiable and

$$\frac{d}{dt}\int_\Omega u(t,\ x)\,dx=\int_\Omega\left[c(t,\ x)-\sum_{i=1}^n\frac{\partial b_i(t,\ x)}{\partial x_i}\right]u(t,\ x)\,dx$$

$$+\int_\Gamma\left[-q(t,\ x)+\sum_{i=1}^n b_i(t,\ x)\nu_i(x)\right]u(t,\ x)\,dS_x$$

for all $t>0$, while from condition (5) we get

$$\frac{d}{dt}\int_\Omega u(t,\ x)\,dx\leqslant 0,$$

which implies

$$\int_\Omega u(t,\ x)\,dx\leqslant\int_\Omega\varphi(x)\,dx. \tag{2.7}$$

Multiplying equation (1) by $u(t,\ x)\chi(|x|-R)$ and integrating over Ω for arbitrary $t>0$, we get

$$\int_\Omega\chi(|x|-R)u(t,x)\frac{\partial u(t,\ x)}{\partial t}dx=\int_\Omega\chi(|x|-R)u(t,\ x)L[u(t,\ x)]\,dx$$

or, integrating by parts,

$$\frac{1}{2}\frac{d}{dt}\int_\Omega \chi(|x|-R)u^2(t,x)\,dx = -\int_\Omega\Big[\chi\sum_{i,j=1}^n a_{ij}\frac{\partial u}{\partial x_i}\frac{\partial u}{\partial x_j}$$

$$+u\sum_{i,j=1}^n a_{ij}\frac{\partial u}{\partial x_j}\frac{\partial \chi}{\partial x_i}\Big]dx + \int_\Gamma \chi u\Big[\frac{\partial u}{\partial N}+\frac{u}{2}\sum_{i=1}^n b_i\nu_i\Big]dS$$

$$-\frac{1}{2}\int_\Omega u^2\Big[\chi\sum_{i=1}^n\frac{\partial b_i}{\partial x_i}+\sum_{i=1}^n b_i\frac{\partial \chi}{\partial x_i}\Big]dx+\int_\Omega \chi c u^2 dx$$

$$= -\int_\Omega \chi\sum_{i,j=1}^n a_{ij}\frac{\partial u}{\partial x_i}\frac{\partial u}{\partial x_j} - \int_\Omega u^2\chi\Big(\frac{1}{2}\sum\frac{\partial b_i}{\partial x_i}-c\Big)dx$$

$$-\int_{\Omega\cap(R<|x|<R+1)}\Big[u\sum_{i,j=1}^n a_{ij}\frac{\partial u}{\partial x_j}\frac{\partial \chi}{\partial x_i}+u^2\sum_{i=1}^n b_i\frac{\partial \chi}{\partial x_i}\Big]dx$$

$$+\int_\Gamma u^2\chi\Big[\frac{1}{2}\Big(\sum_{i=1}^n b_i\nu_i\Big)-q\Big]dS.$$

It follows from the uniform convergence of the integrals (2.1), the maximum principle, the boundedness of the coefficients of equation (1) and of the function $q(t, x)$ and conditions (4) and (5) that as $R\to\infty$

$$\int_\Omega \chi(|x|-R)u^2(t,x)\,dx \rightrightarrows \int_\Omega u^2(t,x)\,dx, \tag{2.8}$$

$$\int_\Omega\Big\{\chi(|x|-R)\Big[\sum_{i,j=1}^n a_{ij}(t,x)\frac{\partial u(t,x)}{\partial x_i}\frac{\partial u(t,x)}{\partial x_j}+u^2(t,x)\Big(\frac{1}{2}\sum_{i=1}^n\frac{\partial b_i(t,x)}{\partial x_i}$$

$$-c(t,x)\Big)\Big]+\Big[u\sum_{i,j=1}^n a_{ij}\frac{\partial u}{\partial x_j}\frac{\partial \chi}{\partial x_i}+u^2\sum_{i=1}^n b_i\frac{\partial\chi}{\partial x_i}\Big]\Big\}dx$$

$$\rightrightarrows \int_\Omega\Big[\sum_{i,j=1}^n a_{ij}(t,x)\frac{\partial u(t,x)}{\partial x_i}\frac{\partial u(t,x)}{\partial x_j}+$$

$$+u^2(t,x)\Big(\frac{1}{2}\sum_{i=1}^n\frac{\partial b_i(t,x)}{\partial x_i}-c(t,x)\Big)\Big]dx$$

and

$$\int_\Gamma u^2(t,x)\chi(|x|-R)\Big[-q(t,x)+\frac{1}{2}\sum_{i=1}^n b_i(t,x)\nu_i(x)\Big]dS$$

$$\rightrightarrows \int_\Gamma\Big[-q(t,x)+\frac{1}{2}\sum_{i=1}^n b_i(t,x)\nu_i(x)\Big]u^2(t,x)\,dS$$

uniformly with respect to $t\in[\delta, T]$. Therefore, as $R\to\infty$,

$$\frac{d}{dt}\int_\Omega \chi(|x|-R)u^2(t, x)\,dx \rightrightarrows -\int_\Omega \Big\{2\sum_{i,j=1}^n a_{ij}\frac{\partial u}{\partial x_i}\frac{\partial u}{\partial x_j} + u^2\Big[\sum_{i=1}^n \frac{\partial b_i}{\partial x_i} - 2c\Big]\Big\}dx + \int_\Gamma \Big[-2q + \sum_{i=1}^n b_i\nu_i\Big]u^2\,dS \tag{2.9}$$

uniformly with respect to $t \in [\delta, T]$.

From relations (2.8) and (2.9) and conditions (4) and (5) we get that for any $t > 0$

$$-\frac{d}{dt}\int_\Omega u^2(t, x)\,dx \geqslant 2\gamma\int_\Omega \sum_{i=1}^n\Big(\frac{\partial u(t, x)}{\partial x_i}\Big)^2 dx. \tag{2.10}$$

We note that estimate (2.10) was obtained without making use of the nonnegativeness of the solution $u(t, x)$.

Since, according to Lemma 3, the solution $u(t, x) \in W_2^1(\Omega) \cap L_1(\Omega)$ for any $t > 0$, inequalities (1.8), (2.7) and (2.10) imply

$$-\frac{d}{dt}\int_\Omega u^2(t, x)\,dx \geqslant \frac{\gamma}{c_2}\frac{\int_\Omega u^2(t, x)\,dx}{\mathscr{P}\Big[\Big(2\int_\Omega u(t, x)\,dx\Big)^2\Big/\int_\Omega u^2(t, x)\,dx\Big]} \geqslant \gamma_2\frac{\int_\Omega u^2(t, x)\,dx}{\mathscr{P}\Big[\Big(2\int_\Omega \varphi(x)\,dx\Big)^2\Big/\int_\Omega u^2(t, x)\,dx\Big]}. \tag{2.11}$$

Let $w(t)$ denote the function

$$-\frac{d}{dt}\Big(\frac{1}{w(t)}\Big)\mathscr{P}(w) \geqslant \gamma_2\frac{1}{w(t)},$$

Then (2.11) can be rewritten in the form

$$\frac{\mathscr{P}(w)}{w}\frac{dw}{dt} \geqslant \gamma_2.$$

or

$$\mathscr{I}(w) = \int_0^w \frac{\mathscr{P}(\theta)}{\theta}\,d\theta \geqslant \gamma_2 t,$$

Hence

$$w(t) \geqslant \mathscr{I}_{-1}(\gamma_2 t),$$

i.e. $w(t) \geqslant \mathfrak{I}_{-1}(\gamma_2 t)$ or

$$\int_\Omega u^2(t, x)\,dx \leqslant \frac{\Big(2\int_\Omega \varphi(x)\,dx\Big)^2}{\mathscr{I}_{-1}(\gamma_2 t)}. \tag{2.12}$$

Now suppose $\varphi(x)$ is an arbitrary continuous initial function with compact support, and let $u_1(t, x)$ denote the solution of problem (1)–(3) with the initial function $|\varphi(x)|$. Since the maximum principle implies $|u(t, x)| \leqslant u_1(t, x)$, by virtue of (2.12) we have

$$\int_\Omega u^2(t, x)\,dx \leqslant \int_\Omega u_1^2(t, x)\,dx \leqslant \frac{4\,\|\varphi\|^2_{L_1(\Omega)}}{\mathcal{J}_{-1}(\gamma_2 t)}.$$

The lemma is proved.

Since the solution $u(t, x)$ can be replaced by the Green function everywhere in the above proof, the following assertion is valid.

COROLLARY. *Suppose $\Omega \in \mathfrak{A}(g)$. Then the Green function $G(t, x; \tau, \xi)$ of problem* (1), (2) *satisfies the inequality*

$$\int_\Omega G^2(t, x; 0, \xi)\,dx \leqslant \frac{4}{\mathcal{J}_{-1}(\gamma_2 t)}, \tag{2.13}$$

where the function $\mathfrak{J}(v)$ is defined by equalities (2.4) *and* (8).

LEMMA 5. *Suppose $\Omega \in \mathfrak{A}(g)$. Then the Green function $G(t, x; \tau, \xi)$ of problem* (1), (2) *satisfies the inequality*

$$G(t, x; 0, \xi) \leqslant 4/\mathcal{J}_{-1}(\gamma_1 t), \tag{2.14}$$

for all $t > 0$ and $x, \xi \in \Omega$, where $\mathfrak{J}_{-1}(t)$ is the inverse of the function $\mathfrak{J}(v)$ defined by (8) *and $\gamma_1 = \gamma_2/2 = \gamma \ln^2 2/32$.*

PROOF. From the formula

$$G(t, x; 0, \xi) = \int_\Omega G(t, x; t/2, \eta)\, G(t/2, \eta; 0, \xi)\,d\eta$$

it follows that

$$G(t, x; 0, \xi) \leqslant \Big(\int_\Omega G^2(t, x; t/2, \eta)\,d\eta\Big)^{1/2} \Big(\int_\Omega G^2(t/2, \eta; 0, \xi)\,d\eta\Big)^{1/2}. \tag{2.15}$$

Fix an arbitrary $t > 0$ and consider the function $\widetilde{G}_t(\tau, \eta; 0, x) = G(t, x; t-\tau, \eta)$. With respect to the variables τ and η it satisfies the equation

$$\frac{\partial \widetilde{G}}{\partial \tau} = \widetilde{L}[\widetilde{G}] = \sum_{i,j=1}^n \frac{\partial}{\partial \eta_i} \tilde{a}_{ij} \frac{\partial}{\partial \eta_j} \widetilde{G} + \sum_{i=1}^n \tilde{b}_i \frac{\partial \widetilde{G}}{\partial \eta_i} + \tilde{c}\widetilde{G}, \tag{2.16}$$

where

$$\tilde{a}_{ij}(\tau, \eta) = a_{ji}(t-\tau, \eta),\quad \tilde{b}_i(\tau, \eta) = -b_i(t-\tau, \eta),$$

$$\tilde{c}(\tau, \eta) = c(t-\tau, \eta) - \sum_{i=1}^n \frac{\partial b_i(t-\tau, \eta)}{\partial \eta_i},$$

and the boundary condition

$$\frac{\partial \tilde{G}}{\partial N_\eta} + \tilde{q}(\tau, \eta)\tilde{G}\big|_{\eta\in\Gamma} = 0, \tag{2.17}$$

where $\tilde{q}(\tau, \eta) = q(t-\tau, \eta) - \Sigma_1^n b_i(t-\tau, \eta)\nu_i(\eta)$, $\eta \in \Gamma$.

From the fact that the coefficients of equation (1) satisfy conditions (4) and (5), while $q(\tau, \eta) \geqslant 0$, it follows that the coefficients of equation (2.16) satisfy condition (4) with the same constant γ and condition (5), while $\tilde{q}(\tau, \eta) \geqslant 0$.

In fact, the satisfaction of (4) is obvious, while from (5) and the fact that $q(\tau, \eta) \geqslant 0$, $\eta \in \Gamma$, we have

$$\tilde{c}(\tau, \eta) = c(t-\tau, \eta) - \sum_{i=1}^n \frac{\partial b_i(t-\tau, \eta)}{\partial \eta_i} \leqslant 0;$$

$$\tilde{c}(\tau, \eta) - \sum_{i=1}^n \frac{\partial \tilde{b}_i(\tau, \eta)}{\partial \eta_i} = c(t-\tau, \eta) - \sum_{i=1}^n \frac{\partial b_i(t-\tau, \eta)}{\partial \eta_i}$$

$$+ \sum_{i=1}^n \frac{\partial b_i(t-\tau, \eta)}{\partial \eta_i} = c(t-\tau, \eta) \leqslant 0;$$

$$\tilde{q}(\tau, \eta) = q(t-\tau, \eta) - \sum_{i=1}^n b_i(t-\tau, \eta)\nu_i(\eta) \geqslant 0, \qquad \eta \in \Gamma,$$

$$-\tilde{q}(\tau, \eta) + \sum_{i=1}^n \tilde{b}_i(\tau, \eta)\nu_i(\eta) = -q(t-\tau, \eta) + \sum_{i=1}^n b_i(t-\tau, \eta)\nu_i(\eta)$$

$$-\sum_{i=1}^n b_i(t-\tau, \eta)\nu_i(\eta) = -q(t-\tau, \eta) \leqslant 0.$$

Consequently the Green function $\tilde{G}_t(\tau, \eta; 0, x)$ of problem (2.16), (2.17) satisfies estimate (2.13). Since $\tilde{G}_t(t/2, \eta; 0, x) = G(t, x; t/2, \eta)$, inequality (2.14) can be obtained from (2.15) with the use of (2.13).

Lemma 5 implies the first assertion (estimate (7)) of Theorem 1. In addition, Lemma 5 immediately implies

LEMMA 6. *Suppose* $\Omega \in \mathfrak{A}(g)$. *Then the solution* $u(t, x)$ *of problem* (1)–(3) *with a continuous initial function* $\varphi(x)$ *satisfies the inequality*

$$|u(t, x)| \leqslant \frac{4\|\varphi\|_{L_1(\Omega)}}{\mathcal{J}_{-1}(\gamma_1 t)}, \tag{2.18}$$

for all $t > 0$ *and* $x \in \Omega$, *where* $\mathcal{J}_{-1}(t)$ *is the inverse of the function* $\mathcal{J}(v)$ *defined by equality* (8) (*and* (2.4)), *and* $\gamma_1 = \gamma \ln^2 2/32$.

In fact,

$$|u(t, x)| = \left|\int_\Omega G(t, x; 0, \xi)\varphi(\xi)\,d\xi\right| \leqslant \sup_{\xi\in\Omega} G(t, x; 0, \xi)\int_\Omega |\varphi(\xi)|\,d\xi$$

and estimate (2.18) follows from inequality (2.14).

Now suppose $\varphi(x)$ is an arbitrary function from $L_1(\Omega)$, let $\varphi_m(x)$ be a sequence of continuous initial functions that converges to $\varphi(x)$ in the norm of $L_1(\Omega)$ and let $u_m(t, x)$ denote the solution of problem (1)–(3) with the initial function $\varphi_m(x)$. Inequality (2.18) implies that the sequence $u_m(t, x)$ converges uniformly with respect to $t \in (0, \delta]$ and $x \in \Omega$ to a continuous function $u(t, x)$. As in [1], we call this limit a weak solution of problem (1)–(3).

LEMMA 7. *For any* $t > 0$ *a weak solution of problem* (1)–(3) *with* $\varphi(x) \in L_1(\Omega)$ *satisfies equation* (1) *and the boundary condition* (2) *in the classical sense.*

The proof of Lemma 7, like the proof of Lemma 5 in [1], immediately follows from inequality (2.18).

Estimate (11) of Theorem 2 immediately follows from Lemma 6 and the definition of a weak solution (for $\varphi \in L_1(\Omega)$).

Suppose now $\Omega \in \mathfrak{A}(g)$, where the function $g(v)$ satisfies in addition to condition a) the stronger condition b), i.e. the function $v^{1-\epsilon_0}/g(v)$ $(0 < \epsilon_0 \leqslant 1/n)$ is monotone nondecreasing. Then the function $\mathfrak{F}(v) = v/g(v)$ is monotone increasing and $\mathfrak{F}(0) = 0$.

It follows from condition b) that $g(\lambda v) \leqslant \lambda^{1-\epsilon_0} g(v)$ for any numbers $\lambda \geqslant 1$ and $v > 0$. On the other hand, the function $g(v)$ is monotone nondecreasing, i.e. $g(\lambda v) \geqslant g(v)$. Thus we have

$$g(v) \leqslant g(\lambda v) \leqslant \lambda^{1-\varepsilon_0} g(v) \tag{2.19}$$

for any $\lambda \geqslant 1$ and $v > 0$. From (2.19) we get

$$\lambda v/g(v) \geqslant \lambda v/g(\lambda v) \geqslant \lambda^{\varepsilon_0} v/g(v)$$

or

$$\lambda^{\varepsilon_0} \mathcal{F}(v) \leqslant \mathcal{F}(\lambda v) \leqslant \lambda \mathcal{F}(v). \tag{2.20}$$

Since the function $\mathfrak{F}_{-1}(t)$ $(t > 0)$ is monotone increasing, it follows from (2.20) that $\mathfrak{F}_{-1}(\lambda^{\epsilon_0} t) \leqslant \lambda\, \mathfrak{F}_{-1}(t) \leqslant \mathfrak{F}_{-1}(\lambda t)$, which implies

$$\lambda \mathcal{F}_{-1}(t) \leqslant \mathcal{F}_{-1}(\lambda t) \leqslant \lambda^{1/\varepsilon_0} \mathcal{F}_{-1}(t). \tag{2.21}$$

Let $\mathcal{J}$ denote the operator defined by the formula

$$\mathcal{J}[f] = \int_0^v \frac{f(\zeta)}{\zeta}\, d\zeta. \tag{2.22}$$

The operator $\mathcal{J}$ is defined on those functions $f(v)$, $v > 0$, for which expression (2.22) is meaningful for all $v > 0$, i.e. for which the integral in the right side of (2.22) converges for all $v > 0$. We note that the operator $\mathcal{J}$ is defined on the functions $\mathfrak{F}(v)$ satisfying condition (2.20).

The function $\mathfrak{P}(v)$ defined by (1.2) can now be written in the form

$$\mathcal{P}(v) = \mathcal{J}^2[\mathcal{F}^2] = \mathcal{J}[\mathcal{J}[\mathcal{F}^2]],$$

while the function $\mathfrak{J}(v)$ defined by (2.4) can be written in the form

$$\mathcal{J}(v) = \mathcal{J}[\mathcal{P}] = \mathcal{J}^3[\mathcal{F}^2]. \tag{2.23}$$

Let $f(y)$ be a function satisfying the inequalities

$$C_1\lambda^\varepsilon f(y) \leqslant f(\lambda y) \leqslant C_2\lambda^\alpha f(y). \tag{2.24}$$

for all $\lambda \geqslant 1$ and $y > 0$, where $0 < \epsilon \leqslant \alpha$, and consider the function

$$h(y) = \mathcal{J}[f] = \int_0^y \frac{f(\zeta)}{\zeta}\, d\zeta.$$

After making the change of variable of integration $\zeta = \kappa y$ $(\kappa \leqslant 1)$, we get

$$h(y) = \int_0^1 \frac{f(\varkappa y)}{\varkappa}\, d\varkappa.$$

Since (2.24) can be rewritten in the form

$$\frac{\varkappa^\alpha}{C_2} f(y) \leqslant f(\varkappa y) \leqslant \frac{\varkappa^\varepsilon}{C_1} f(y), \quad \text{where} \quad \varkappa \leqslant 1,$$

it follows that

$$h(y) \geqslant \frac{f(y)}{C_2} \int_0^1 \varkappa^{\alpha-1} dx = \frac{f(y)}{C_2\alpha}$$

and

$$h(y) \leqslant \frac{f(y)}{C_1} \int_0^1 \varkappa^{\varepsilon-1} d\varkappa = \frac{f(y)}{C_1\varepsilon},$$

i.e.

$$\frac{f(y)}{\alpha C_2} \leqslant h(y) \leqslant \frac{f(y)}{\varepsilon C_1}. \tag{2.25}$$

Moreover, from (2.24) we have

$$h(\lambda y) = \int_0^1 \frac{f(\lambda \varkappa y)}{\varkappa}\, d\varkappa \leqslant C_2\lambda^\alpha \int_0^1 \frac{f(\varkappa y)}{\varkappa}\, d\varkappa = C_2\lambda^\alpha h(y)$$

and

$$h(\lambda y) \geqslant C_1\lambda^\varepsilon \int_0^1 \frac{f(\varkappa y)}{\varkappa}\, d\varkappa = C_1\lambda^\varepsilon h(y),$$

i.e.

$$C_1\lambda^\varepsilon h(y) \leqslant h(\lambda y) \leqslant C_2\lambda^\alpha h(y) \tag{2.26}$$

for any $\lambda \geqslant 1$ and $y > 0$. We have consequently proved

LEMMA 8. *If a function $f(y)$ satisfies inequalities* (2.24), *then the function $h(y) = \mathcal{J}[f]$ satisfies* (2.24) *with the same exponents α and ϵ and constants C_1 and C_2, and the functions $f(y)$ and $h(y)$ are connected by relation* (2.25).

We note that the function $\mathfrak{F}^2(v)$ satisfies inequalities (2.24) with $\epsilon = 2\epsilon_0$, $\alpha = 2$ and $C_1 = C_2 = 1$. It therefore follows from Lemma 8 that

$$\frac{1}{8}\mathcal{F}^2(v) \leqslant \frac{\mathcal{G}(v)}{4} = \frac{\mathfrak{J}[\mathcal{F}^2]}{4} \leqslant \frac{1}{2}\mathcal{P}(v)$$
$$\leqslant \mathcal{J}(v) \leqslant \frac{1}{2\varepsilon_0}\mathcal{P}(v) \leqslant \frac{1}{4\varepsilon_0^2}\mathcal{G}(v) \leqslant \frac{1}{8\varepsilon_0^3}\mathcal{F}^2(v).$$

Hence for the inverse functions we have

$$\mathcal{F}_{-1}(\sqrt{\varepsilon_0^3 8t}) \leqslant \mathcal{G}_{-1}(4\varepsilon_0^2 t) \leqslant \mathcal{P}_{-1}(2\varepsilon_0 t) \leqslant \mathcal{J}_{-1}(t) \leqslant \mathcal{F}_{-1}(\sqrt{8t})$$

and from (2.21)

$$C_3\mathcal{F}_{-1}(\sqrt{t}) \leqslant \mathcal{J}_{-1}(\gamma_1 t) \leqslant C_4\mathcal{F}_{-1}(\sqrt{t}),$$

where the constants C_3 and C_4 depend only on ϵ_0 and γ. For example, $C_3 = (\epsilon_0^3 8\gamma_1)^{1/2\epsilon_0} = 4/C_0$. We have thus proved

LEMMA 9. *If the function* $g(v)$ *satisfies condition* b), *then estimates* (7) *and* (11) *in Theorems* 1 *and* 2 *can be continued*:

$$G(t, x; 0, \xi) \leqslant \frac{4}{\mathcal{J}_{-1}(\gamma_1 t)} \leqslant \frac{C_0'''}{\mathcal{J}_{-1}(t)} \leqslant \frac{C_0''}{\mathcal{P}_{-1}(t)} \leqslant \frac{C_0'}{\mathcal{G}_{-1}(t)} \leqslant \frac{C_0}{\mathcal{F}_{-1}(\sqrt{t})}, \tag{2.27}$$

$$|u(t, x)| \leqslant \frac{4\|\varphi(x)\|_{L_1(\Omega)}}{\mathcal{J}_{-1}(\gamma_{-1}t)} \leqslant \frac{C_0'''\|\varphi(x)\|_{L_1(\Omega)}}{\mathcal{J}_1(t)} \leqslant \frac{C_0''\|\varphi\|_{L_1(\Omega)}}{\mathcal{P}_{-1}(t)}$$
$$\leqslant \frac{C_0'\|\varphi(x)\|_{L_1(\Omega)}}{\mathcal{G}_{-1}(t)} \leqslant \frac{C_0\|\varphi\|_{L_1(\Omega)}}{\mathcal{F}_{-1}(\sqrt{t})},$$

where the constants C_0''', C_0'', C_0' *and* C_0 *depend only on the constants* γ *and* ϵ_0 *in conditions* (4) *and* b).

Theorems 1 and 2 are proved.

§3. Generalized solution

We note that the right side of estimate (11) as well as the constants C_0, C_0', C_0'' and C_0''' in (2.27) depend only on the constant γ in condition (4) (in the second case they also depend on ϵ_0) and do not depend on the coefficients of equation (1). This permits one to drop the condition of smoothness of the coefficients (condition A)). For the sake of simplicity we will prove this assertion in the case of the second boundary value problem for an equation without minor terms ($b_i = c = q = 0$), which will be of interest to us in subsequent sections, where the exactness of our estimates will be established under these conditions.

Consider the second boundary value problem for equation (1) without minor terms

$$\frac{\partial u(t, x)}{\partial t} = \sum_{i, j=1}^{n} \frac{\partial}{\partial x_i}\left(a_{ij}(t, x)\frac{\partial u(t, x)}{\partial x_j}\right), \tag{3.1}$$

$$\left.\frac{\partial u(t, x)}{\partial N}\right|_{x\in\Gamma}=0, \tag{3.2}$$

where the symmetric matrix $a_{ij}(t, x)$, whose elements are measurable functions, satisfies condition (4).

By the generalized solution of problem (3.1), (3.2), (3) with an initial function from $L_2(\Omega)$ we will mean the function which for any $T > 0$ is the generalized solution of this problem from the space $V_2^{1,0}(D_T)$ (see [4]), where $D_T = \Omega \times (0, T)$. As in [4], we denote by $W_2^{1,0}(D_T)$ the Hilbert space with scalar product

$$(u, v)_{W_2^{1,0}(\Omega)} = \int_{D_T} [u(t, x)v(t, x) + (\nabla_x u(t, x), \nabla_x v(t, x))]\,dx\,dt,$$

and by $V_2^{1,0}(D_T)$ the Banach space consisting of all of the elements of $W_2^{1,0}(D_T)$ that are continuous in $L_2(\Omega)$ with respect to t and having the finite norm

$$\|u\|_{V_2^{1,0}(D_T)} = \max_{0\leqslant t\leqslant T} \|u(t, x)\|_{L_2(\Omega)} + \|\nabla_x u(t, x)\|_{L_2(D_T)}.$$

Such a solution of problem (3.1), (3.2), (3) exists and is unique (see [4]), and the following assertion, whose proof is a repetition of a proof presented in [4] for the analogous assertion for the first boundary value problem, is valid.

PROPOSITION 1. *Let $u^{(m)}(t, x)$ be the generalized solutions (from $V_2^{1,0}$) of problem* (3.1), (3.2), (3) *with the coefficients $a_{ij}^{(m)}(t, x)$ of equation* (3.1) *satisfying condition* (4). *If the $a_{ij}^{(m)}(t, x)$ converge almost everywhere in $\Omega \times (t > 0)$ to the $a_{ij}(t, x)$, then for any $T > 0$ the solutions $u^{(m)}(t, x)$ converge strongly in $V_2^{1,0}(D_T)$ to the generalized solution $u(t, x)$ of the limit problem.*

PROPOSITION 2. *Suppose the initial function $\varphi(x)$ is continuous and has compact support while the coefficients of equation* (3.1) *satisfy condition* A). *Then the classical solution of problem* (3.1), (3.2), (3) *is the generalized solution.*

PROOF. From the uniform in $t \in [0, T]$ convergence of the integral $\int_\Omega |u(t, x)|\,dx$ (Lemma 3) and the maximum principle it follows that $u(t, x)$ is continuous in $L_2(\Omega)$ with respect to $t \in [0, \infty)$. Consequently $u(t, x) \xrightarrow{L_2(\Omega)} \varphi(x)$ and

$$\|u(t, x)\|^2_{L_2(\Omega)} \to \|\varphi(x)\|^2_{L_2(\Omega)} \tag{3.3}$$

as $t \to 0$.

Integrating (2.10) over the segment $[\delta, T]$, where $\delta > 0$, we get

$$\int_\delta^T \int_\Omega |\nabla_x u(t, x)|^2\,dx \leqslant -\frac{1}{2\gamma}\int_\delta^T \left[\frac{d}{dt}\int_\Omega u^2(t, x)\,dx\right]dt$$

$$= \frac{1}{2\gamma}\left[\int_\Omega u^2(\delta, x)\,dx - \int_\Omega u^2(T, x)\,dx\right],$$

which together with relation (3.3) implies the existence of a limit of the integral

$$\int_\delta^T \int_\Omega |\nabla_x u(t, x)|^2 \, dx$$

as $\delta \to 0$. Consequently $u(t, x) \in V_2^{1,0}(D_T)$ for any $T > 0$. Multiplying equation (3.1) by an arbitrary smooth function $\psi(t, x)$ with compact support and integrating over $\Omega \times (0, t)$, we get for any $t > 0$ the identity

$$\int_\Omega u(t, x)\psi(t, x)\,dx + \int_0^t \int_\Omega [-u(t, x)\psi_t(t, x)]\,dxdt$$
$$+ \int_0^t \int_\Omega \sum_{i,j=1}^n a_{ij}(t, x) \frac{\partial u(t, x)}{\partial x_i} \frac{\partial \psi(t, x)}{\partial x_j} dxdt = \int_\Omega \varphi(x)\psi(0, x)\,dx,$$

which by the density in $W_2^1(D_t)$ of smooth functions with compact support is valid for any $\psi(t, x) \in W_2^1(D_t)$. The proposition is proved.

Suppose $u(t, x)$ is the generalized solution of problem (3.1), (3.2), (3) with an initial function $\varphi(x)$ that is continuous and has compact support, suppose $a_{ij}(t, x) = \delta_{ij}$ outside the domain $\Omega \times (t > 0)$ and consider the problem composed of (3.2), (3) and the equation

$$\frac{\partial u^{(m)}(t, x)}{\partial t} = L^{(m)}[u^{(m)}] = \sum_{i,j=1}^n \frac{\partial}{\partial x_i}\left(a_{ij}^{(m)}(t, x) \frac{\partial u^{(m)}(t, x)}{\partial x_j}\right), \tag{3.4}$$

where $a_{ij}^{(m)}(t, x)$ is an average of the function $\tilde{a}_{ij}^{(m)}(t, x)$ equal to $a_{ij}(t, x)$ for $|x| \leqslant m$ and δ_{ij} for $|x| > m$, i.e.

$$a_{ij}^{(m)}(t, x) = \text{const}\, m^{n+1} \int_{|\zeta|^2+\tau^2 \leqslant 1/m^2} \exp\left[\frac{|\zeta|^2 + \tau^2}{|\zeta|^2 + \tau^2 - 1/m^2}\right] \tilde{a}_{ij}^{(m)}(t + \tau, x + \zeta)\,d\zeta d\tau.$$

The constant is chosen so that the average of unity will be unity. Since

$$\sum_{i,j=1}^n a_{ij}^{(m)}(t, x) y_i y_j = \text{const}\, m^{n+1} \int_{|\zeta|^2+\tau^2 \leqslant 1/m^2} \exp\left[\frac{|\zeta|^2 + \tau^2}{|\zeta|^2 + \tau^2 - 1/m^2}\right]$$
$$\times \sum_{i,j=1}^n \tilde{a}_{ij}^{(m)}(t + \tau, x + \zeta) y_i y_j \,d\tau d\zeta,$$

the satisfaction of condition (4) by the matrix $\tilde{a}_{ij}^{(m)}(t, x)$ (the unit matrix also satisfies condition (4)) implies the satisfaction of condition (4) with the same constant γ by the matrix $a_{ij}^{(m)}(t, x)$. The classical solutions $u^{(m)}(t, x)$ of problems (3.4), (3.2), (3) satisfy estimates (2.27) with the same right sides. Since, according to Proposition 2, the $u^{(m)}(t, x)$ are also the generalized solutions while, according to Proposition 1, the convergence almost everywhere of the $a_{ij}^{(m)}(t, x)$ to $a_{ij}(t, x)$ implies the convergence almost everywhere in $V_2^{1,0}$ of the $u^{(m)}(t, x)$ to $u(t, x)$, inequalities (2.27) hold almost everywhere in $\Omega \times (t > 0)$ for the generalized solution $u(t, x)$ of problem (3.1), (3.2), (3). But $u(t, x)$ is continuous for $t > 0$ and

$x \in \Omega$ (see [5], [6]), and hence estimates (2.27) are valid for all $t > 0$ and $x \in \Omega$.

Now, as in §2, by approximating any function $\varphi(x) \in L_1(\Omega)$ by continuous functions $\varphi^{(m)}(x)$ with compact supports, we can define a weak solution as the uniform limit of the corresponding generalized solutions. Such a solution will satisfy estimates (2.27).

We have thus proved the right inequality of inequalities (18) in Theorem 4.

§4. Estimation of the moment

In this and the following section we will assume that the coefficients of equation (3.1) are sufficiently smooth (satisfying condition A)) functions.

Let $G(t, x; \tau, \xi)$ be the Green function of problem (3.1), (3.2), let ξ be a fixed point of Ω and let

$$Q_\xi(t) = -\int_\Omega G(t, x; 0, \xi) \ln G(t, x; 0, \xi)\, dx \tag{4.1}$$

and

$$\mu_\xi(t) = \int_\Omega |x - \xi|\, G(t, x; 0, \xi)\, dx. \tag{4.2}$$

Following [5], we will call these functions the *entropy* and *moment*, respectively.

LEMMA 10. *For any point* $\xi \in \Omega$ *the functions* $Q_\xi(t)$ *and* $\mu_\xi(t)$ *are absolutely continuous for* $t \geqslant \delta > 0$,

$$\frac{dQ_\xi(t)}{dt} = \int_\Omega \sum_{i, j=1}^{n} a_{ij}(t, x) \frac{\partial G(t, x; 0, \xi)}{\partial x_i} \frac{\partial G(t, x; 0, \xi)}{\partial x_j} \frac{dx}{G(t, x; 0, \xi)}, \tag{4.3}$$

$$\left| \frac{d\mu_\xi(t)}{dt} \right| \leqslant \int_\Omega |A \nabla_x G|\, dx, \quad A(t, x) = \{a_{ij}(t, x)\} \tag{4.4}$$

and $\lim_{\delta \to 0} \mu_\xi(\delta) = 0$.

The proof of Lemma 10 will be given in §7. In the case when $\Omega = E_n$, i.e. $G(t, x; \tau, \xi) = \Gamma(t, x; \tau, \xi)$ is a fundamental solution of equation (3.1), J. Nash proved in [5] that

$$k_0 \sqrt{t} \leqslant \int_{E_n} |x - \xi|\, \Gamma(t, x; 0, \xi)\, dx \leqslant k \sqrt{t}$$

with the constants k_0 and k depending only on the constant γ in condition (4). Using a method similar to Nash's method, we obtain analogous inequalities from estimates (2.27).

THEOREM 5. *Suppose* $\Omega \in \mathfrak{B}(g)$. *Then for all* $t > 0$

$$k_0 \sqrt{t} \leqslant \mu_\xi(t) \leqslant k \sqrt{t}, \tag{4.5}$$

where the constants k_0 *and* k *depend only on* γ *(in condition* (4)), ϵ_0 *(in condition* b)) *and* $d = d(\xi)$ *(in inequality* (13)). *Moreover,* $k \leqslant d_1 \ln(d + 1)$, *where* d_1 *depends only on* ϵ_0 *and* γ.

PROOF. From estimates (2.27) we have (recall that $\Omega \in \mathfrak{B}(g)$ implies $\Omega \in \mathfrak{A}(g)$)

$$G(t,\ x;\ 0,\ \xi) \leqslant \frac{C_0'}{\mathscr{G}_{-1}(t)}.$$

Substituting this result into (4.1), we get

$$Q_\xi(t) \geqslant \min_{x \in \Omega}[-\ln G] \geqslant k_1 + \ln \mathscr{G}_{-1}(t). \tag{4.6}$$

1) Since for any real λ we have

$$Z[\ln Z + \lambda] \geqslant -e^{-1-\lambda} \tag{4.7}$$

for all $Z > 0$, it follows upon integrating (4.7) over Ω when $Z = G$ and $\lambda = ar + b$, where a and b do not depend on x while $r = |x - \xi|$, that

$$-Q_\xi(t) + a\mu_\xi(t) + b \geqslant -e^{-1-b} \int_\Omega e^{-a|x-\xi|}\,dx.$$

Let $\mathfrak{D}_\xi(a)$ denote the function $\int_\Omega e^{-a|x-\xi|}dx$, and set $a = 1/\mu_\xi(t)$ and $e^{-b} = e/\mathfrak{D}_\xi(a)$. Then the latter inequality takes the form

$$-Q_\xi(t) + 1 + \ln \mathscr{D}_\xi(a) \geqslant 0$$

and hence the form

$$\mathscr{D}_\xi(1/\mu_\xi(t)) \geqslant e^{Q_\xi(t)-1}. \tag{4.8}$$

Consider the function

$$\mathscr{D}_\xi(a) = \int_\Omega e^{-a|x-\xi|}dx = \int_0^\infty e^{-ar}s_\xi(r)\,dr.$$

Using estimate (13) ($\Omega \in \mathfrak{B}(g)$), we get

$$\mathscr{D}_\xi(a) \leqslant d(\xi) \int_0^\infty e^{-ar} \frac{\mathscr{F}_{-1}(r)}{r}\,dr$$

and hence, making the change of variable $ar = \rho$,

$$\mathscr{D}_\xi(a) \leqslant d(\xi) \int_0^\infty e^{-\rho} \frac{\mathscr{F}_{-1}(\rho/a)}{\rho}\,d\rho$$

$$= d\left\{\int_0^1 e^{-\rho} \frac{\mathscr{F}_{-1}(\rho/a)}{\rho}\,d\rho + \int_1^\infty e^{-\rho} \frac{\mathscr{F}_{-1}(\rho/a)}{\rho}\,d\rho\right\}. \tag{4.9}$$

From estimate (2.21) ($\Omega \in \mathfrak{B}(g)$, so that $g(v)$ satisfies condition b)) it follows that

$$\mathscr{F}_{-1}(\rho/a) \leqslant \rho^{1/\varepsilon_0}\mathscr{F}_{-1}(1/a) \qquad \text{for } \rho \geqslant 1 \tag{4.10}$$

$$\mathscr{F}_{-1}(\rho/a) \leqslant \rho\mathscr{F}_{-1}(1/a) \qquad \text{for } \rho \leqslant 1. \tag{4.11}$$

Substituting (4.10) and (4.11) into (4.9), we get

$$\mathscr{D}_\xi(a) \leqslant \mathscr{F}_{-1}(1/a)\,d(\xi)\left\{\int_0^1 e^{-\rho}d\rho + \int_1^\infty e^{-\rho}\rho^{1/\varepsilon_0-1}d\rho\right\} = d_2 d(\xi)\,\mathscr{F}_{-1}(1/a). \tag{4.12}$$

From (4.8) and (4.12) we have

$$\mathcal{F}_{-1}(1/a) \geqslant \frac{1}{d_2 d(\xi)} e^{Q_\xi(t)-1}$$

or

$$\mu_\xi(t) \geqslant \mathcal{F}\left(\frac{1}{d_2 d(\xi)} e^{Q_\xi(t)-1}\right)$$

and hence, using (2.20),

$$\mu_\xi(t) \geqslant d_3 \mathcal{F}(e^{Q_\xi(t)}), \tag{4.13}$$

where $d_3 = (ed_2 d)^{-\epsilon_0}$ with the constant d_2 depending only on ϵ_0.

2) Consider the derivative of the entropy $Q_\xi(t)$ with respect to t. According to Lemma 10, for almost all $t > 0$ this derivative exists and satisfies the equality

$$\frac{dQ_\xi(t)}{dt} = \int_\Omega \frac{\nabla_x G A \nabla_x G}{G} dx$$

$$= \int_\Omega (\nabla_x \ln G(t, x; 0, \xi)) A(t, x) (\nabla_x \ln G(t, x; 0, \xi)) (G(t, x; 0, \xi) dx). \tag{4.14}$$

Since the matrix $A = \{a_{ij}(t, x)\}$ satisfies condition (4), the following scalar product can be introduced for n-dimensional vectors y and z:

$$[y, z] = (y, Az) = yAz = \sum_{i, j=1}^{n} y_i a_{ij}(t, x) z_j.$$

Inequalities (4) then give

$$\gamma |y|^2 \leqslant [y, y] \leqslant \gamma^{-1} |y|^2,$$

where $|y|^2 = \Sigma_1^n y_i^2$. We have

$$(Ay, Ay) = [Ay, y] \leqslant [y, y]^{1/2} [Ay, Ay]^{1/2}$$
$$= (y, Ay)^{1/2} (Ay, A^2 y)^{1/2} \leqslant (y, Ay)^{1/2} \gamma^{-1/2} (Ay, Ay)^{1/2},$$

so that $(Ay, Ay) \leqslant \gamma^{-1}(y, Ay)$ for any vector y. But then from (4.14) we get

$$\gamma^{-1} \frac{dQ_\xi(t)}{dt} \geqslant \int_\Omega |A(\nabla_x \ln G)|^2 G dx.$$

And since according to Hölder's inequality

$$\int_\Omega |A(\nabla_x \ln G)| G dx \leqslant \left(\int_\Omega |A(\nabla_x \ln G)|^2 G dx\right)^{1/2} \times \left(\int_\Omega G dx\right)^{1/2}$$
$$= \left(\int_\Omega |A(\nabla_x \ln G)|^2 G dx\right)^{1/2},$$

it follows that

$$\gamma^{-1} \frac{dQ_\xi(t)}{dt} \geqslant \left(\int_\Omega |A(\nabla_x \ln G)| G dx\right)^2 = \left(\int_\Omega |A(\nabla_x G)| dx\right)^2.$$

On the other hand, according to Lemma 10 the moment $\mu_\xi(t)$ is an absolutely continuous function of t and inequality (4.4) holds. Consequently

$$\gamma^{-1}\frac{dQ_\xi(t)}{dt}\geqslant\left(\frac{d\mu_\xi(t)}{dt}\right)^2. \tag{4.15}$$

3) Let

$$\mathcal{R}_\xi(t)=Q_\xi(t)-k_1-\ln\mathcal{G}_{-1}(t).$$

Estimate (4.6) implies that $\mathcal{R}_\xi(t)\geqslant 0$ for all $t\geqslant 0$. Since the function

$$\mathcal{G}(v)=\mathcal{J}[\mathcal{F}^2](v)=\int_0^v\frac{\mathcal{F}^2(\zeta)}{\zeta}\,d\zeta$$

together with its first derivative is continuous and $\mathcal{G}'_v=\mathcal{F}^2(v)/v\neq 0$ for $v>0$, it follows that $\mathcal{G}_{-1}(t)$ together with its first derivative is also continuous for $t>0$ and

$$\frac{d\mathcal{G}_{-1}(t)}{dt}=\frac{1}{\mathcal{G}'_v(\mathcal{G}_{-1}(t))}=\frac{\mathcal{G}_{-1}(t)}{\mathcal{F}^2(\mathcal{G}_{-1}(t))}.$$

Consequently from Lemma 10 we get that $\mathcal{R}_\xi(t)$ is absolutely continuous and that for $t>0$

$$\frac{d\mathcal{R}_\xi(t)}{dt}=\frac{dQ_\xi(t)}{dt}-\frac{1}{\mathcal{F}^2(\mathcal{G}_{-1}(t))}.$$

And from the inequalities immediately following Lemma 8 we get that

$$\frac{dQ_\xi(t)}{dt}-\frac{d\mathcal{R}_\xi(t)}{dt}=\frac{1}{\mathcal{F}^2_\xi(\mathcal{G}_{-1})(t))}\leqslant\frac{1}{2\varepsilon_0\mathcal{G}(\mathcal{G}_{-1}(t))}=\frac{1}{2\varepsilon_0 t}. \tag{4.16}$$

Since $\mu_\xi(t)$ is absolutely continuous for $t\geqslant\delta>0$, we have for any $\delta>0$

$$\mu_\xi(t)-\mu_\xi(\delta)=\int_\delta^t\frac{d\mu_\xi(t)}{dt}dt\leqslant\int_\delta^t\left(\gamma^{-1}\frac{dQ_\xi(t)}{dt}\right)^{1/2}dt$$

$$\leqslant\gamma^{-1/2}\int_\delta^t\left[\frac{d\mathcal{R}_\xi(t)}{dt}+\left(\frac{dQ_\xi(t)}{dt}-\frac{d\mathcal{R}_\xi(t)}{dt}\right)\right]^{1/2}dt=\gamma^{-1/2}\int_\delta^t\left[\frac{d\mathcal{R}_\xi(t)}{dt}+\frac{1}{2\varepsilon_0 t}\right]^{1/2}dt.$$

Here we have used (4.15) and (4.16). From the fact that the inequality

$$a+b\leqslant a+b+\frac{a^2}{4b}=\left(\sqrt{b}+\frac{a}{2\sqrt{b}}\right)^2$$

holds for any positive number b and any a it follows that

$$(a+b)^{1/2}\leqslant\sqrt{b}+\frac{a}{2\sqrt{b}},$$

i.e.

$$\mu_\xi(t)-\mu_\xi(\delta)\leqslant\gamma^{-1/2}\int_\delta^t\left[\frac{1}{\sqrt{2\varepsilon_0 t}}+\frac{d\mathcal{R}_\xi(t)}{dt}\sqrt{\frac{\varepsilon_0 t}{2}}\right]dt$$

$$=\sqrt{\frac{2t}{\gamma\varepsilon_0}}+\sqrt{\frac{\varepsilon_0}{2\gamma}}\int_\delta^t\frac{d\mathcal{R}_\xi(t)}{dt}\sqrt{t}\,dt. \tag{4.17}$$

Consider

$$\int_\delta^t \frac{d\mathcal{R}_\xi(t)}{dt}\sqrt{t}\,dt = \mathcal{R}_\xi(t)\sqrt{t} - \mathcal{R}_\xi(\delta)\sqrt{\delta} - \frac{1}{2}\int_\delta^t \frac{\mathcal{R}_\xi(t)}{\sqrt{t}}\,dt.$$

Since $\mathfrak{R}_\xi(t) \geqslant 0$ for all $t > 0$, it follows that

$$\int_\delta^t \frac{d\mathcal{R}_\xi(t)}{dt}\sqrt{t}\,dt \leqslant \sqrt{t}\,\mathcal{R}_\xi(t)$$

for all $\delta > 0$. Consequently from (4.17), taking into account the fact that $\mu_\xi(\delta) \to 0$ as $\delta \to 0$ (Lemma 10), we obtain the inequality

$$\mu_\xi(t) \leqslant \sqrt{t}\left(\sqrt{\frac{2}{\gamma\varepsilon_0}} + \sqrt{\frac{\varepsilon_0}{2\gamma}}\,\mathcal{R}_\xi(t)\right). \tag{4.18}$$

At the same time it follows from (4.13) that

$$\mu_\xi(t) \geqslant d_3\mathcal{F}\,[\exp[\mathcal{R}_\xi(t) + k + \ln \mathcal{G}_{-1}(t)]] \geqslant d_4\mathcal{F}\,[\mathcal{G}_{-1}(t)\exp(\mathcal{R}_\xi(t))].$$

From inequality (2.20) we have

$$\mathcal{F}\,(\mathcal{G}_{-1}(t)\exp(\mathcal{R}_\xi(t))) \geqslant \exp[\varepsilon_0\mathcal{R}_\xi(t)]\,\mathcal{F}\,(\mathcal{G}_{-1}(t)) = \exp[\varepsilon_0\mathcal{R}_\xi(t)]\,[\mathcal{F}^2(\mathcal{G}_{-1}(t))]^{1/2}$$

and from (4.16)

$$\mathcal{F}^2[\mathcal{G}_{-1}(t)] \geqslant 2\varepsilon_0 t.$$

Consequently

$$\mu_\xi(t) \geqslant \frac{d_5}{d(\xi)}\sqrt{t}\exp[\varepsilon_0\mathcal{R}_\xi(t)], \tag{4.19}$$

where d_5 depends on ϵ_0 and γ. Comparing (4.19) with (4.18), we get that

$$d_6\exp[\varepsilon_0\mathcal{R}_\xi(t)] \leqslant d(\xi)\,(2/\varepsilon_0 + \mathcal{R}_\xi(t)),$$

i.e.

$$\mathcal{R}_\xi(t) \leqslant d_7\ln(d(\xi) + 1), \tag{4.20}$$

where d_7 depends only on ϵ_0 and γ (recall that $d(\xi) \geqslant \epsilon_0^{(1-\epsilon_0)/\epsilon_0} > 0$).

By virtue of (4.18), (4.20) and (4.19), (4.6), we obtain inequality (4.5), where $k_0 = d_5/d$ and $k = d_1\ln(d(\xi) + 1)$, with d_1 depending only on ϵ_0 and γ.

§5. Estimates from below

PROOF OF THEOREM 3. Take an arbitrary $t > 0$ and an arbitrary $\xi \in \Omega$. Let

$$\sup_{x\in\Omega\cap(|x-\xi|<R)} G(t, x; 0, \xi),$$

where $G(t, x; \tau, \xi)$ is the Green function of problem (3.1), (3.2). Then from the identity $\int_\Omega G(t, x; 0, \xi)dx \equiv 1$ we get that for any $R > 0$

$$1 - \int_\Omega G(t, x; 0, \xi)\,dx \leqslant M_{R,\xi}(t)\,v_\xi(R) + \mu_\xi(t)/R, \tag{5.1}$$

Recall that $v_\xi(R) = \operatorname{meas}_n[\Omega \cap (|x-\xi| < R)]$. Substituting into (5.1) the estimate from above in Theorem 5 of the moment $\mu_\xi(t)$ and taking $R = 2k\sqrt{t}$, we get

$$1 \leqslant M_{2k\sqrt{t},\,\xi}(t)\, v_\xi(2k\sqrt{t}) + {}^1/_2,$$

or

$$M_{2k\sqrt{t},\,\xi}(t) \geqslant \frac{1}{2v_\xi(2k\sqrt{t})}. \tag{5.2}$$

Consider the function $v_\xi(R)$. Since $\Omega \in \mathfrak{B}(g)$, from inequality (13) we have

$$\frac{dv_\xi(R)}{dR} = s_\xi(R) \leqslant d(\xi)\frac{\mathscr{T}_{-1}(R)}{R} \quad \text{and} \quad v_\xi(0) = 0,$$

which implies

$$v_\xi(R) \leqslant d(\xi)\int_0^R \frac{\mathscr{T}_{-1}(\zeta)}{\zeta}\, d\zeta = d(\xi)\,\mathcal{J}[\mathscr{T}_{-1}](R). \tag{5.3}$$

The function $g(v)$ satisfies condition b) $(\Omega \in \mathfrak{B}(g))$, and hence (2.21) holds. Therefore from (2.21) and (2.25) we get

$$\int_0^R \frac{\mathscr{T}_{-1}(\zeta)}{\zeta}\, d\zeta = \mathcal{J}[\mathscr{T}_{-1}](R) \leqslant \mathscr{T}_{-1}(R)$$

and from (5.3)

$$v_\xi(R) \leqslant d(\xi)\,\mathscr{T}_{-1}(R). \tag{5.4}$$

Substituting (5.4) into (5.2), we have

$$\sup_{x\in\Omega} G(t, x; 0, \xi) \geqslant M_{2k\sqrt{t},\,\xi}(t) \geqslant \frac{1}{2d(\xi)\,\mathscr{T}_{-1}(2k\sqrt{t})}. \tag{5.5}$$

Estimate (14) can be obtained from (5.5) with the use of inequalities (22) and (2.21). Thus Theorem 3 is proved, the constant k in the statement of the theorem being taken from inequalities (4.5).

Proof of Theorem 4. Suppose to begin with that the coefficients of equation (3.1) satisfy condition A) and that the initial function $\varphi(x)$ is continuous and has compact support. Consider the expression

$$\mu(t) = \int_\Omega |x|\,|u(t, x)|\, dx.$$

From Theorem 5 we get

$$\begin{aligned}
\mu(t) &= \int_\Omega |x| \left| \int_\Omega G(t, x; 0, \xi)\,\varphi(\xi)\, d\xi \right| dx \\
&\leqslant \int_\Omega |\varphi(\xi)| \int_\Omega (|x-\xi| + |\xi|)\, G(t, x; 0, \xi)\, dx\, d\xi \\
&= \int_\Omega |\varphi(\xi)|\,(\mu_\xi(t) + |\xi|)\, d\xi \leqslant d_1\sqrt{t}\int_\Omega |\varphi(\xi)| \ln(d(\xi)+1)\, d\xi \\
&+ \int_\Omega |\xi|\,|\varphi(\xi)|\, d\xi = a\sqrt{t} + b.
\end{aligned} \tag{5.6}$$

Since $\int_\Omega \varphi(x)\,dx > 0$, as in the proof of Theorem 3 we have

$$0 < \int_\Omega \varphi(\xi)\,d\xi = \int_\Omega u(t, x)\,dx \leqslant M_R(t)\,v(R) + \mu(t)/R,$$

where

$$M_R(t) = \sup_{x\in\Omega\cap(|x|<R)} u(t, x).$$

Using estimate (5.6) and setting

$$R = \frac{2(a\sqrt{t} + b)}{\int_\Omega \varphi(x)\,dx}$$

we get

$$\int_\Omega \varphi(x)\,dx \leqslant M_R(t)\,v(R) + \frac{1}{2}\int_\Omega \varphi(x)\,dx,$$

or, using (5.4),

$$\sup_{x\in\Omega} u(t, x) \geqslant M_R(t) \geqslant \frac{1}{2}\int_\Omega \varphi(x)\,dx \Big/ v\left(\frac{2(a\sqrt{t}+b)}{\int_\Omega \varphi(x)\,dx}\right)$$

$$\geqslant \frac{1}{2d(0)}\int_\Omega \varphi(x)\,dx \Big/ \mathcal{F}_{-1}\left(\frac{2(a\sqrt{t}+b)}{\int_\Omega \varphi(x)\,dx}\right). \tag{5.7}$$

Since the constant d_1 in Theorem 5, and hence the constants a and b in (5.7), do not depend on the coefficients of the equation (they are one and the same for all equations of form (3.1) satisfying condition (4) with a given constant of ellipticity γ), it follows as in §3 that the condition of smoothness of the coefficients can be dropped in (5.7). It is sufficient that the coefficients $a_{ij}(t, x)$ be bounded measurable functions satisfying condition (4).

Suppose now that $\varphi(x)$ is an arbitrary function from $L_1(\Omega)$ with finite integrals (17) and such that $\int_\Omega \varphi(x)\,dx > 0$. Let $\varphi_m(x)$ be a sequence of continuous functions with compact supports that converges to $\varphi(x)$ in the norm of $L_1(\Omega)$ and is such that as $m \to \infty$

$$\int_\Omega |x|\,|\varphi_m(x)|\,dx \to \int_\Omega |x|\,|\varphi(x)|\,dx, \tag{5.8}$$

$$\int_\Omega |\varphi_m(x)| \ln(d(x)+1)\,dx \to \int_\Omega |\varphi(x)| \ln(d(x)+1)\,dx, \tag{5.9}$$

and

$$\int_\Omega \varphi_m(x)\,dx > 0. \tag{5.10}$$

It is easy to see that such a sequence can be chosen. In fact, let

$$\varphi_{R_m}(x)=\varphi(x) \quad \text{for} \quad |x|<R_m, \quad x\in\Omega$$

and

$$\varphi_{R_m}(x)=0 \quad \text{for} \quad |x|>R_m, \quad x\in\Omega,$$

where $R_m \to \infty$ as $m\to\infty$. The finiteness of the integrals (17) and condition (16) imply relations (5.8)–(5.10) for sufficiently large m. We now approximate each of the functions $\varphi_{R_m}(x)$ in $L_1(\Omega)$ by a continuous function $\varphi_m(x)$ with the same support.

Let $u_m(t, x)$ denote the solution of problem (3.1), (3.2), (3) with the initial function $\varphi_m(x)$. It follows from Theorem 1 that the sequence of solutions $u_m(t, x)$ converges uniformly in $x\in\Omega$ and $t\geqslant\delta>0$ to a weak solution $u(t, x)$. Each of the functions $u_m(t, x)$ satisfies (5.7). And since $\mathcal{F}_{-1}(t)$ is a continuous function, we can pass to the limit as $m\to\infty$ and conclude that (5.7) also holds for the solution $u(t, x)$.

The left inequality in (18) is obtained from (5.7) with the use of (2.21). Theorem 4 is proved.

We note that we have in fact proved the stronger estimate

$$\sup_{x\in\Omega\cap(|x|<R)} u(t, x)\geqslant\frac{c_1}{\mathcal{F}_{-1}(\sqrt{t}+a_1)},$$

where

$$R=2(a\sqrt{t}+b)\Big/\int_\Omega\varphi(x)\,dx,$$

$$a=d_1\int_\Omega|\varphi(x)|\ln(d(x)+1)\,dx,$$

$$b=\int_\Omega|x||\varphi(x)|\,dx.$$

Estimates (9′) of the Corollary can be directly obtained from estimates (9) and (12) by virtue of (5.4).

§6. Examples

Suppose $n=2$ and $\Omega=\{(x, y)\colon y>f(x)\}$, where $f(x)$ is a continuously differentiable function such that

$$f(-x)=f(x); \qquad f(0)=0; \tag{6.1}$$
$$f'(x_1)\leqslant f'(x_2) \quad \text{for} \quad x_1\leqslant x_2.$$

We wish to estimate the function $l(v)$. Recall that

$$l(v)=\inf \operatorname{meas}_1[\partial Q\cap\Omega],$$
$$\operatorname{meas}_2 Q=v,$$

where Q is an otherwise arbitrary open subset of Ω and ∂Q is the boundary of Q. We will denote $\operatorname{meas}_1[\partial Q\cap\Omega]$ by $l(Q)$, and $\operatorname{meas}_2 Q$ by $v(Q)$.

Suppose first that Q is a subdomain of Ω. If $\partial Q \cap \Gamma = \varnothing$ or $\partial Q \cap \Gamma = \mathfrak{M}$, where $\mathfrak{M}$ is a point, then $l(Q) = \text{meas}_1 \partial Q$ and

$$v(Q) \leqslant \frac{l^2(Q)}{4\pi}. \tag{6.2}$$

When $\partial Q \cap \Gamma \neq \varnothing$ and $\partial Q \cap \Gamma \neq \mathfrak{M}$, we let $\mathfrak{M}_1 = (x_1, y_1)$ and $\mathfrak{M}_2 = (x_2, y_2)$ denote the points of the intersection of ∂Q and Γ such that the inequality $x_1 \geqslant x \geqslant x_2$ is satisfied for any point $(x, y) \in \partial Q \cap \Gamma$. If at least one of the points $\mathfrak{M}_1$ and $\mathfrak{M}_2$ does not exist (i.e. $\partial Q \cap \Gamma$ is not bounded), then $\partial Q \cap \Omega$ has infinite length and we can assume that estimate (6.2) is satisfied. We denote the part of Γ included between the points $\mathfrak{M}_1$ and $\mathfrak{M}_2$ by $\Gamma_{\mathfrak{M}_1\mathfrak{M}_2}$, and the part of Ω bounded by the curve $\Gamma_{\mathfrak{M}_1\mathfrak{M}_2}$ and the straight line $\mathfrak{M}_1\mathfrak{M}_2$ by $\Omega_{\mathfrak{M}_1\mathfrak{M}_2}$. When $l(Q)$ is finite, the domain Q is bounded and there exists a connected curve S_1 belonging to $\partial Q \cap \Omega$ and joining the points $\mathfrak{M}_1$ and $\mathfrak{M}_2$ such that the subdomain Q_1 of Ω bounded by the curves S_1 and $\Gamma_{\mathfrak{M}_1\mathfrak{M}_2}$ contains Q. Consequently

$$l(Q_1) = \text{meas}_1 S_1 \leqslant l(Q), \quad \text{while} \quad v(Q_1) \geqslant v(Q).$$

Suppose $v(Q_1) \geqslant v(\Omega_{\mathfrak{M}_1\mathfrak{M}_2})$, and let S_2 denote the circular arc in Ω joining $\mathfrak{M}_1$ and $\mathfrak{M}_2$ such that the measure of the domain Q_2 bounded by S_2 and $\Gamma_{\mathfrak{M}_1\mathfrak{M}_2}$ is equal to $v(Q_1)$. Then $Q_2 \supset \Omega_{\mathfrak{M}_1\mathfrak{M}_2}$ and $l(Q_2) = \text{meas}_1 S_2 \leqslant l(Q_1)$. If $v(Q_1) \leqslant v(\Omega_{\mathfrak{M}_1\mathfrak{M}_2})$, we take the line segment $\mathfrak{M}_1\mathfrak{M}_2$ as S_2 (then $Q_2 = \Omega_{\mathfrak{M}_1\mathfrak{M}_2}$). Consequently, in either case,

$$l(Q_2) \leqslant l(Q_1) \leqslant l(Q),$$

while

$$v(Q_2) \geqslant v(Q_1) \geqslant v(Q).$$

Let R denote the radius of the circle on which the circular arc S_2 lies and let α denote the central angle subtended by S_2. When $S_2 = \mathfrak{M}_1\mathfrak{M}_2$, we take $\alpha = 0$. Then $l(Q_2) = \alpha R$ and

$$v(Q_2) = \frac{\alpha R^2}{2} - \frac{R^2 \sin \alpha}{2} + v(\Omega_{\mathcal{M}_1\mathcal{M}_2}). \tag{6.3}$$

Suppose $\alpha > \pi$ and consider the circular segment $Q_2 \backslash \Omega_{\mathfrak{M}_1\mathfrak{M}_2}$. The area

$$v(Q_2 \setminus \Omega_{\mathcal{M}_1\mathcal{M}_2}) = \frac{R^2}{2}(\alpha - \sin \alpha) = \frac{l^2(Q_2)}{2\alpha^2}(\alpha - \sin \alpha).$$

For a fixed arc length $l(Q_2) = l_0$ the function $l_0^2(\alpha - \sin \alpha)/2\alpha^2$ monotone decreases with increasing $\alpha \in [\pi, 2\pi]$. In fact,

$$\frac{d}{d\alpha}\left[\frac{l_0^2}{2}\left(\frac{1}{\alpha} - \frac{\sin \alpha}{\alpha^2}\right)\right] = \frac{l_0^2}{2}\left(-\frac{1}{\alpha^2} + \frac{2 \sin \alpha}{\alpha^3} - \frac{\cos \alpha}{\alpha^2}\right)$$

$$= \frac{l_0^2}{2\alpha^3}[-\alpha(1 + \cos \alpha) + 2 \sin \alpha)] \leqslant 0$$

for $\alpha \in [\pi, 2\pi]$. Consequently

$$v(Q_2 \setminus \Omega_{\mathcal{M}_1\mathcal{M}_2}) = \frac{l^2(Q_2)}{2\alpha^2}(\alpha - \sin\alpha) \leqslant \frac{l^2(Q_2)}{2\pi}. \tag{6.4}$$

Let $\mathfrak{M}_3 = (x_3, y_3)$ and $\mathfrak{M}_4 = (x_4, y_4)$ denote the points on Γ such that $x_3 > x_1$, $x_4 < x_2$, the straight line $\mathfrak{M}_3\mathfrak{M}_4$ is parallel to $\mathfrak{M}_1\mathfrak{M}_2$ and its length is equal to $2l(Q_2)/\pi$. Then $\Omega_{\mathfrak{M}_3\mathfrak{M}_4} \supset \Omega_{\mathfrak{M}_1\mathfrak{M}_2}$. Let Q_3 denote the subdomain of Ω bounded by $\Gamma_{\mathfrak{M}_3\mathfrak{M}_4}$ and the semicircle S_3 in Ω joining $\mathfrak{M}_3$ and $\mathfrak{M}_4$. We have

$$l(Q_3) = \operatorname{meas}_1(S_3) = l(Q_2),$$

$$v(Q_3) = \frac{l^2(Q_2)}{2\pi} + v(\Omega_{\mathcal{M}_3\mathcal{M}_4}) \geqslant \frac{l^2(Q_2)}{2\pi} + v(\Omega_{\mathcal{M}_1\mathcal{M}_2})$$

and from (6.3) and (6.4) we get $v(Q_3) \geqslant v(Q_2)$.

If $\alpha \leqslant \pi$, we take $Q_3 = Q_2$ ($\mathfrak{M}_3 = \mathfrak{M}_1, \mathfrak{M}_4 = \mathfrak{M}_2, S_3 = S_2$), so that

$$v(Q_3) \geqslant v(Q), \quad l(Q_3) \leqslant l(Q). \tag{6.5}$$

Let $2h$ denote the length of the line segment $\mathfrak{M}_3\mathfrak{M}_4$, let $\mathfrak{M}$ denote the point of intersection of the tangents to Γ at $\mathfrak{M}_3$ and $\mathfrak{M}_4$ and let β denote the angle $\mathfrak{M}_3\mathfrak{M}\mathfrak{M}_4$. Clearly $\beta \leqslant \pi$. We consider two cases.

1) $\beta \geqslant \pi/2$. The area of the triangle $\mathfrak{M}_3\mathfrak{M}\mathfrak{M}_4$ is not greater than $h_2/\tan(\beta/2)$. From the convexity of Γ we get that $\Omega_{\mathfrak{M}_3\mathfrak{M}_4}$ is contained in the triangle $\mathfrak{M}_3\mathfrak{M}\mathfrak{M}_4$, and hence

$$v(Q_3) < \pi h^2/2 + h^2/\tan(\beta/2) \leqslant h^2\left(\frac{\pi}{2}+1\right) \leqslant \frac{1}{4}\left(\frac{\pi}{2}+1\right) l^2(Q_3). \tag{6.6}$$

2) $\beta \leqslant \pi/2$. In this case condition (6.1) directly implies that $x_4 < 0$. Since Γ is symmetric with respect to the y axis, we can suppose that $|x_4| < x_3$. It follows from the inequalities $x_3 - x_4 \leqslant 2h$ and $f(x_3) - f(x_4) \leqslant 2h$ and condition (6.1) that $|x_4| \leqslant h$, $f(x_4) \leqslant f(h)$ and

$$f(x_3) \leqslant f(x_4) + 2h \leqslant f(h) + 2h.$$

Therefore

$$v(Q_3) \leqslant \frac{\pi h^2}{2} + [f(x_3) + f(x_4)]\,h \leqslant \frac{\pi h^2}{2} + 2h(f(h) + h) = 2hf(h) + h^2\left(\frac{\pi}{2}+2\right). \tag{6.7}$$

From the inequalities (6.2) and (6.5)–(6.7) we get that for an arbitrary domain $Q \subset \Omega$

$$v(Q) \leqslant l^2(Q)\frac{1}{4}\left(\frac{\pi}{2}+2\right) + l(Q) f\left(\frac{l(Q)}{2}\right) \leqslant \left(\frac{\pi}{4}+1\right)\max\left\{l^2(Q),\ l(Q) f\left(\frac{l(Q)}{2}\right)\right\}$$
$$\leqslant \left(\frac{\pi}{4}+1\right)\max\{l^2(Q),\ l(Q) f(l(Q))\}. \tag{6.8}$$

Let $\widetilde{f}(x) = xf(x)$ and suppose Q is an arbitrary open subset of Ω. Then $Q = \bigcup_i Q^i$, where the Q^i are connected. Since (6.8) holds for each Q^i and $\widetilde{f}'(x)$ is a monotone increasing function for $x > 0$ by virtue of the fact that $f'(x)$ is monotone increasing, we get

$$v(Q)=\sum_i v(Q^i)\leqslant\left(\frac{\pi}{4}+1\right)\sum_i \max\{l^2(Q^i),\ \tilde f(l(Q^i))\}$$
$$\leqslant\left(\frac{\pi}{4}+1\right)\max\{l^2(Q),\ \tilde f(l(Q))\},$$

which implies

$$l(Q)\geqslant\min\left\{2\sqrt{\frac{v(Q)}{\pi+4}},\ \tilde f_{-1}\left(\frac{v(Q)}{1+\pi/4}\right)\right\},\tag{6.8'}$$

where $\tilde f_{-1}(v)$ is the inverse of the function $\tilde f(l)$. From condition (6.1) we see that either the function $\max\{x^2, \tilde f(x)\}$ is equal to x^2 for all $x>0$ or there exists a point $x_0>0$ such that $\max\{x^2, \tilde f(x)\}=x^2$ for $x\leqslant x_0$ and $\max\{x^2, \tilde f(x)\}=\tilde f(x)$ for $x\geqslant x_0$. It follows that the function

$$\min\left\{\sqrt{\frac{v}{1+\pi/4}},\ \tilde f\left(\frac{v}{1+\pi/4}\right)\right\}$$

is equal to $\sqrt{v/(1+\pi/4)}$ for sufficiently small v and therefore satisfies condition a).

From (6.8') we get

$$l(v)=\inf_{\mathrm{meas}_2 Q=v}\mathrm{meas}_1[\partial Q\cap\Omega]\geqslant\min\left\{\sqrt{\frac{v}{1+\pi/4}},\ \tilde f_{-1}\left(\frac{v}{1+\pi/4}\right)\right\},$$

i.e. $\Omega\in\mathfrak{A}(g)$, where

$$g(v)=\min\left\{\sqrt{\frac{v}{1+\pi/4}},\ \tilde f_{-1}\left(\frac{v}{1+\pi/4}\right)\right\}.$$

Suppose $\lambda\geqslant 1$ and $x>0$ are arbitrary numbers. By the convexity of $f(x)$

$$f(\lambda x)\geqslant\lambda f(x).\tag{6.10}$$

Therefore

$$\tilde f(\lambda x)=\lambda x f(\lambda x)\geqslant\lambda^2\tilde f(x).$$

Since

$$g_{-1}(x)=(1+\pi/4)\begin{cases}x^2 & \text{for } x\leqslant x_0\\ \tilde f(x) & \text{for } x\geqslant x_0\end{cases}$$

(if $g_{-1}(x)=(1+\pi/4)x^2$ for all $x>0$, we take $x_0=\infty$), it follows that $g_{-1}(\lambda x)\geqslant\lambda^2 g_{-1}(x)$, which implies

$$g(v)=\begin{cases}\sqrt{\dfrac{v}{1+\pi/4}} & \text{for } \dfrac{v}{1+\pi/4}\leqslant x_0^2,\\ \tilde f_{-1}\left(\dfrac{v}{1+\pi/4}\right) & \text{for } \dfrac{v}{1+\pi/4}\geqslant x_0^2\end{cases}$$

and $g(\lambda v)\leqslant\lambda^{1/2}g(v)$.

Consequently $\sqrt{v/g(v)}$ is a monotonely nondecreasing function and $g(v)$ satisfies condition b) with $\epsilon_0=1/2$.

Consider the monotone increasing function

$$\mathcal{F}(v)=\frac{v}{g(v)}=\begin{cases}\sqrt{1+\pi/4}\sqrt{v} & \text{for } \dfrac{v}{1+\pi/4}\leqslant x_0^2,\\ \dfrac{v}{\tilde{f}_{-1}\left(\dfrac{v}{1+\pi/4}\right)} & \text{for } \dfrac{v}{1+\pi/4}\geqslant x_0^2.\end{cases}$$

Then the function

$$\mathcal{F}_{-1}(R)=\begin{cases}\dfrac{R^2}{1+\pi/4} & \text{for } R\leqslant(1+\pi/4)\,x_0,\\ \left(\dfrac{v}{\tilde{f}_{-1}\left(\dfrac{v}{1+\pi/4}\right)}\right)_{-1}(R) & \text{for } R\geqslant(1+\pi/4)\,x_0\end{cases}$$

is also monotone increasing, and hence

$$g(\mathcal{F}_{-1}(R))=\frac{\mathcal{F}_{-1}(R)}{R}=\begin{cases}\dfrac{R}{1+\pi/4} & \text{for } R\leqslant(1+\pi/4)\,x_0,\\ \dfrac{1}{R}\left(\dfrac{v}{\tilde{f}_{-1}\left(\dfrac{v}{1+\pi/4}\right)}\right)_{-1}(R) & \text{for } R\geqslant(1+\pi/4)\,x_0.\end{cases}$$

Suppose $R \geqslant (\pi/4+1)x_0$. Then $\mathcal{F}_{-1}(R)\geqslant(1+\pi/4)x_0^2$ and the function

$$t=g(\mathcal{F}_{-1}(R))=\tilde{f}_{-1}\left(\frac{\mathcal{F}_{-1}(R)}{1+\pi/4}\right)=\tilde{f}_{-1}\left(\frac{\left(\dfrac{v}{\tilde{f}_{-1}\left(\dfrac{v}{1+\pi/4}\right)}\right)_{-1}(R)}{1+\pi/4}\right),$$

so that

$$(1+\pi/4)\tilde{f}(t)=\left(\frac{v}{\tilde{f}_{-1}\left(\dfrac{v}{1+\pi/4}\right)}\right)_{-1}(R)$$

and

$$R=\frac{(1+\pi/4)\tilde{f}(t)}{\tilde{f}_{-1}(\tilde{f}(t))}=\frac{(1+\pi/4)\tilde{f}(t)}{t}=(1+\pi/4)f(t),$$

which implies

$$g(\mathcal{F}_{-1}(R))=\min\left\{\frac{R}{1+\pi/4},\ f_{-1}\left(\frac{R}{1+\pi/4}\right)\right\}=\begin{cases}\dfrac{R}{1+\pi/4} & \text{for } R\leqslant\left(\dfrac{\pi}{4}+1\right)x_0,\\ f_{-1}\left(\dfrac{R}{1+\pi/4}\right) & \text{for } R\geqslant\left(\dfrac{\pi}{4}+1\right)x_0.\end{cases}\tag{6.11}$$

Let $\xi=(x_\xi, y_\xi)$ denote an arbitrary point of Ω and let $S_\xi(R)$ denote the circle of radius R with center at ξ. We have

$$s_\xi(R)=\operatorname{meas}_1\{S_\xi(R)\cap\Omega\}\leqslant\operatorname{meas}_1 S_\xi(R)=2\pi R.\tag{6.12}$$

Suppose

$$g(v)=\sqrt{\frac{v}{1+\pi/4}}$$

(i.e. $x_0=\infty$). Then

$$\frac{\mathcal{F}_{-1}(R)}{R}=\frac{R}{1+\pi/4}$$

and hence

$$s_\xi(R) \leqslant 2\pi(1+\pi/4)\frac{\mathcal{F}_{-1}(R)}{R},$$

so that

$$\Omega \in \mathfrak{B}\left(\sqrt{\frac{v}{1+\pi/4}}\right),$$

with the constant $d(\xi) = \pi(2 + \pi/2)$ and not depending on ξ. Suppose now $x_0 < \infty$. From the evenness of $f(x)$ we can suppose that $x_\xi \geqslant 0$. Take $R \geqslant 2f_{-1}(|\xi|)$. Then the points $(x_\xi \pm R, y_\xi)$ are not contained in Ω, and the monotonicity of $f(x)$ implies that the boundary Γ intersects the semicircle $S_\xi(R) \cap (y > y_\xi)$ at exactly two points, which we denote by $\mathfrak{M}_1 = (x_1, y_1)$ and $\mathfrak{M}_2 = (x_2, y_2)$ $(x_2 < 0 < x_1, y_\xi < y_2 < y_1)$. Since

$$s_\xi(R) \leqslant 2\,\mathrm{meas}_1\{S_\xi(R) \cap (y > y_\xi) \cap \Omega\} < \pi\,|\mathcal{M}_1\mathcal{M}_2|,$$

while the angle between the vector $\overrightarrow{\mathfrak{M}_2\mathfrak{M}_1}$ and the positive direction of the x axis is less than $\pi/4$, it follows that

$$s_\xi(R) \leqslant \pi\frac{2x_1}{\cos\pi/4} < 2\sqrt{2}\,\pi f_{-1}(R+|\xi|) < 4\pi f_{-1}(R+|\xi|).$$

Since

$$4\pi f_{-1}(R+|\xi|) > 4\pi f_{-1}(|\xi|) = 2\pi R$$

for $R = 2f_{-1}(|\xi|) > 0$, it follows from estimate (6.12) and the convexity of the function $f_{-1}(y)$ that

$$s_\xi(R) < 4\pi f_{-1}(R+|\xi|) \tag{6.13}$$

for all $R > 0$. From inequality (6.10) we get that

$$f_{-1}(\lambda y) \leqslant \lambda f_{-1}(y) \tag{6.14}$$

for any $y > 0$ and $\lambda \geqslant 1$. Suppose $R \geqslant (1+\pi/4)x_0$. It follows from (6.14) that

$$\begin{aligned} f_{-1}(R+|\xi|) &= f_{-1}\left((1+\pi/4)(1+|\xi|/R)\frac{R}{1+\pi/4}\right) \\ &\leqslant (1+\pi/4)(1+|\xi|/R)f_{-1}\left(\frac{R}{1+\pi/4}\right) \leqslant (1+\pi/4+|\xi|/x_0)f_{-1}\left(\frac{R}{1+\pi/4}\right), \end{aligned}$$

and from (6.13) that

$$s_\xi(R) \leqslant 4\pi(1+\pi/4+|\xi|/x_0)f_{-1}\left(\frac{R}{1+\pi/4}\right). \tag{6.15}$$

Making use of (6.12) for $R \leqslant (1+\pi/4)x_0$ and (6.15) for $R \geqslant (1+\pi/4)x_0$, we get

$$s_\xi(R) \leqslant \begin{cases} 2\pi(1+\pi/4)\dfrac{R}{1+\pi/4} & \text{for } R \leqslant (1+\pi/4)x_0 \\ 4\pi(1+\pi/4+|\xi|/x_0)f_{-1}\left(\dfrac{R}{1+\pi/4}\right) & \text{for } R \geqslant (1+\pi/4)x_0. \end{cases}$$

Therefore, taking (6.11) into account,

$$s_\xi(R) \leqslant 4\pi(1+\pi/4+|\xi|/x_0)\frac{\mathcal{F}_{-1}(R)}{R}.$$

We have thus proved the following assertion.

LEMMA 11. *A domain* $\Omega = \{(x, y):\ y > f(x)\}$, *where the function* $f(x)$ *together with its first derivative is continuous and satisfies condition* (6.1), *belongs to the class* $\mathfrak{B}(g)$, *the function* $g(v)$ *being given by relation* (6.9) *and*

$$d(\xi) = 4\pi(1 + \pi/4 + |\xi|/x_0).$$

Theorems 1 and 3 together with Lemma 11 imply that the Green function $G(t, x; 0, \xi)$ of problem (3.1), (3.2) satisfies the inequalities

$$c_0 \max\{t^{-1}, (\sqrt{t} f_{-1}(\sqrt{t}))^{-1}\} \leqslant c_0 \max\left\{\frac{1+\pi/4}{t}, \left(\sqrt{t} f_{-1}\left(\frac{\sqrt{t}}{1+\pi/4}\right)\right)^{-1}\right\}$$
$$\leqslant \sup_{x\in\Omega} G(t, x; 0, \xi) \leqslant C_0 \max\left\{\frac{1+\pi/4}{t}, \left(\sqrt{t} f_{-1}\left(\frac{\sqrt{t}}{1+\pi/4}\right)\right)^{-1}\right\}$$
$$\leqslant C_0(1+\pi/4)\max\{t^{-1}, (\sqrt{t} f_{-1}(\sqrt{t}))^{-1}\}. \tag{6.16}$$

Here we have made use of formulas (6.11) and (6.14).

EXAMPLE 1. Suppose $\alpha \geqslant 1$ and $f(x) = x^\alpha$ for $x > 0$. Then estimates (6.16) take the form

$$c_0 \max\left\{t^{-1}, t^{-\frac{\alpha+1}{2\alpha}}\right\} \leqslant \sup_{x\in\Omega} G(t, x; 0, \xi) \leqslant C_0(1+\pi/4)\max\left\{t^{-1}, t^{-\frac{\alpha+1}{2\alpha}}\right\}.$$

EXAMPLE 2. Suppose $\alpha \geqslant 1$ and a function $f(x)$ satisfying condition (6.1) has the form

$$y = f(x) = x^\alpha(1 + \delta(x))\ln^\beta x \tag{6.17}$$

for $x > 1$, where $\delta(x) \to 0$ as $x \to \infty$, while $\beta > 0$ when $\alpha = 1$. Then the function $f_{-1}(y)$ has the form

$$f_{-1}(y) = \alpha^{\beta/\alpha} y^{1/\alpha}(1 + \epsilon(y))\ln^{-\beta/\alpha} y \tag{6.18}$$

for $y > 1$, where $\epsilon(y) \to 0$ as $y \to \infty$. In fact, substituting (6.18) into (6.17), we get

$$y = \alpha^\beta y \ln^{-\beta} y(1+\varepsilon(y))^\alpha\left[\frac{\beta}{\alpha}\ln\alpha + \frac{1}{\alpha}\ln y - \frac{\beta}{\alpha}\ln\ln y + \ln(1+\varepsilon(y))\right]^\beta[1+\delta(f_{-1}(y))]$$
$$= y(1+\varepsilon(y))^\alpha[1 + o(1) + \alpha\ln^{-1} y\ln(1+\varepsilon(y))]^\beta[1+\delta(f_{-1}(y))].$$

Since $y = f(x) \to \infty$ as $x \to \infty$, $f_{-1}(y) \to \infty$ as $y \to \infty$. Consequently

$$(1+\varepsilon(y))^{\alpha/\beta}[1 + o(1) + \alpha\ln^{-1} y\ln(1+\varepsilon(y))] \to 1 \quad \text{as} \quad y \to \infty,$$

which implies $\epsilon(y) \to 0$ as $y \to \infty$. Thus in this case estimates (6.16) take the form

$$\text{const}\, t^{-\frac{\alpha+1}{2\alpha}} \ln^{\beta/\alpha} t \leqslant \sup_{x\in\Omega} G(t, x; 0, \xi) \leqslant \text{const}\, t^{-\frac{\alpha+1}{2\alpha}} \ln^{\beta/\alpha} t$$

for sufficiently large t.

EXAMPLE 3. Suppose a function $f(x)$ satisfying condition (6.1) has the form

$$f(x) = e^x(1 + o(1)).$$

for sufficiently large x. (For example, $f(x) = \sinh x$.) Then

$$\frac{\text{const}}{\sqrt{t}\ln t} \leqslant \sup_{x\in\Omega} G(t, x; 0, \xi) \leqslant \frac{\text{const}}{\sqrt{t}\ln t}$$

for sufficiently large t.

§7. Appendix

PROOF OF LEMMAS 3 AND 10. The proof of Lemma 3 is almost indistinguishable from the proof of Lemma 3 in [1]. The conditions on the domain Ω, the coefficients of equation (1) and the function $q(t, x)$ ensure (see [7]) the possibility of representing the Green function $G(t, x; \tau, \xi)$ of problem (1), (2) in the form

$$G(t, x; \tau, \xi) = \Gamma(t, x; \tau, \xi) + \int_0^t \int_\Gamma \Gamma(t, x; \theta, \eta)\psi(\theta, \eta; \tau, \xi)\, dS_\eta\, d\theta, \tag{7.1}$$

where $\Gamma(t, x; \tau, \xi)$ is a fundamental solution of equation (1) and the density $\psi(t, x; \tau, \xi)$ is determined from the integral equation

$$-\frac{1}{2}\psi(t, x; \tau, \xi) = \int_0^t \int_\Gamma \left[\frac{\partial}{\partial N_x} + q(t, x)\right]\Gamma(t, x; \theta, \eta)\psi(\theta, \eta; \tau, \xi)\, dS_\eta d\theta + \left[\frac{\partial}{\partial N_x} + q(t, x)\right]\Gamma(t, x; \tau, \xi).$$

Suppose $0 < t \leqslant T$ and $0 \leqslant \tau < t$. Then (see [4], [7]) the following estimates of the fundamental solution and its derivatives are valid:

$$\Gamma(t, x; \tau, \xi) \leqslant \frac{\text{const}}{(t-\tau)^{n/2}} \exp\left[-\varkappa_0 \frac{|x-\xi|^2}{t-\tau}\right], \tag{7.2}$$

$$|\nabla_x \Gamma(t, x; \tau, \xi)| \leqslant \frac{\text{const}}{(t-\tau)^{\frac{n+1}{2}}} \exp\left[-\varkappa_0 \frac{|x-\xi|^2}{t-\tau}\right], \tag{7.3}$$

$$\left|\frac{\partial \Gamma(t, x; \tau, \xi)}{\partial N_x}\right| \leqslant \frac{\text{const}}{|t-\tau|^{1-\nu}\,|x-\eta|^{n-1+2\nu-\mu_0}}, \tag{7.4}$$

the latter estimate holding for $x, \eta \in \Gamma$ and $0 < 2\nu < \mu_0$, while in each estimate const and κ_0 depend on T. The density $\psi(t, x; \tau, \xi)$ is given by the formula

$$\psi(t, x; \tau, \xi) = \text{const}\left\{\int_0^t \int_\Gamma \mathfrak{N}(t, x; \theta, \eta) \times \left[\frac{\partial}{\partial N_\eta} + q(\theta, \eta)\right]\Gamma(\theta, \eta; \tau, \xi)\, dS_\eta d\theta + \left[\frac{\partial}{\partial N_x} + q(t, x)\right]\Gamma(t, x; \tau, \xi)\right\},$$

where $\mathfrak{N}(t, x; \theta, \eta)$ is the sum of the Neumann series.

Let K denote a compactum contained in Ω, let R_0 denote a positive number such that $K \subset \{|x| \leqslant R_0\}$ and let ρ_0 denote the distance from K to Γ. From estimates (7.3) and (7.4) we get that the following inequalities hold for $\xi \in K$ and $|x| > R_0 + 2$:

$$|\psi(t, x; 0, \xi)| \leqslant \text{const}\left\{\frac{\exp\left[-\varkappa_0 \frac{|x-\xi|^2}{t}\right]}{t^{(n+1)/2}}\right.$$

$$\left.+\int_0^t\int_\Gamma |\mathfrak{N}(t, x; \theta, \eta)| \frac{\exp\left[-\varkappa_0\frac{|\eta-\xi|^2}{\theta}\right]}{\theta^{(n+1)/2}}\, dS_\eta d\theta\right\}$$

$$\leqslant \text{const}\left\{\exp\left[-\varkappa_1\frac{|x|^2}{t}\right]+\int_0^t\int_{\Gamma\cap(|x-\eta|<1)}\frac{\exp\left[-\varkappa\frac{(|\eta|-R)^2}{\theta}\right]}{|t-\theta|^{1-\nu}|x-\eta|^{(n-1+2\nu-\mu_0)}}\, dS_\eta d\theta\right.$$

$$+\int_0^t\int_{\Gamma\cap(|x-\eta|\geqslant 1)\cap\left(|\eta|\geqslant\frac{|x|}{2}\right)}\frac{\exp\left[-\varkappa\frac{|\eta|^2}{\theta}\right]}{|x-\eta|^{\mu_1}\rho_0^{n+1}}\, dS_\eta d\theta$$

$$\left.+\int_0^t\int_{\Gamma\cap(|x-\eta|\geqslant 1)\cap\left(|\eta|\leqslant\frac{|x|}{2}\right)}\frac{\exp\left[-\varkappa\frac{(\rho_0^2+|\eta|^2)}{\theta}\right]}{|x-\eta|^{\mu_1}\rho_0^{n+1}}\, dS_\eta d\theta\right\}$$

$$\leqslant\text{const}\left\{\exp\left[-\varkappa_1\frac{|x|^2}{t}\right]+\exp\left[-\varkappa\frac{(|x|-R-1)^2}{t}\right]\right.$$

$$+\exp\left[-\varkappa\frac{|x|^2}{4t}\right]\int_{\Gamma\cap(|x-\eta|\geqslant 1)}\frac{dS_\eta}{|x-\eta|^{\mu_1}}$$

$$\left.+\exp\left[-\varkappa_1\frac{\rho_0^2}{t}\right]\frac{1}{|x|^{\mu_1}}\int_\Gamma\exp\left[-\varkappa\frac{|\eta|^2}{T}\right]dS_\eta\right\}\leqslant\text{const}\frac{\exp\left[-\varkappa_1\frac{\rho_0^2}{t}\right]}{|x|^{\mu_1}},$$

where μ_1 is a sufficiently large positive number and const and the exponents κ and κ_1 depend on T, ρ_0 and R_0.

Analogously, when $|x| \leqslant R_0 + 2$ we obtain the estimate

$$|\psi(t, x; 0, \xi)| \leqslant \text{const}\exp\left[-\varkappa_1\frac{\rho_0^2}{t}\right].$$

Consequently for any $x \in \Gamma$ $(\xi \in K)$

$$|\psi(t, x; 0, \xi)| \leqslant \frac{\text{const}}{(1+|x|)^{\mu_1}}\exp\left[-\varkappa_1\frac{\rho_0^2}{t}\right]. \tag{7.5}$$

From (7.2), (7.3) and (7.5) we get, as in [1], that (i) the integrals (2.1) converge uniformly in $t \in [\delta, T]$, $\delta > 0$, and in $\xi \in K$, and (ii) the integral

$$\int_{\Omega}\left[\int_{\Omega} G(t, x; 0, \xi)\varphi(\xi)\,d\xi\right]dx$$

converges uniformly in $t \in [0, T]$ for any function $\varphi(\xi)$ that is continuous and has compact support. We note that an additional condition on Ω that was used in [1] has not been used here. Consider the integral

$$I = \int_{\Omega\cap(R<|x|<R+1)} \sum_{i=1}^{n}\left|\frac{\partial G(t, x; 0, \xi)}{\partial x_i}\right| dx.$$

From the representation (7.1) and estimates (7.3) and (7.5) we get that for $0 < \delta \leqslant t \leqslant T$ and sufficiently large $R > 2$

$$I \leqslant \text{const} \int_{\Omega\cap(R<|x|<R+1)} \left\{\frac{1}{t^{\frac{n+1}{2}}}\exp\left[-\varkappa_0\frac{|x-\xi|^2}{t}\right]\right.$$
$$\left.+\int_0^t\int_{\Gamma}(t-\theta)^{-\frac{n+1}{2}}(|\eta|+1)^{-\mu_1}\exp\left[-\varkappa_0\frac{|x-\eta|^2}{t-\theta}\right]dS_\eta d\theta\right\}dx$$
$$\leqslant \text{const}\left\{[\exp(-\varkappa_1 R^2)]R^n + \int_{\Gamma}(|\eta|+1)^{-\mu_1}\int_{\Omega\cap(R<|x|<R+1)}\int_0^t \tau^{-\frac{n+1}{2}}\right.$$
$$\times\exp\left(-\varkappa_0\frac{|x-\eta|^2}{\tau}\right)d\tau\,dx\,dS_\eta.$$

Since

$$\int_0^t \tau^{-\frac{n+1}{2}}e^{-a^2/\tau}d\tau = \frac{1}{a^{n-1}}\int_{a^2/t}^{\infty}\zeta^{\frac{n-3}{2}}e^{-\zeta}d\zeta$$
$$\leqslant \frac{\exp(-a^2/2t)}{a^{n-1}}\int_0^{\infty}\zeta^{\frac{n-3}{2}}e^{-\zeta/2}d\zeta \qquad (n \geqslant 2), \tag{7.6}$$

it follows that

$$\int_{\Omega\cap(R<|x|<R+1)}\int_0^t \tau^{-\frac{n+1}{2}}\exp\left[-\varkappa_0\frac{|x-\eta|^2}{\tau}\right]d\tau\,dx$$
$$\leqslant \int_{\Omega\cap(R<|x|<R+1)} \text{const}\,|x-\eta|^{-(n-1)}\exp\left[-\varkappa_0\frac{|x-\eta|^2}{2t}\right]dx$$

and, when $R/2 < |\eta|$,

$$\int_{\Omega\cap(R<|x|<R+1)}|x-\eta|^{-(n-1)}dx \leqslant \text{const}\,R^n,$$

while when $|\eta| \leqslant R/2$ and $|x-\eta| \geqslant |x| - |\eta| \geqslant R/2$,

$$\int\limits_{\Omega\cap(R<|x|<R+1)} \frac{1}{|x-\eta|^{n-1}} \exp\left(-\varkappa_0 \frac{|x-\eta|^2}{2t}\right) dx \leqslant \text{const}\, R \exp\left(-\varkappa_0 \frac{R^2}{8T}\right).$$

Consequently

$$I \leqslant \text{const}\left\{\left(\int\limits_{\Gamma\cap(|\eta|\leqslant R/2)} dS_\eta\right) R\,[\exp(-\varkappa_1 R^2)] + \frac{R^n}{R^{\mu_1-\mu_2}} \int\limits_{\Gamma\cap(|\eta|>R/2)} \frac{dS_\eta}{(1+|\eta|)^{\mu_2}}\right. \tag{7.7}$$

and hence, choosing μ_1 sufficiently large, we conclude that (2.2) holds uniformly in $t \in [\delta, T]$.

Thus Lemma 3 is proved.

Suppose now that $G(t, x; 0, \xi)$ is the Green function of problem (3.1), (3.2), that ξ is a fixed point of Ω and that ρ_0 is the distance from ξ to Γ. Using the representation (7.1) and estimates (7.2) and (7.5), we obtain the inequality

$$G(t, x; 0, \xi) \leqslant \text{const}\left\{\frac{1}{t^{n/2}}\left[\exp\left(-\varkappa_0 \frac{|x-\xi|^2}{t}\right)\right]\right.$$
$$\left. + \int\limits_0^t \int\limits_\Gamma \frac{1}{(t-\theta)^{n/2}} \left[\exp\left(-\varkappa_0 \frac{|x-\eta|^2}{t-\theta}\right)\right] \frac{1}{(|\eta|+1)^{\mu_1}} \exp\left(-\varkappa_1 \frac{\rho_0^2}{\theta}\right) dS_\eta\, d\theta.\right.$$

Hence, placing the origin at ξ (we denote $x - \xi$ by y) and using (7.6), we get that for $|y| = |x - \xi| > 0$ and $0 < t \leqslant T$

$$G(t, x; 0, \xi) \leqslant \text{const}\left\{\frac{1}{t^{n/2}}\left[\exp\left(-\varkappa_0 \frac{|y|^2}{t}\right)\right]\right.$$
$$\left. + \int\limits_0^t \int\limits_\Gamma \frac{1}{(t-\theta)^{n/2}} \left[\exp\left(-\varkappa_0 \frac{|y-(\eta-\xi)|^2}{t-\theta} - \varkappa_1 \frac{\rho_0^2}{t}\right)\right] \frac{dS_\eta\, d\theta}{(1+|\eta-\xi|)^{\mu_1}}\right\}$$
$$\leqslant \text{const}\left\{\frac{1}{t^{n/2}}\left[\exp\left(-\varkappa_0 \frac{|y|^2}{t}\right)\right] + \left[\exp\left(-\varkappa_1 \frac{\rho_0^2}{t}\right)\right]\right.$$
$$\left. \times \int\limits_\Gamma \frac{\exp\left(-\varkappa_0 \dfrac{|y-\zeta|^2}{2t}\right)}{|y-\zeta|^{n-2}} \frac{dS_\eta}{(1+|\zeta|)^{\mu_1}}\right\} \tag{7.8}$$
$$= \text{const}\left\{\frac{1}{t^{n/2}} \exp\left(-\varkappa_0 \frac{|y|^2}{t}\right) + \exp\left(-\varkappa_1 \frac{\rho_0^2}{t}\right) I_1(y, t)\right\}.$$

We estimate the integral I_1 for $|y| \geqslant 2$:

$$\begin{aligned}
I_1(t, y) &= \int_{\Gamma\cap(|y-\zeta|<1)} \exp\left[-\varkappa_0 \frac{|y-\zeta|^2}{2t}\right] \frac{dS_\zeta}{|y-\zeta|^{n-2}(|\zeta|+1)^{\mu_1}} \\
&+ \int_{\Gamma\cap(|y-\zeta|>y/2)} \exp\left[-\varkappa_0 \frac{|y-\zeta|^2}{2t}\right] \frac{dS_\zeta}{|y-\zeta|^{n-2}(|\zeta|+1)^{\mu_1}} \\
&+ \int_{\Gamma\cap(1\leqslant|y-\zeta|\leqslant y/2)} \exp\left[-\varkappa_0 \frac{|y-\zeta|^2}{2t}\right] \frac{dS_\zeta}{|y-\zeta|^{n-2}(|\zeta|+1)^{\mu_1}} \\
&\leqslant \operatorname{const}\left\{\frac{1}{|y|^{\mu_1}} \int_{\Gamma\cap(|y-\zeta|<1)} \frac{dS_\zeta}{|y-\zeta|^{n-2}} + \exp\left[-\varkappa_0 \frac{|y|^2}{8t}\right] \int \frac{dS_\zeta}{(|\zeta|+1)^{\mu_1}}\right. \\
&\left. + \frac{1}{|y|^{\mu_1}} \int_\Gamma \exp\left[-\varkappa_0 \frac{|y-\zeta|^2}{2T}\right] dS_\zeta\right\} \\
&\leqslant \operatorname{const}\left[\frac{1}{|y|^{\mu_1}} + \exp(-\varkappa_1 |y|^2)\right] \leqslant \frac{\operatorname{const}}{|y|^{\mu_1}}.
\end{aligned}$$

Since I_1 is bounded for $|y| \leqslant 2$, it follows upon substituting the above estimates into (7.8) that

$$G(t, x; 0, \xi) \leqslant \operatorname{const}\left\{\frac{1}{t^{n/2}} \exp\left[-\varkappa_0 \frac{|x-\xi|^2}{t}\right] + \exp\left[-\varkappa_1 \frac{\rho_0^2}{t}\right] \frac{1}{(|x-\xi|+1)^{\mu_1}}\right\}. \tag{7.9}$$

Consequently for any $\alpha > 0$ the integral $\int_\Omega G^\alpha(x, t; 0, \xi)\,dx$ converges uniformly in $t \in [\delta, T]$ $(\delta > 0)$. Consider the entropy $Q(t) = -\int_\Omega G \ln G dx$. Since $|G \ln G| < \operatorname{const} G^\alpha$ for $G \leqslant 1$ and $\alpha < 1$, while $|G \ln G| < \operatorname{const} G^\beta$ for $G \geqslant 1$ and $\beta > 1$, we see that $|G \ln G| \leqslant \operatorname{const}(G^\alpha + G^\beta)$ and hence that the integral $\int_\Omega G \ln G dx$ converges absolutely and uniformly in $t \in [\delta, T]$. Thus the function $Q(t)$ is continuous for $t > 0$.

We note that it immediately follows from (7.9) that for any $\epsilon > 0$ there exists a constant $C = C(\epsilon)$ such that $|Q(t)| \leqslant C(\epsilon) t^{-\epsilon}$ for all $t \in (0, T]$.

Also, the integral $\int_\Omega |x-\xi| G(t, x; 0, \xi)\,dx$ converges uniformly in $t \in [\delta, T]$, i.e. the moment

$$\mu_\xi(t) = \int_\Omega |x-\xi|\, G(t, x; 0, \xi)\, dx$$

is a continuous function for $t > 0$. Further,

$$\begin{aligned}
\int_\Omega |x-\xi|\, G(t, x; 0, \xi)\, dx &\leqslant \operatorname{const}\left\{\frac{1}{t^{n/2}} \int_\Omega |x-\xi| \exp\left[-\varkappa_0 \frac{|x-\xi|^2}{t}\right] dx\right. \\
&\left. + \exp\left[-\varkappa_1 \frac{\rho_0^2}{t}\right] \int \frac{|x-\xi|}{(|x-\xi|+1)^{\mu_1}}\, dx\right\} \leqslant \operatorname{const}\left\{\frac{1}{t^{n/2}} \int_{E_n} |y| \exp\left[-\varkappa_0 \frac{|y|^2}{t}\right] dy\right. \\
&\left. + \exp\left[-\varkappa_1 \frac{\rho_0^2}{t}\right] \int_{E_n} \frac{dy}{(|y|+1)^{\mu_1-1}}\right\} \leqslant \operatorname{const}\left\{\sqrt{t} + \exp\left[-\varkappa_1 \frac{\rho_0^2}{t}\right]\right\} \leqslant \operatorname{const}\sqrt{t},
\end{aligned}$$

which implies $\lim_{t\to 0} \mu_\xi(t) = 0$.

Suppose the origin is located at the point ξ, let $\chi(\theta)$ denote a smooth function such that $\chi(\theta) \in [0, 1]$ for all real θ, $\chi(\theta) = 0$ for $0 \leqslant -1$ and $\chi(\theta) = 1$ for $\theta \geqslant 0$, and let $\mu_{R,0}(t)$ denote the integral

$$\int_\Omega \chi(R - |y|)|y|G(t, y; 0, 0)dy.$$

Since the integral $\int_\Omega |y|G(t, y, 0, 0)dy$ converges uniformly in $t \in [\delta, T]$ $(\delta > 0)$, it follows that

$$\mu_{R,0}(t) \rightrightarrows \mu_0(t) \quad \text{as} \quad R \to \infty \tag{7.10}$$

uniformly in $t \in [\delta, T]$.

Consider the derivative

$$\begin{aligned}
\frac{d\mu_R(t)}{dt} &= \int_\Omega \chi(R - |y|)|y|G_t(t, y; 0, 0)\,dy \\
&= \int_\Omega \chi(R - |y|)|y|(\nabla_y A(t, y)\nabla_y)G(t, y; 0, 0)\,dy \\
&= -\int_\Omega \chi(R - |y|)(\nabla_y |y| A(t, y)\nabla_y G(t, y; 0, 0))\,dy \\
&\quad - \int_\Omega |y|(\nabla_y \chi(R - |y|)A(t, y)\nabla_y G(t, y; 0, 0))\,dy \\
&= -\int_{\Omega \cap (|y| < R)} (\nabla_y |y| A(t, y)\nabla_y G(t, y; 0, 0))\,dy \\
&\quad - \int_{\Omega \cap (R < |y| < R+1)} [(|y|\nabla_y \chi(R - |y|) \\
&\quad + \chi(R - |y|)\nabla_y |y|)A(t, y)\nabla_y G(t, y; 0, 0)]\,dy.
\end{aligned}$$

From the boundedness of $|\nabla_y \chi|$ and χ and inequality (4) we get that

$$\begin{aligned}
&\Bigg|\int_{\Omega \cap (R < |y| < R+1)} [(|y|\nabla_y \chi(R - |y|) \\
&\qquad + \chi(R - |y|)\nabla_y |y|)A(t, x)\nabla_y G(t, y; 0, 0)]\,dy\Bigg| \\
&\qquad \leqslant \text{const} \int_{\Omega \cap (R < |y| < R+1)} |y|\gamma^{-1}|\nabla_y G(t, y; 0, 0)|\,dy
\end{aligned}$$

for sufficiently large R, and from (7.7) we get that as $R \to \infty$ the integral in the right side of this inequality tends to 0 uniformly in $t \in [\delta, T]$. Choosing a sufficiently large μ_1 in (7.7), we get that the integral $\int_\Omega |\nabla_y G(t, y; 0, 0)|dy$ converges uniformly in $t \in [\delta, T]$ $(\delta > 0)$. Consequently, as $R \to \infty$,

$$\frac{d\mu_{R,0}(t)}{dt} \rightrightarrows -\int_\Omega (\nabla_y |y| A(t, y)\nabla_y G(t, y; 0, 0))\,dy = f(t) \tag{7.11}$$

uniformly in $t \in [\delta, T]$ $(\delta > 0)$, and $f(t)$ is continuous.

Take arbitrary positive numbers t and δ such that $t \in (\delta, T]$, and consider the equality

$$\mu_{R,0}(t) = \mu_{R,0}(\delta) + \int_\delta^t \frac{d\mu_{R,0}(t)}{dt}\,dt.$$

Passing to the limit as $R \to \infty$ and making use of relations (7.10) and (7.11), we get

$$\mu_0(t) = \mu_0(\delta) + \int_\delta^t f(t)\,dt.$$

Consequently

$$\frac{d\mu_0(t)}{dt} = f(t) = -\int_\Omega (\nabla_y |y| A(t, y) \nabla_y G(t, y; 0, 0)\,dy.$$

Therefore (4.4) follows from condition (4).

Let $Q_\sigma(t)$ denote the function $-\int_\Omega G \ln(G + \sigma) dy$. It is continuous for $t > 0$. Consider

$$Q_{\sigma,R}(t) = -\int_\Omega \chi(R - |y|) G(t, y; 0, 0) \ln(G + \sigma)\,dy.$$

Clearly

$$Q_{\sigma,R}(t) \rightrightarrows Q_\sigma(t) \quad \text{as} \quad R \to \infty \tag{7.12}$$

uniformly in $t \in [\delta, T]$. Differentiating $Q_{\sigma,R}(t)$ with respect to t, we get

$$\frac{dQ_{\sigma,R}(t)}{dt} = -\int_\Omega \chi(R - |y|)\Big[G_t \ln(G+\sigma) + \frac{G}{G+\sigma} G_t\Big] dy$$
$$= -\int_\Omega \chi(R - |y|)(\nabla_y (A \nabla_y G)) \ln(G+\sigma)\,dy$$
$$-\int_\Omega \chi(R - |y|) G_t dy + \sigma \int_\Omega \chi(R - |y|) \frac{(\nabla_y (A \nabla_y G))}{G+\sigma} dy$$
$$= \int_\Omega \frac{(\nabla_y G) A (\nabla_y G)}{G+\sigma} \chi(R - |y|)\,dy - \int_\Omega \chi(R - |y|) G_t\,dy$$
$$+ \sigma \int_\Omega \chi(R - |y|) \frac{(\nabla_y G) A (\nabla_y G)}{(\sigma + G)^2} dy + \int_\Omega (\nabla_y \chi) A (\nabla_y G)\Big[\ln(G+\sigma) - \frac{\sigma}{G+\sigma}\Big] dy.$$

In view of the fact that $\sigma > 0$ it follows from (4), (7.9) and (7.7) that as $R \to \infty$

$$\int_\Omega (\nabla_y \chi) A (\nabla_y G)\Big[\ln(G+\sigma) - \frac{\sigma}{G+\sigma}\Big] dy \rightrightarrows 0$$

uniformly in $t \in [\delta, T]$. From (2.6), since $b_i(t, x) - c(t, x) = 0$ and $q(t, x) = 0$, we get that

$$\int_\Omega \chi(R - |y|) G_t dy \rightrightarrows 0 \qquad \text{as} \qquad R \to \infty$$

uniformly in $t \in [\delta, T]$. While from the uniform convergence of the integral $\int_\Omega |\nabla G|^2 dx$ we have

$$\int_\Omega (\nabla_y G) A (\nabla_y G) \chi (R - |y|) \left[\frac{1}{G+\sigma} + \frac{\sigma}{(G+\sigma)^2}\right] dy$$

$$\rightrightarrows \int_\Omega (\nabla_y G) A (\nabla_y G) \left[\frac{1}{G+\sigma} + \frac{\sigma}{(G+\sigma)^2}\right] dy \quad \text{as} \quad R \to \infty.$$

Consequently, as $R \to \infty$,

$$\frac{dQ_{\sigma,R}(t)}{dt} \rightrightarrows \int_\Omega \frac{(\nabla_y G) A (\nabla_y G)}{G+\sigma} dy + \int_\Omega \frac{\sigma (\nabla_y G) A (\nabla_y G)}{(G+\sigma)^2} dy = f_\sigma(t) \tag{7.13}$$

uniformly in $t \in [\delta, T]$, and $f_\sigma(t)$ is continuous. Hence, from (7.12) and (7.13),

$$\frac{dQ_\sigma(t)}{dt} = f_\sigma(t) = \int_\Omega \frac{(\nabla_y G) A (\nabla_y G)}{G+\sigma} dy + \sigma \int_\Omega \frac{(\nabla_y G) A (\nabla_y G)}{(G+\sigma)^2} dy > 0. \tag{7.14}$$

Consider

$$\frac{\partial f_\sigma(t)}{\partial \sigma} = -\int_\Omega \frac{(\nabla_y G) A (\nabla_y G)}{(G+\sigma)^2} dy + \int_\Omega \frac{(\nabla_y G) A (\nabla_y G)}{(G+\sigma)^2} dy - 2\sigma \int_\Omega \frac{(\nabla_y G) A (\nabla_y G)}{(G+\sigma)^3} dy.$$

This equality holds for $\sigma > 0$ since the integrals in the right side converge uniformly in $\sigma \geqslant \sigma_0$ for an arbitrary number σ_0. Thus

$$\frac{\partial f_\sigma(t)}{\partial \sigma} = -2\sigma \int_\Omega \frac{(\nabla_y G) A (\nabla_y G)}{(G+\sigma)^3} dy < 0$$

and the family of functions $f_\sigma(t)$ monotone increases as $\sigma \to 0^+$.

Take an arbitrary pair of positive numbers t and t_1 such that $t > t_1$. From (7.14) we have

$$Q_\sigma(t) - Q_\sigma(t_1) = \int_{t_1}^{t} f_\sigma(t)\, d\tau. \tag{7.15}$$

The convergence of the integral $\int_\Omega G \ln G dy$ in (7.9) implies that the integral $\int_\Omega G \ln (G+\sigma) dy$ converges uniformly in $\sigma \in [0, \sigma_1]$ $(\sigma_1 > 0)$, and hence one can pass to the limit as $\sigma \to 0$ in (7.15) and write

$$Q(t) - Q(t_1) = \lim_{\sigma \to 0} \int_{t_1}^{t} f_\sigma(\tau)\, d\tau. \tag{7.16}$$

From the existence of a limit for the integral in (7.16) and the monotonicity with respect to σ of the family of positive functions $f_\sigma(t)$ we get that there exists an almost everywhere finite positive function $F(t)$ to which the family $f_\sigma(t)$ converges as $\sigma \to 0$ for almost all $t > 0$. According to a theorem of Levi, $F(t)$ is summable and

$$\int_{t_1}^{t} F(\tau)\, d\tau = \lim_{\sigma \to 0} \int_{t_1}^{t} f_\sigma(\tau)\, d\tau = Q(t) - Q(t_1)$$

for any $t > t_1 > 0$. This means that $Q(t)$ is absolutely continuous. Hence for almost all $t > 0$ the derivative of Q with respect to t exists and satisfies the relation

$$\frac{dQ(t)}{dt} = F(t) = \lim_{\sigma \to 0} \left\{ \int_{\Omega} \frac{(\nabla_y G) A (\nabla_y G)}{G + \sigma} dy + \int_{\Omega} \sigma \frac{(\nabla_y G) A (\nabla_y G)}{(G+\sigma)^2} dy \right\}. \quad (7.17)$$

Since both of the summands in the right side of (7.17) are positive, each of them has a finite limit for $\sigma \to 0$ for almost all $t > 0$.

Consider the function

$$g_\sigma(t, y) = \frac{(\nabla_y G) A (\nabla_y G)}{G + \sigma}.$$

Since

$$\frac{\partial g_\sigma(t, y)}{\partial \sigma} = -\frac{(\nabla_y G) A (\nabla_y G)}{(G+\sigma)^2} < 0,$$

the family of functions $g_\sigma(t, y)$ monotone increases to the function $(\nabla_y G) A (\nabla_y G)/G$ as $\sigma \to 0^+$. Consequently for almost all $t > 0$ (for which the first integral in (7.17) has a finite limit for $\sigma \to 0$) the function $(\nabla_y G) A (\nabla_y G)/G$ is summable with respect to y, and

$$\lim_{\sigma \to 0} \int_{\Omega} \frac{(\nabla_y G) A (\nabla_y G)}{G + \sigma} dy = \int_{\Omega} \frac{(\nabla_y G) A (\nabla_y G)}{G} dy. \quad (7.18)$$

Take any $\epsilon > 0$ and a $t > 0$ such that the integral standing in the right side of (7.18) converges. Then the monotonicity of the family $g_\sigma(t, y)$ with respect to σ implies the existence of a number $R = R(\epsilon) > 0$ such that for all $\sigma > 0$

$$0 < \sigma \int_{\Omega \cap (|y| > R)} \frac{(\nabla_y G) A (\nabla_y G)}{(G+\sigma)^2} dy \leqslant \int_{\Omega \cap (|y| > R)} \frac{(\nabla_y G) A (\nabla_y G)}{G + \sigma} dy < \int_{\Omega \cap (|y| > R)} \frac{(\nabla_y G) A (\nabla_y G)}{G} dy < \varepsilon.$$

Since $G(t, y; 0, 0) > 0$ for $t > 0$,

$$\sigma \int_{\Omega \cap (|y| \leqslant R)} \frac{(\nabla_y G) A (\nabla_y G)}{(G+\sigma)^2} \leqslant \frac{\sigma}{\min\limits_{y \in \Omega, |y| \leqslant R} G^2(t, y; 0, 0)} \int_{\Omega} (\nabla_y G) A (\nabla_y G)\, dy < \varepsilon$$

for sufficiently small σ. Consequently

$$\sigma \int_{\Omega} \frac{(\nabla_y G) A (\nabla_y G)}{(G+\sigma)^2} dy \to 0 \quad \text{as} \quad \sigma \to 0 \quad (7.19)$$

for almost all $t > 0$. Substituting (7.18) and (7.19) into (7.17), we obtain (4.3).

Thus Lemma 10 is proved.

BIBLIOGRAPHY

1. A. K. Guščin, *The rate of stabilization of a solution of a parabolic equation in an unbounded region,* Differencial′nye Uravnenija **6** (1970), 741–761 = Differential Equations **6** (1970), 567–581. MR **42** #4864.

2. ———, *On lower bounds for the solutions of boundary value problems for a parabolic equation,* Dokl. Akad. Nauk SSSR **203** (1972), 274–277 = Soviet Math. Dokl. **13** (1972), 368–372.

3. E. De Giorgi, *Sulla differenziabilità e l'analiticità delle estremali degli integrali multipli regolari,* Mem. Accad. Sci. Torino. Cl. Sci. Fis. Mat. (3) **3** (1957), 25–43; Russian transl., Matematika **4** (1960), no. 6, 23–38. MR **20** #172.

4. O. A. Ladyženskaja, V. A. Solonnikov and N. N. Ural′ceva, *Linear and quasilinear equations of parabolic type,* "Nauka", Moscow, 1967; English transl., Transl. Math. Monographs, vol. 23, Amer. Math. Soc., Providence, R. I., 1968. MR **39** #3159a, b.

5. J. Nash, *Continuity of solutions of parabolic and elliptic equations,* Amer. J. Math. **80** (1958), 931–954; Russian transl., Matematika **4** (1960), no. 1, 31–52. MR **20** #6592.

6. J. Moser, *A Harnack inequality for parabolic differential equations,* Comm. Pure Appl. Math. **17** (1964), 101–134. MR **28** #2357.

7. V. Pogožel′skiĭ, *Investigation of integrals of a parabolic equation and of boundary problems in an unbounded domain,* Mat. Sb. **47 (89)** (1959), 397–430. (Russian) MR **22** #2791.

Trudy Mat. Inst. Steklov.
126 (1973)

Proc. Steklov Inst. Math.
126 (1973)

ON THE RATE OF DECAY OF A VORTEX IN A VISCOUS FLUID

UDC 517.947.4

V. N. MASLENNIKOVA

ABSTRACT. In this paper the asymptotic behavior for large time is studied of the solution of the Cauchy problem for the linearized Navier-Stokes system describing the motion of a rotating viscous fluid. A theorem is proved giving the rate of decrease of the solution as $t \longrightarrow \infty$, uniformly with respect to all x in E_3.

Bibliography: 16 items.

In the present paper we study the solution of the linearized Navier-Stokes system

$$\begin{cases} \dfrac{\partial \mathbf{v}}{\partial t} - [\mathbf{v}, \ \omega] - \nu \Delta \mathbf{v} + \operatorname{grad} P = \mathbf{F}(x, \ t) \\ \operatorname{div} \mathbf{v} = 0, \end{cases} \tag{1}$$

describing the motion of a rotating viscous fluid in the region

$$\{x \in E_3, \quad t \geqslant 0\}$$

with the initial conditions

$$\mathbf{v}(x, \ t)|_{t=0} = \mathbf{v}^0(x), \qquad \operatorname{div} \mathbf{v}^0 = 0, \tag{2}$$

where $\mathbf{v}(x, t) = (v_1, v_2, v_3)$, $\boldsymbol{\omega}$ is a given constant rotation vector, which (see [1]) can be reduced to the form $\boldsymbol{\omega} = (0, 0, \omega)$ since the Laplace operator is invariant relative to an orthogonal linear transformation of the independent variables, and $\nu (= \text{const} > 0)$ is the kinematic viscosity. It will be assumed without loss of generality (as will be seen below) that $\omega > 0$. System (1) without the Coriolis terms $[\mathbf{v}, \boldsymbol{\omega}]$ was extensively studied from the mathematical point of view by Oseen [2] and Leray [3].

In particular, Oseen was the first to construct a fundamental solution of system (1) without taking rotation into account, the so-called Oseen tensor.

We will give an explicit representation of the solution of problem (1), (2) (i.e. we will construct a fundamental solution for system (1)) and then use this representation to study the asymptotic behavior of the solution as $t \longrightarrow \infty$.

Thus our purpose is to investigate the influence of the vector product $[\mathbf{v}, \boldsymbol{\omega}]$ on both the structure of a fundamental solution and the behavior of the solution of the Cauchy problem for $t \longrightarrow \infty$.

AMS (MOS) subject classifications (1970). Primary 35Q10, 76D05; Secondary 35B40.

We note that even in a system with viscosity, as well as in a Sobolev system [4], the presence of the terms $[\mathbf{v}, \omega]$ causes the solutions to have an oscillatory character and leads to the emergence of a vortex, whereas in the absence of these terms a solution of system (1) is in a definite sense analogous to a solution of the heat equation. We observe, however, that system (1) is not a parabolic system.

The following basic theorem will be proved.

THEOREM 1. *If the initial data* $\mathbf{v}_0(x) \in L_1(E_3)$ *and satisfy the natural consistency condition* $\operatorname{div} \mathbf{v}^0 = 0$, *the solution of the Cauchy problem for the homogeneous system* (1) $(\mathbf{F} \equiv 0)$ *satisfies the following asymptotic estimates for large* $t > 0$:

$$|\mathbf{v}(x, t)| \leqslant \frac{C_1}{(\nu t)^{3/2}\omega t} \|\mathbf{v}^0\|_{L_1(E_3)} \tag{3}$$

$$|P(x, t)| \leqslant \frac{C_2}{(\nu t)(\omega t)} \|\mathbf{v}^0\|_{L_1(E_3)}, \tag{4}$$

where the constants C_1 *and* C_2 *do not depend on* x *or* t.

Theorem 1 describes the rate of decay of a vortex in an incompressible viscous fluid.

In the course of proving Theorem 1 we prove two lemmas on the uniform (with respect to an unrestricted family of strongly oscillating functions) boundedness of some definite integrals. These lemmas have been set aside in a separate section since they are of independent mathematical interest.

The results of the present article were briefly announced in the note [5].

§1. Construction of the solution of the Cauchy problem

We assume that system (1) and the initial data (2) are such that they admit a Fourier transform with respect to x and a Laplace transform with respect to t.

In order to avoid encumbering with unnecessary calculations arguments that are analytically quite complicated as it is, we will consider the homogeneous case of system (1):

$$\begin{cases} \dfrac{\partial \mathbf{v}}{\partial t} - [\mathbf{v}, \boldsymbol{\omega}] - \nu \Delta \mathbf{v} + \operatorname{grad} P = 0, \\ \operatorname{div} \mathbf{v} = 0 \end{cases} \tag{1*}$$

with the nonhomogeneous initial data of (2). Duhamel's principle can later be used to pass over to the nonhomogeneous system (1).

The Fourier-Laplace transform of system (1*) with the initial conditions (2) has the form

$$\begin{cases} \lambda \tilde{\mathbf{v}}(\alpha, \lambda) - [\tilde{\mathbf{v}}, \boldsymbol{\omega}] + \nu |\alpha|^2 \tilde{\mathbf{v}}(\alpha, \lambda) + i\boldsymbol{\alpha} \tilde{P}(\alpha, \lambda) = \tilde{\mathbf{v}}(\boldsymbol{\alpha}), \\ i\alpha_1 \tilde{v}_1 + i\alpha_2 \tilde{v}_2 + i\alpha_3 \tilde{v}_3 = 0, \end{cases} \tag{5}$$

where $\boldsymbol{\alpha} = (\alpha_1, \alpha_2, \alpha_3)$ and $|\boldsymbol{\alpha}|^2 = \Sigma_1^3 \alpha_i^2$. Solving the algebraic system (5) and making use of the solenoidalness of the initial data of (2), we obtain a representation of the Fourier-Laplace transform of the solution in the form

$$\tilde{\mathbf{v}}(\alpha, \lambda) = \frac{|\alpha|^2 (\lambda + \nu |\alpha|^2)}{|\alpha|^2 (\lambda + \nu |\alpha|^2)^2 + \omega^2 \alpha_3^2} \tilde{\mathbf{v}}(\alpha) + \frac{\omega \alpha_3 [\tilde{\mathbf{v}}^0, \boldsymbol{\alpha}]}{|\alpha|^2 (\lambda + \nu |\alpha|^2)^2 + \omega^2 \alpha_3^2};$$

$$\tilde{P}(\alpha, \lambda) = \frac{i\omega\alpha_2 (\lambda + \nu |\alpha|^2)}{|\alpha|^2 (\lambda + \nu |\alpha|^2)^2 + \omega^2 \alpha_3^2} \tilde{v}_1^0(\alpha) - \frac{i\omega\alpha_1 (\lambda + \nu |\alpha|^2)}{|\alpha|^2 (\lambda + \nu |\alpha|^2)^2 + \omega^2 \alpha_3^2} \tilde{v}_2^0(\alpha)$$
$$- \frac{i\omega^2 \alpha_3}{|\alpha|^2 (\lambda + \nu |\alpha|^2)^2 + \omega^2 \alpha_3^2} \tilde{v}_3^0(\alpha). \qquad (6)$$

The inverse Laplace transform can be found quite easily, since (6) contains only rational expressions in λ.

Choosing $\operatorname{Re} \lambda \geqslant \lambda_0 > 0$ and using formulas for the inverse Laplace transforms of the rational functions appearing in (6):

$$L^{-1}\left\{\frac{1}{|\alpha|^2 (\lambda + \nu |\alpha|^2)^2 + \omega^2 \alpha_3^2}\right\} = \frac{1}{|\alpha| \omega \alpha_3} e^{-\nu|\alpha|^2 t} \sin \frac{\omega \alpha_3 t}{|\alpha|}, \qquad (7)$$
$$L^{-1}\left\{\frac{\lambda + \nu |\alpha|^2}{|\alpha|^2 (\lambda + \nu |\alpha|^2)^2 + \omega^2 \alpha_3^2}\right\} = \frac{1}{|\alpha|^2} e^{-\nu|\alpha|^2 t} \cos \frac{\omega \alpha_3}{|\alpha|} t,$$

we obtain a representation of the Fourier transform of the solution of the Cauchy problem in the form

$$\tilde{\mathbf{v}}(\alpha, t) = \tilde{\mathbf{v}}^0(\alpha) e^{-\nu|\alpha|^2 t} \cos \frac{\omega \alpha_3}{|\alpha|} t + [\tilde{\mathbf{v}}^0, \boldsymbol{\alpha}] e^{-\nu|\alpha|^2 t} \frac{1}{|\alpha|} \sin \frac{\omega \alpha_3}{|\alpha|} t.$$
$$\tilde{P}(\alpha, t) = (\tilde{v}_1^0(\alpha) \cdot \alpha_2 - \tilde{v}_2^0(\alpha) \alpha_1) \frac{i\omega}{|\alpha|^2} e^{-\nu|\alpha|^2 t} \cos \frac{\omega \alpha_3 t}{|\alpha|} - \tilde{v}_3^0(\alpha) \frac{i\omega}{|\alpha|} e^{-\nu|\alpha|^2 t} \sin \frac{\omega \alpha_3 t}{|\alpha|}. \qquad (8)$$

In order to determine $\mathbf{v}$ and P it is necessary to calculate the following kernels under the assumption $t > 0$:

$$K_1(x, t) = \frac{1}{(2\pi)^3} \int_{-\infty}^{+\infty} e^{i(\alpha, \mathbf{x})} e^{-\nu|\alpha|^2 t} \cos \frac{\omega \alpha_3 t}{|\alpha|} d\alpha, \qquad (9)$$

$$\mathbf{K}_2(x, t) = \frac{1}{(2\pi)^3} \int_{-\infty}^{+\infty} e^{i(\alpha, \mathbf{x}) - \nu|\alpha|^2 t} \frac{\boldsymbol{\alpha}}{|\alpha|} \sin \frac{\omega \alpha_3 t}{|\alpha|} d\alpha, \qquad (10)$$

$$K_{3_j}(x, t) = \frac{1}{(2\pi)^3} \int_{-\infty}^{+\infty} e^{i(\alpha, \mathbf{x}) - \nu|\alpha|^2 t} \frac{i\alpha_j \omega}{|\alpha|^2} \cos \frac{\omega \alpha_3 t}{|\alpha|} d\alpha, \qquad (11)$$

$$K_4(x, t) = \frac{1}{(2\pi)^3} \int_{-\infty}^{+\infty} e^{i(\alpha, \mathbf{x}) - \nu|\alpha|^2 t} \frac{i\omega}{|\alpha|} \sin \frac{\omega \alpha_3 t}{|\alpha|} d\alpha. \qquad (12)$$

Let us find $K_1(x, t)$. We use the following formula of Bochner [7] for multiple Fourier integrals:

$$\int_{-\infty}^{+\infty} \varphi(|\alpha|)\, e^{i(\alpha, x)} d\alpha_1 \dots d\alpha_n = \frac{(2\pi)^{n/2}}{r^{\frac{n-2}{2}}} \int_0^{\infty} \varphi(|\alpha|)\, |\alpha|^{n/2} J_{\frac{n-2}{2}}(|\alpha| r)\, d|\alpha|, \qquad (13)$$

where $r = (\Sigma_1^n x_i^2)^{1/2}$ and $J_{(n-2)/2}$ is the Bessel function of order $(n-2)/2$. Then

$$K_1(x, t) = \frac{1}{(2\pi)^2} \int_{-\infty}^{+\infty} e^{i\alpha_3 x_3 - \nu t \alpha_3^2} d\alpha_3 \int_0^{\infty} e^{-\nu t \beta^2} \beta J_0(\beta\rho) \cos \frac{\omega \alpha_3 t}{\sqrt{\beta^2 + \alpha_3^2}}\, d\beta, \qquad (14)$$

where $\rho^2 = x_1^2 + x_2^2$.

Now in (14) we pass over to polar coordinates by means of the relations $\alpha_3 = \lambda \cos\theta$ and $\beta = \lambda \sin\theta$, we replace $\cos(\omega t \cos\theta)$ by its exponential value

$$\cos(\omega t \cos\theta) = \frac{\exp(i\omega t \cos\theta) + \exp(-i\omega t \cos\theta)}{2}$$

and we use the following formula of Sonin ([8], §35) for cylindrical functions:

$$\int_0^{\pi} e^{iZ\cos\theta} J_{\frac{n-3}{2}}(z \sin\theta) \sin^{\frac{n-1}{2}}\theta\, d\theta = \sqrt{2\pi z}\; z^{\frac{n-3}{2}} \frac{J_{\frac{n-2}{2}}(\sqrt{z^2 + Z^2})}{(\sqrt{z^2 + Z^2})^{(n-2)/2}}. \qquad (15)$$

This gives us

$$K_1(x, t) = \frac{1}{(2\pi)^2} \int_0^{\infty} \lambda^2 e^{-\nu t \lambda^2} d\lambda \int_0^{\pi} e^{i x_3 \lambda \cos\theta} \sin\theta \cos(\omega t \cos\theta)\, J_0(\rho\lambda \sin\theta)\, d\theta$$

$$= \frac{1}{2(2\pi)^{3/2}} \int_0^{\infty} \lambda^2 e^{-\nu t \lambda^2} \left(\frac{J_{1/2}(\sqrt{\lambda^2 r^2 + 2\lambda x_3 \omega t + \omega^2 t^2})}{(\sqrt{\lambda^2 r^2 + 2\lambda x_3 \omega t + \omega^2 t^2})^{1/2}} + \frac{J_{1/2}(\sqrt{\lambda^2 r^2 - 2\lambda x_3 \omega t + \omega^2 t^2})}{(\sqrt{\lambda^2 r^2 - 2\lambda x_3 \omega t + \omega^2 t^2})^{1/2}} \right) d\lambda,$$

where $r^2 = \rho^2 + x_3^2$.

Making the substitution $\lambda = -\lambda_1$ in the second summand and introducing the notation

$$\Omega = \sqrt{\lambda^2 r^2 + 2\lambda x_3 \omega t + \omega^2 t^2}, \qquad (16)$$

we get

$$K_1(x, t) = \frac{1}{2(2\pi)^{3/2}} \int_{-\infty}^{+\infty} \lambda^2 e^{-\nu t \lambda^2} \frac{J_{1/2}(\Omega)}{\Omega^{1/2}}\, d\lambda. \qquad (17)$$

Before attacking the integral in (17), let us first reduce the other kernels $\mathbf{K}_2$, K_{3j} and K_4 to the same form. We calculate each component of the vector kernel $\mathbf{K}_2$ separately. We have

$$K_{21}(x,\ t)=\frac{1}{(2\pi)^3}\int\limits_{-\infty}^{+\infty} e^{i(\alpha,\, x)-\nu t|\alpha|^2}\cdot\frac{\alpha_1}{|\alpha|}\sin\frac{\omega\alpha_3 t}{|\alpha|}\,d\alpha.$$

We pass over to polar coordinates by means of the relations $\alpha_1 = \beta\cos\varphi$ and $\alpha_2 = \beta\sin\varphi$:

$$K_{21}=\frac{1}{(2\pi)^3}\int\limits_{-\infty}^{+\infty} e^{i\alpha_3 x_3-\nu t\alpha_3^2}\int\limits_0^{\infty} e^{-\nu t\beta^2}\cdot\frac{\beta^2}{\sqrt{\alpha_3^2+\beta^2}}\sin\frac{\omega t\alpha_3}{\sqrt{\alpha_3^2+\beta^2}}$$
$$\times\left(\int\limits_0^{2\pi} e^{i(x_1\beta\cos\varphi+x_2\beta\sin\varphi)}\cos\varphi d\varphi\right)d\beta d\alpha_3.$$

But

$$\int\limits_0^{2\pi} e^{i(x_1\beta\cos\varphi+x_2\beta\sin\varphi)}\cos\varphi d\varphi=2\pi i\frac{x_1}{\rho}J_1(\beta\rho), \tag{18}$$

where again $\rho = \sqrt{x_1^2 + x_2^2}$.

Taking into account (18), making the substitution $\alpha_3 = \lambda\cos\theta$, $\beta = \lambda\sin\theta$ once more and replacing $\sin(\omega t\cos\theta)$ by its exponential valuc

$$\sin(\omega t\cos\theta)=\frac{\exp(i\omega t\cos\theta)-\exp(-i\omega t\cos\theta)}{2i},$$

we get

$$K_{21}(x,t)=\frac{x_1}{2(2\pi)^2\rho}\int\limits_0^{\infty}\lambda^2 e^{-\nu t\lambda^2}d\lambda\int\limits_0^{\pi}\sin^2\theta$$
$$\times\left\{e^{i(x_3\lambda+\omega t)\cos\theta}-e^{i(x_3\lambda-\omega t)\cos\theta}\right\}J_1(\lambda\rho\sin\theta)\,d\theta.$$

We again apply Sonin's formula (15) for $n = 5$:

$$K_{21}(x,t)=\frac{x_1}{2(2\pi)^{3/2}}\int\limits_0^{\infty}\lambda^3 e^{-\nu t\lambda^2}\left[\frac{J_{3/2}\left(\sqrt{\lambda^2\rho^2+(\lambda x_3+\omega t)^2}\right)}{\left(\sqrt{\lambda^2\rho^2+(\lambda x_3+\omega t)^2}\right)^{3/2}}\right.$$
$$\left.-\frac{J_{3/2}\left(\sqrt{\lambda^2\rho^2+(\lambda x_3-\omega t)^2}\right)}{\left(\sqrt{\lambda^2\rho^2+(\lambda x_3-\omega t)^2}\right)^{3/2}}\right]d\lambda.$$

Making the substitution $\lambda = \lambda_1$ in the second summand and using the notation (16), we get

$$K_{21}(x,t)=\frac{x_1}{2(2\pi)^{3/2}}\int\limits_{-\infty}^{+\infty}\lambda^3 e^{-\nu t\lambda^2}\frac{J_{3/2}(\Omega)}{\Omega^{3/2}}\,d\lambda. \tag{19}$$

Analogously

$$K_{22}(x,t)=\frac{x_2}{2(2\pi)^{3/2}}\int\limits_{-\infty}^{+\infty}\lambda^3 e^{-\nu t\lambda^2}\frac{J_{3/2}(\Omega)}{\Omega^{3/2}}\,d\lambda. \tag{20}$$

The kernel K_{23} has a somewhat different structure. In order to calculate it we first apply Bochner's formula (13):

$$K_{23}(x,t)=\frac{1}{(2\pi)^3}\int_{-\infty}^{+\infty} e^{i(\alpha,x)-\nu t|\alpha|^2}\frac{\alpha_3}{|\alpha|}\sin\frac{\omega\alpha_3 t}{|\alpha|}\,d\alpha$$

$$=\frac{1}{(2\pi)^2}\int_{-\infty}^{+\infty}\alpha_3 e^{i\alpha_3 x_3-\nu t\alpha_3^2}d\alpha_3\int_0^\infty \frac{\beta J_0(\beta\rho)}{\sqrt{\beta^2+\alpha_3^2}}e^{-\nu t\beta^2}\sin\frac{\omega\alpha_3 t}{\sqrt{\beta^2+\alpha_3^2}}\,d\beta.$$

We again pass over to polar coordinates by means of the relations $\alpha_3 = \lambda\cos\theta$, $\beta = \lambda\sin\theta$ and we replace $\sin(\omega t\cos\theta)$ by its exponential value:

$$K_{23}(x,t)=\frac{1}{(2\pi)^2}\int_0^\infty \lambda^2 e^{-\nu t\lambda^2}d\lambda\int_0^\pi e^{ix_3\lambda\cos\theta}\sin\theta\cos\theta\sin(\omega t\cos\theta)\,J_0(\lambda\rho\sin\theta)\,d\theta$$

$$=\frac{1}{2i(2\pi)^2}\int_0^\infty \lambda^2 e^{-\nu t\lambda^2}d\lambda\int_0^\pi\left[e^{i(x_3\lambda+\omega t)\cos\theta}-e^{i(x_3\lambda-\omega t)\cos\theta}\right]\sin\theta\cos\theta J_0(\rho\lambda\sin\theta)\,d\theta.$$

Now, having added the arguments, we pass back from exponential functions to trigonometric functions:

$$K_{23}(x,t)=\frac{1}{2i(2\pi)^2}\int_0^\infty \lambda^2 e^{-\nu t\lambda^2}d\lambda\int_0^\pi\{\sin\theta\cos\theta J_0(\rho\lambda\sin\theta)[\cos[(x_3\lambda+\omega t)\cos\theta]$$
$$+i\sin[(x_3\lambda+\omega t)\cos\theta]-\cos[(x_3\lambda-\omega t)\cos\theta]$$
$$-i\sin[(x_3\lambda-\omega t)\cos\theta]]\}\,d\theta. \tag{21}$$

The integrals with respect to θ of the first and third summands are equal to zero since these summands are odd functions of $\gamma=\theta-\pi/2$. In fact, consider the integral of the first summand, letting $x_3\lambda+\omega t=z$. We have

$$\int_0^\pi \sin\theta\cos\theta J_0(\rho\lambda\sin\theta)\cos(z\cos\theta)\,d\theta$$

$$=-\int_{-\pi/2}^{\pi/2}\cos\gamma\sin\gamma J_0(\rho\lambda\cos\gamma)\cos(z\sin\gamma)\,d\gamma=0,$$

since J_0 is an even function of γ.

In the fourth summand in (21) we make the substitution $\lambda=-\lambda_1$. Then

$$K_{23}(x,t)=\frac{1}{(2\pi)^2}\int_{-\infty}^{+\infty}\lambda^2 e^{-\nu t\lambda^2}\int_0^{\pi/2}\sin\theta\cos\theta\sin[(x_3\lambda+\omega t)\cos\theta]J_0(\rho\lambda\sin\theta)\,d\theta\,d\lambda. \tag{22}$$

We now replace $\sin[(x_3\lambda+\omega t)\cos\theta]$ in (22) by the Bessel function of order 1/2 according to the formula

$$\sin\alpha=\sqrt{\frac{\pi\alpha}{2}}\,J_{1/2}(\alpha) \tag{23}$$

and use the following formula of Sonin ([8], §33) for integrals of Bessel functions:

$$\int_0^{\pi/2} J_\mu(z \sin\theta) J_\nu(Z\cos\theta) \sin^{\mu+1}\theta \cos^{\nu+1}\theta d\theta = \frac{z^\mu Z^\nu J_{\mu+\nu+1}(\sqrt{z^2+Z^2})}{(z^2+Z^2)^{\frac{\mu+\nu+1}{2}}},$$

$$\operatorname{Re}\mu,\ \operatorname{Re}\nu > -1, \tag{24}$$

when $\mu = 0$ and $\nu = 1/2$. This gives us

$$K_{23}(x, t) = \frac{1}{2(2\pi)^{3/2}} \int_{-\infty}^{+\infty} \lambda^2 (x_3\lambda + \omega t) e^{-\nu t\lambda^2} \frac{J_{3/2}(\Omega)}{\Omega^{3/2}} d\lambda. \tag{25}$$

Thus the components of $\mathbf{K}_2$ have the forms (19), (20) and (25).

In order to find $P(x, t)$ in the Cauchy problem we must also calculate the kernels K_{3j} and K_4 for $j = 1, 2$.

In formula (11) for K_{31} we pass over to polar coordinates by means of the relations $\alpha_1 = \beta\cos\varphi$, $\alpha_2 = \beta\sin\varphi$ and use formula (18):

$$K_{31}(x, t) = \frac{i\omega}{(2\pi)^3} \int_{-\infty}^{+\infty} e^{i\alpha_3 x_3 - \nu t\alpha_3^2} \int_0^\infty e^{-\nu t\beta^2} \cdot \frac{\beta^2}{\beta^2+\alpha_3^2}$$

$$\times \cos\frac{\omega t\alpha_3}{\sqrt{\beta^2+\alpha_3^2}} \int_0^{2\pi} e^{i\beta(x_1\cos\varphi + x_2\sin\varphi)} \cos\varphi\, d\varphi\, d\beta\, d\alpha_3$$

$$= -\frac{\omega x_1}{(2\pi)^2\rho} \int_{-\infty}^{+\infty} e^{i\alpha_3 x_3 - \nu t\alpha_3^2} \int_0^\infty J_1(\beta\rho) e^{-\nu t\beta^2} \frac{\beta^2}{\alpha_3^2+\beta^2} \cos\frac{\omega t\alpha_3}{\sqrt{\alpha_3^2+\beta^2}} d\beta\, d\alpha_3. \tag{26}$$

In (26) we pass over once more to polar coordinates by means of the relations $\alpha_3 = \lambda\cos\theta$, $\beta = \lambda\sin\theta$ and replace $\cos(\omega t\cos\theta)$ by its exponential value:

$$K_{31}(x, t) = -\frac{\omega x_1}{(2\pi)^2\rho} \int_0^\infty \lambda e^{-\nu t\lambda^2} d\lambda \int_0^\pi e^{ix_3\lambda\cos\theta} \sin^2\theta \cdot J_1(\lambda\rho\sin\theta)$$

$$\times \cos(\omega t\cos\theta)\, d\theta = -\frac{\omega x_1}{2(2\pi)^2\rho} \int_0^\infty \lambda e^{-\nu t\lambda^2} d\lambda \int_0^\pi \sin^2\theta J_1(\rho\lambda\sin\theta)$$

$$\times [e^{i(x_3\lambda+\omega t)\cos\theta} + e^{i(x_3\lambda-\omega t)\cos\theta}]\, d\theta. \tag{27}$$

We apply Sonin's formula (15) for $n = 5$ to (27):

$$K_{31}(x, t) = -\frac{\omega x_1}{2(2\pi)^{3/2}} \int_{-\infty}^{+\infty} \lambda^2 e^{-\nu t\lambda^2} \frac{J_{3/2}(\Omega)}{\Omega^{3/2}} d\lambda. \tag{28}$$

Analogously

$$K_{32}(x, t) = -\frac{\omega x_2}{2(2\pi)^{3/2}} \int_{-\infty}^{+\infty} \lambda^2 e^{-\nu t\lambda^2} \frac{J_{3/2}(\Omega)}{\Omega^{3/2}} d\lambda. \tag{29}$$

We calculate $K_4(x, t)$ by first using Bochner's formula (11) and then going over to polar coordinates and using (15):

$$K_4(x, t) = \frac{i\omega}{(2\pi)^2} \int_{-\infty}^{+\infty} e^{i\alpha_3 x_3 - \nu t\alpha_3^2} d\alpha_3 \int_0^{\infty} \frac{e^{-\nu t\beta^2}}{\sqrt{\beta^2+\alpha_3^2}} \beta J_0(\beta\rho) \sin \frac{\omega\alpha_3 t}{\sqrt{\beta^2+\alpha_3^2}} d\beta$$

$$= \frac{i\omega}{(2\pi)^2} \int_0^{\infty} \lambda e^{-\nu t\lambda^2} d\lambda \int_0^{\pi} \sin\theta e^{ix_3\lambda\cos\theta} J_0(\lambda\rho\sin\theta) \sin(\omega t\cos\theta)\, d\theta$$

$$= \frac{\omega}{2(2\pi)^2} \int_0^{\infty} \lambda e^{-\nu t\lambda^2} d\lambda \int_0^{\pi} [e^{i(x_3\lambda+\omega t)\cos\theta} - e^{i(x_3\lambda-\omega t)\cos\theta}] \sin\theta \cdot J_0(\lambda\rho\sin\theta)\, d\theta$$

$$= \frac{\omega}{2(2\pi)^{3/2}} \int_{-\infty}^{+\infty} \lambda e^{-\nu t\lambda^2} \frac{J_{1/2}(\Omega)}{\Omega^{1/2}} d\lambda. \tag{30}$$

Thus the solution of the Cauchy problem (1*), (2) has the form

$$\mathbf{v}(x, t) = \int_{E_3} \{\mathbf{v}^0(y) K_1(x-y, t) + [\mathbf{v}^0(y), \mathbf{K}_2(x-y, t)]\}\, dy,$$

$$P(x, t) = \int_{E_3} \{v_1^0(y) K_{32}(x-y, t) - v_2^0(y) K_{31}(x-y, t) - v_3^0(y) K_4(x-y, t)\}\, dy, \tag{31}$$

where

$$K_1(x-y, t) = \frac{1}{2(2\pi)^{3/2}} \int_{-\infty}^{+\infty} \lambda^2 e^{-\nu t\lambda^2} \frac{J_{1/2}(\Omega)}{\Omega^{1/2}} d\lambda,$$

$$K_{2i}(x-y, t) = \frac{(x_i - y_i)}{2(2\pi)^{3/2}} \int_{-\infty}^{+\infty} \lambda^3 e^{-\nu t\lambda^2} \frac{J_{3/2}(\Omega)}{\Omega^{3/2}} d\lambda, \qquad i = 1, 2 \tag{32}$$

$$K_{23}(x-y, t) = \frac{1}{2(2\pi)^{3/2}} \int_{-\infty}^{+\infty} \lambda^2 [(x_3 - y_3)\lambda + \omega t] e^{-\nu t\lambda^2} \frac{J_{3/2}(\Omega)}{\Omega^{3/2}} d\lambda.$$

From the last formula we see that the kernel K_{23} consists of two summands: $K_{23} = K'_{23} + K''_{23}$, where

$$K'_{23}(x-y, t) = \frac{x_3 - y_3}{2(2\pi)^{3/2}} \int_{-\infty}^{+\infty} \lambda^3 e^{-\nu t\lambda^2} \frac{J_{3/2}(\Omega)}{\Omega^{3/2}} d\lambda,$$

$$K''_{23}(x-y, t) = \frac{\omega t}{2(2\pi)^{3/2}} \int_{-\infty}^{+\infty} \lambda^2 e^{-\nu t\lambda^2} \frac{J_{3/2}(\Omega)}{\Omega^{3/2}} d\lambda. \tag{32'}$$

Thus $\mathbf{K}_2 = \mathbf{K}'_2 + \mathbf{K}''_2$, where $\mathbf{K}'_2 = (K'_{21}, K'_{22}, K'_{23})$ and $\mathbf{K}''_2 = (0, 0, K''_{23})$. Further,

$$K_{3j}(x-y,\ t) = -\frac{\omega(x_j-y_j)}{2(2\pi)^{3/2}}\int_{-\infty}^{+\infty}\lambda^2 e^{-\nu t\lambda^2}\frac{J_{3/2}(\Omega)}{\Omega^{3/2}}\,d\lambda, \qquad j=1,\ 2,$$

$$K_4(x-y,\ t) = \frac{\omega}{2(2\pi)^{3/2}}\int_{-\infty}^{+\infty}\lambda e^{-\nu t\lambda^2}\frac{J_{1/2}(\Omega)}{\Omega^{1/2}}\,d\lambda, \tag{32''}$$

where

$$\Omega = \sqrt{\lambda^2 r^2 + 2\lambda(x_3-y_3)\,\omega t + \omega^2 t^2},$$

$$r^2 = \sum_{i=1}^{3}(x_i-y_i)^2.$$

In studying the asymptotic behavior of the solution of the Cauchy problem for $t \longrightarrow \infty$ we will use the expressions (32) as well as a somewhat different set of expressions, which we now proceed to derive from (32), going into detail in the case of K_1. We first express Ω as a square root of a sum of squares by introducing new parameters. In (17) we make the change of variable $\sqrt{\nu t}\lambda = \lambda_1$ under the assumption $t \geqslant T > 0$. Then

$$K_1(x,\ t) = \frac{1}{2(2\pi)^{3/2}(\nu t)^{3/2}\sqrt{\omega t}}\int_{-\infty}^{+\infty}\lambda^2 e^{-\lambda^2}\frac{J_{1/2}\left[\omega t\sqrt{a^2(\lambda+\beta/a)^2+\alpha^2}\right]}{\left(\sqrt{a^2(\lambda+\beta/a)^2+\alpha^2}\right)^{1/2}}\,d\lambda, \tag{33}$$

where $a^2 = r^2/\nu t(\omega t)^2$, $\beta = x_3/r$ and $\alpha = \rho/r$ (here $\alpha^2 + \beta^2 = 1$). We now make the substitution $\lambda + \beta/a = \lambda_2$:

$$K_1(x,\ t) = \frac{1}{2(2\pi)^{3/2}(\nu t)^{3/2}\sqrt{\omega t}}\int_{-\infty}^{+\infty}(\lambda-\beta/a)^2\exp\{-(\lambda-\beta/a)^2\}$$

$$\times\frac{J_{1/2}\left(\omega t\sqrt{a^2\lambda^2+\alpha^2}\right)}{\left(\sqrt{a^2\lambda^2+\alpha^2}\right)^{1/2}}\,d\lambda.$$

Substituting the following Sonin-Poisson integral representation for Bessel functions ([8], §33):

$$\frac{J_{\mu+\sigma+1}\left(t\sqrt{q^2+z^2}\right)}{\left(\sqrt{q^2+z^2}\right)^{\mu+\sigma+1}} = \frac{q^{-\mu}z^{-\sigma}}{t^{\mu+\sigma+1}}\int_0^t J_\mu(qy)\,y^{\mu+1}J_\sigma\left(z\sqrt{t^2-y^2}\right)\left(\sqrt{t^2-y^2}\right)^{\sigma}dy, \tag{34}$$

$$\operatorname{Re}\mu,\ \operatorname{Re}\sigma > -1,$$

when $\mu = -1/2$, $\sigma = 0$, $q = a\lambda$ and $z = \alpha$ and replacing $J_{-1/2}(qy)$ by a trigonometric function according to the formula

$$J_{-1/2}(qy)=\sqrt{\frac{2}{\pi qy}}\cos(qy),$$

we get

$$K_1(x, t)=\sqrt{\frac{2}{\pi}}\frac{1}{2(2\pi\nu t)^{3/2}\omega t}\int_{-\infty}^{+\infty}\Big\{(\lambda-\beta/a)^2$$

$$\times\exp\{-(\lambda-\beta/a)^2\}\,d\lambda\int_0^{\omega t}\cos(a\lambda y)\,J_0\big(\alpha\sqrt{(\omega t)^2-y^2}\big)dy\Big\}.$$

In view of the absolute convergence of the integrals on the right, we can change the order of integration and integrate with respect to λ. We then obtain the desired expression for $K_1(x, t)$ in the form

$$K_1(x, t)=\frac{1}{4\sqrt{2}\,(2\pi\nu t)^{3/2}\omega t}\int_0^{\omega t}J_0\big(\alpha\sqrt{(\omega t)^2-y^2}\big)$$

$$\times\exp\Big\{-\frac{a^2y^2}{4}\Big\}\cos(\beta y)(2-a^2y^2)\,dy. \tag{35}$$

We have made use here of the well-known formula

$$\int_{-\infty}^{+\infty}x^n e^{-(ax^2+bx+c)}\cos(px+q)\,dx=\Big(-\frac{1}{2a}\Big)^n\sqrt{\frac{\pi}{a}}\exp\Big\{\frac{b^2-p^2}{4a}-c\Big\}$$

$$\times\sum_{k=0}^{E\left(\frac{n}{2}\right)}\frac{n!}{(n-2k)!\,k!}a^k\sum_{j=0}^{n-2k}\binom{n-2k}{j}b^{n-2k-j}p^j\cos\Big(\frac{pb}{2a}-q+\frac{\pi}{2}j\Big),\quad a>0. \tag{36}$$

Formula (34) and (36) are used in exactly the same way to derive analogous expressions for the other kernels.

In the case of $\mathbf{K}_2(x, t)$ we use formula (34) when $\mu=-1/2$, $\sigma=1$, $q=a\lambda$ and $z=\alpha$. Here

$$K''_{23}(x, t)=\frac{1}{2(2\pi)^{3/2}(\nu t)^{3/2}\sqrt{\omega t}}\int_{-\infty}^{+\infty}\lambda^2e^{-\lambda^2}\frac{J_{3/2}\big[\omega t\sqrt{a^2(\lambda+\beta/a)^2+\alpha^2}\big]}{\big(\sqrt{a^2(\lambda+\beta/a)^2+\alpha^2}\big)^{3/2}}\,d\lambda$$

$$=\frac{1}{(2\pi)^2(\nu t)^{3/2}(\omega t)^2\alpha}\int_{-\infty}^{+\infty}(\lambda-\beta/a)^2e^{-(\lambda-\beta/a)^2}\Big(\int_0^{\omega t}\cos(a\lambda y)\,J_1\big(\alpha\sqrt{(\omega t)^2-y^2}\big)$$

$$\times\sqrt{(\omega t)^2-y^2}dy\Big)d\lambda=\frac{1}{4\sqrt{2}\,(2\pi\nu t)^{3/2}\omega t}\int_0^{\omega t}J_1\big(\alpha\sqrt{(\omega t)^2-y^2}\big)$$

$$\times\frac{\sqrt{(\omega t)^2-y^2}}{\omega t\sigma}e^{-a^2y^2/4}\cos(\beta y)(2-a^2y^2)\,dy. \tag{37}$$

Analogously

$$K'_{2i}(x, t) = \frac{x_i}{\rho} \frac{a}{4\sqrt{2}\,(2\pi\nu t)^{3/2}(\omega t)^2} \int_0^{\omega t} \sin(\beta y)$$
$$\times J_1\left(\alpha\sqrt{(\omega t)^2 - y^2}\right)\sqrt{(\omega t)^2 - y^2}\, e^{-a^2y^2/4}\left(\frac{a^3y^3}{2} - 3ay\right)dy \qquad (38)$$

for $i = 1, 2, 3$. Further, when $j = 1, 2$ we have

$$K_{3j}(x, t) = -\frac{x_j}{\rho} \frac{\omega a}{4\sqrt{2}\,(2\pi)^{3/2}(\nu t)(\omega t)^2}$$
$$\times \int_0^{\omega t} J_1\left(\alpha\sqrt{(\omega t)^2 - y^2}\right)\left(\sqrt{(\omega t)^2 - y^2}\right)e^{-a^2y^2/4}\cos(\beta y)(2 - a^2y^2)\,dy, \qquad (39)$$

$$K_4(x, t) = -\frac{\omega}{2\sqrt{2}\,(2\pi)^{3/2}(\nu t)(\omega t)} \int_0^{\omega t} J_0\left(\alpha\sqrt{(\omega t)^2 - y^2}\right)\sin(\beta y)\, ay \exp\left\{-\frac{a^2y^2}{4}\right\}dy.$$

We note that if we set $y = \omega t \cos\varphi$, formulas (35) and (37)–(39) go over to the corresponding formulas of our note [5], which we obtained in another way.

Let us now consider the nonhomogeneous system (1).

We consider only the case of solenoidal right sides, since we can use the Sobolev-Weyl expansion [9], [10], as we did in [4], to reduce an arbitrary **F** to a solenoidal **F**. In this case the solution of the Cauchy problem (1), (2) can be obtained by Duhamel's principle:

$$\mathbf{v}(x, t) = \int_{E_3} \{\mathbf{v}^0(y)K_1(x - y, t) + [\mathbf{v}^0(y), \mathbf{K}_2(x - y, t)]\}\,dy$$
$$+ \int_0^t \int_{E_3} \{\mathbf{F}(y, \tau)K_1(x - y, t - \tau) + [\mathbf{F}(y, \tau), \mathbf{K}_2(x - y, t - \tau)]\}\,dy\,d\tau,$$

$$P(x, t) = \int_{E_3} \{v_1^0(y)K_{32}(x - y, t) - v_2^0(y)K_{31}(x - y, t) \qquad (40)$$
$$- v_3^0(y)K_4(x - y, t)\}\,dy + \int_0^t \int_{E_3} \{F_1(y, \tau)K_{31}(x - y, t - \tau)$$
$$- F_2(y, \tau)K_{31}(x - y, t - \tau) - F_3(y, \tau)K_4(x - y, t - \tau)\}\,dy\,d\tau.$$

If, on the other hand, the solution for an arbitrary **F** is written out directly without using the Weyl-Sobolev expansion, the solution (40) can be expressed in a more complicated way in tensor form.

We will not do this here, since below, when investigating the asymptotic behavior of the solution of the Cauchy problem as $t \to \infty$, we will use only the form (31) of the solution. It can be proved that the solution of the Cauchy problem (1), (2) is unique in the function class $W^{1,0}_{2,x,t}(E_3 \times (0 \leqslant t < \infty))$. A corresponding theorem for the first boundary value problem for system (1) was proved by us in [11]. The proof of the Cauchy problem is analogous and will not be given here.

The solution constructed by us will belong to this uniqueness class under some weak assumptions on $\mathbf{v}^0(x)$ and $\mathbf{F}(x, t)$ (e.g. the membership of $\mathbf{F}$ and $\mathbf{v}^0$ in L_2).

This follows from the estimates developed by us in [12] concerning the membership of the solution (40) in the Sobolev space $W^{l,2l}_{p,t,x}$. The corresponding theorem proved in [12] is analogous to

THEOREM 2. *Suppose* $\mathbf{v}^0(x) \in W_p^{2l+2-2/p}$, $\operatorname{div} \mathbf{v}^0 = 0$, *and* $\mathbf{F} \in W^{l,2l}_{p,t,x}$, *where* $p > 1$ *and* l *is a nonnegative integer. Then the solution* (40) *of problem* (1), (2) *is such that*

$$\mathbf{v}(x, t) \in W^{l+1,\,2l+2}_{p,\,t,\,x}(E_4^T), \qquad \operatorname{grad} P \in W^{l,\,2l}_{p,\,t,\,x}(E_4^T),$$

where $E_4^T = \{x \in E_3,\ 0 \leqslant t \leqslant T\}$, *and satisfies the estimate*

$$\|\mathbf{v}\|_{W^{l+1,\,2l+2}_{p,\,t,\,x}(E_4^T)} + \|\operatorname{grad} P\|_{W^{l,\,2l}_{p,\,t,\,x}(E_4^T)} \leqslant C\left(\|\mathbf{v}^0\|_{W_p^{2l+2-2/p}(E_3)} + \|\mathbf{F}\|_{W^{l,\,2l}_{p,\,t,\,x}(E_4^T)}\right),$$

the norm of a function $u(x)$ *in the Banach space* $W_p^m(E_n)$ *being defined for nonintegral positive* m *by the formula*

$$\|u\|_{W_p^m(E_n)} = \sum_{j=0}^{[m]} \sum_{(j)} \|D_x^j u\|_{L_p(E_n)}$$
$$+ \sum_{(j=[m])} \left(\int_{E_n} dx \int_{E_n} |D_x^j u(x) - D_y^j u(y)|^p \frac{dy}{|x-y|^{n+p(m-[m])}} \right)^{1/p},$$

in which $[m]$ *is the integral part of* m *and* $D_x^j = D_{x_1}^{j_1} \cdots D_{x_n}^{j_n}$.

§2. The uniform boundedness of some definite integrals containing Bessel functions

The present section is devoted to the proof of Lemmas 1 and 2, which will be used in §3 to study the asymptotic behavior as $t \to \infty$ of the solution of the Cauchy problem (1*), (2).

LEMMA 1. *If* $0 \leqslant y \leqslant t$ *and* $\alpha \leqslant \beta$ $(\alpha^2 + \beta^2 = 1)$, *the integral*

$$Q_1(x, t, y) = \int_0^y J_0\left(\alpha\sqrt{t^2 - z^2}\right) \cos(\beta z)\, dz \tag{41}$$

satisfies the estimate

$$|Q_1(x, t, y)| \leqslant C_1. \tag{42}$$

Here $\alpha = \rho/r$, $\beta = x_3/r$, $\rho^2 = x_1^2 + x_2^2$, $r^2 = \rho^2 + x_3^2$, $t > 0$ *and* C_1 *is an absolute constant. It can clearly be assumed without loss of generality that* $\beta \geqslant 0$, *i.e.* $x_3 \geqslant 0$, *in* (41).

PROOF. Substituting the following representation of Sonin [8] (see also [13], Chapter VII, §3.95) of a Bessel function of a real argument:

$$J_0(\alpha\sqrt{t^2-z^2})=\frac{1}{2\pi i}\int_{|\zeta|=1}\exp\left\{\frac{\alpha t}{2}\left(\zeta-\frac{1}{\zeta}\right)+\frac{\alpha z}{2}\left(\zeta+\frac{1}{\zeta}\right)\right\}\frac{d\zeta}{\zeta}, \tag{43}$$

into (41) and integrating with respect to z, we get $Q_1 = \Sigma_1^3 Q_{1i}$, where

$$Q_{11}=\frac{\cos(\beta y)}{2\pi i}\int_{|\zeta|=1}\exp\left\{\frac{\alpha t}{2}\left(\zeta-\frac{1}{\zeta}\right)+\frac{\alpha y}{2}\left(\zeta+\frac{1}{\zeta}\right)\right\}\frac{\alpha/2\,(\zeta^2+1)}{\alpha^2/4\,(\zeta^2+1)^2+\zeta^2\beta^2}\,d\zeta,$$

$$Q_{12}=\frac{\sin(\beta y)}{2\pi i}\int_{|\zeta|=1}\exp\left\{\frac{\alpha t}{2}\left(\zeta-\frac{1}{\zeta}\right)+\frac{\alpha y}{2}\left(\zeta+\frac{1}{\zeta}\right)\right\}\frac{\beta\zeta\,d\zeta}{\alpha^2/4\,(\zeta^2+1)^2+\zeta^2\beta^2}, \tag{44}$$

$$Q_{13}=-\frac{1}{2\pi i}\int_{|\zeta|=1}\exp\left\{\frac{\alpha t}{2}\left(\zeta-\frac{1}{\zeta}\right)\right\}\frac{\alpha/2\,(\zeta^2+1)}{\alpha^2/4\,(\zeta^2+1)^2+\zeta^2\beta^2}\,d\zeta.$$

Since Q_{13} is $-Q_{11}$ for $y = 0$, it suffices to estimate Q_{11} and Q_{12}.

Let us first consider Q_{11}. We transform the argument of the exponential function:

$$\frac{\alpha t}{2}\left(\zeta-\frac{1}{\zeta}\right)+\frac{\alpha y}{2}\left(\zeta+\frac{1}{\zeta}\right)=\frac{\alpha}{2}\sqrt{t^2-y^2}\left(\frac{\zeta}{\sigma}-\frac{\sigma}{\zeta}\right), \tag{45}$$

where $\sigma = ((t-y)/(t+y))^{1/2}$. We note that the parameter σ can again vary from 0 to 1 (we will always take the positive root).

We factor the denominator of the rational part of Q_{11} into a product of polynomials of the second degree in ζ:

$$\frac{\alpha^2}{4}(\zeta^2+1)^2+\zeta^2\beta^2=\frac{1}{4\alpha^2}[\alpha^2\zeta^2+(1+\beta)^2][\alpha^2\zeta^2+(1-\beta)^2].$$

The rational part of Q_{11} can then be expressed as a sum of two partial fractions:

$$\frac{\alpha/2\,(\zeta^2+1)}{\alpha^2/4\,(\zeta^2+1)^2+\zeta^2\beta^2}=\frac{\alpha(1+\beta)}{\alpha^2\zeta^2+(1+\beta)^2}+\frac{\alpha(1-\beta)}{\alpha^2\zeta^2+(1-\beta)^2},$$

and Q_{11} can be written in the form

$$Q_{11}=\frac{\cos(\beta y)}{2\pi i}\int_{|\zeta|=1}\exp\left\{\frac{\alpha}{2}\sqrt{t^2-y^2}\left(\frac{\zeta}{\sigma}-\frac{\sigma}{\zeta}\right)\right\}\times\left[\frac{\alpha(1+\beta)}{\alpha^2\zeta^2+(1+\beta)^2}+\frac{\alpha(1-\beta)}{\alpha^2\zeta^2+(1-\beta)^2}\right]d\zeta. \tag{46}$$

Making the substitution $\zeta = z\sigma$ in (46), so that the circle $|\zeta| = 1$ goes over into the circle $|z| = 1/\sigma$, and letting $(1+\beta)/\alpha\sigma = \theta_1$ and $(1-\beta)/\alpha\sigma = \theta_2$, we get

$$Q_{11}=\frac{\cos(\beta y)}{2\pi i}\int_{|z|=1/\sigma}\exp\left\{\frac{\alpha}{2}\sqrt{t^2-y^2}\left(z-\frac{1}{z}\right)\right\}\times\left[\frac{\theta_1}{z^2+\theta_1^2}+\frac{\theta_2}{z^2+\theta_2^2}\right]dz=Q'_{11}+Q''_{11}. \tag{47}$$

Consider Q'_{11}. The integrand of Q'_{11} has two first order poles at $z = \pm i\theta_1$. Since the inequality $\theta_1 = (1+\beta)/\alpha\sigma > 1/\sigma$ is always satisfied under the assumptions of the lemma, it follows from Cauchy's integral theorem that the circle $|z| = 1/\sigma$ can be replaced by the circle $|z| = 1$. Thus

$$Q'_{11} = \frac{\cos(\beta y)}{2\pi i} \int\limits_{|z|=1} \exp\left\{\frac{\alpha}{2}\sqrt{t^2-y^2}\left(z-\frac{1}{z}\right)\right\}\frac{\theta_1}{z^2+\theta_1^2}\,dz.$$

Since $z = e^{i\varphi}$ (φ a real variable) on the unit circle, the expression

$$\exp\left\{\frac{\alpha}{2}\sqrt{t^2-y^2}\left(z-\frac{1}{z}\right)\right\} = \exp\left\{i\alpha\sqrt{t^2-y^2}\sin\varphi\right\}$$
$$= \cos\left(\alpha\sqrt{t^2-y^2}\sin\varphi\right) + i\sin\left(\alpha\sqrt{t^2-y^2}\sin\varphi\right)$$

is obviously bounded in modulus when φ varies from $-\pi$ to π and $y \leqslant t$. Also,

$$\left|\frac{\theta_1}{z^2+\theta_1^2}\right| \leqslant \left|\frac{1}{z-i\theta_1}\right| + \left|\frac{1}{z+i\theta_1}\right|.$$

Clearly the denominator $|z - i\theta_1|$ achieves its minimum value at $z = i$, and hence

$$\left|\frac{1}{z-i\theta_1}\right| \leqslant \frac{1}{\left|1-\dfrac{1+\beta}{\sigma\alpha}\right|} = \frac{\alpha\sigma}{|1+\beta-\alpha\sigma|} \leqslant \frac{\alpha\sigma}{\beta} \leqslant 1,$$

since, by assumption, $\alpha\sigma \leqslant 1$ and $\alpha \leqslant \beta$. Analogously $|1/(z + i\theta_2)| \leqslant 1$, and hence $|\theta_1/(z^2 + \theta_1^2)| \leqslant 2$, which implies

$$|Q'_{11}| \leqslant 4. \tag{48}$$

Consider Q''_{11}. Here the integrand has two simple poles at $z = \pm i\theta_2$. But now $\theta_2 = (1-\beta)/\alpha\sigma \leqslant 1/\sigma$, with equality being achieved when $1 - \beta = \alpha$, i.e. when $\alpha = 0$, since, by assumption, $\alpha \leqslant \beta$. We note that the integral Q_1 is bounded uniformly in t and y when $\alpha = 0$.

For then $J_0(0) \equiv 1$ and we have

$$|Q_1| = \left|\int\limits_0^y \cos z\,dz\right| = |\sin y| \leqslant 1.$$

Also, it is useful to note that $\lim_{\alpha\to 0}(1-\beta)/\alpha = 0$. It will be assumed below that an integral with a pole lying on its path of integration should be understood in the sense of the Cauchy principal value. The sequel will clearly substantiate this assumption. Since

$$\frac{\theta_2}{z^2+\theta_2^2} = \frac{1}{2i}\,\frac{1}{z-i\theta_2} - \frac{1}{2i}\,\frac{1}{z+i\theta_2},$$

it follows that

$$Q''_{11} = -\frac{\cos(\beta y)}{4\pi} \int\limits_{|z|=1/\sigma} \exp\left\{\frac{\alpha}{2}\sqrt{t^2-y^2}\left(z-\frac{1}{z}\right)\right\}\left(\frac{1}{z-i\theta_2}-\frac{1}{z+i\theta_2}\right)dz. \quad (49)$$

We consider two cases: 1) $\theta_2 \geqslant 1$ and 2) $\theta_2 \leqslant 1$.

1) Suppose $\theta_2 \geqslant 1$. In general, we can facilitate the discussion by calculating or estimating our integrals (49) for any fixed $\theta_2 > 1$ inasmuch as the final result is an estimate that does not depend on θ_2.

Applying the residue theorem to the integral of (49) and letting $z_1 = i\theta_2$ and $z_2 = -i\theta_2$, we get

$$\int\limits_{|z|=1/\sigma} f(z)\,dz = 2\pi i \sum_{k=1}^{2} \operatorname{res} f(z_k) + \int\limits_{|z|=1} f(z)\,dz, \quad (50)$$

where

$$f(z) = \exp\left\{\frac{\alpha}{2}\sqrt{t^2-y^2}\left(z-\frac{1}{z}\right)\right\}\left(\frac{1}{z-i\theta_2}-\frac{1}{z+i\theta_2}\right).$$

It is easily seen that

$$2\pi i \sum_{k=1}^{2} \operatorname{res} f(z_k) = -4\pi \sin\left[\frac{\alpha}{2}\sqrt{t^2-y^2}\left(\theta_2+\frac{1}{\theta_2}\right)\right]$$
$$= -4\pi \sin(t-\beta y), \quad (51)$$

and hence the sum of the residues is bounded in modulus for any values of the parameters satisfying the condition of the lemma.

Let us estimate the modulus of the integral around the unit circle in (50). It clearly suffices to carry out the estimate for a single summand, since the analogous estimate for the other summand can then be obtained from the first estimate by simply replacing i by $-i$.

Thus consider

$$I = \int\limits_{|z|=1} \exp\left\{\frac{\alpha}{2}\sqrt{t^2-y^2}\left(z-\frac{1}{z}\right)\right\}\frac{dz}{z-i\theta_2}.$$

Here one should consider the cases of unbounded and bounded θ_2.

a) Suppose $\theta_2 \geqslant 3/2$ (we note that, by assumption, θ_2 is always nonnegative). Then on the unit circle

$$\max\frac{1}{|z-i\theta_2|} \leqslant \frac{1}{|1-\theta_2|} \leqslant \frac{1}{\left|1-\frac{3}{2}\right|} = 2$$

and hence

$$|I| \leqslant 8\pi. \quad (52)$$

b) Suppose $1 \leqslant \theta_2 \leqslant 3/2$. Then $z = e^{i\varphi}$ (φ a real variable) on the unit circle, and hence

$$I = i \int_{-\pi}^{\pi} e^{i\alpha\sqrt{t^2-y^2}\sin\varphi} \frac{\cos\varphi + i\sin\varphi}{\cos\varphi + i\sin\varphi - i\theta_2} d\varphi. \tag{53}$$

Take a small *fixed* $\delta > 0$ and divide the integral in (53) into three integrals as follows:

$$\int_{-\pi}^{\pi} = \int_{-\pi}^{\frac{\pi}{2}-\delta} + \int_{\frac{\pi}{2}-\delta}^{\frac{\pi}{2}+\delta} + \int_{\frac{\pi}{2}+\delta}^{\pi}. \tag{54}$$

The moduli of the first and third integrals on the right side are easily estimated, since their integrands are absolutely integrable functions. In fact,

$$\int_{\frac{\pi}{2}+\delta}^{\pi} e^{i\alpha\sqrt{t^2-y^2}\sin\varphi} \frac{(\cos\varphi + i\sin\varphi)[\cos\varphi - i(\sin\varphi - \theta_2)]}{[\cos\varphi + i(\sin\varphi - \theta_2)][\cos\varphi - i(\sin\varphi - \theta_2)]} d\varphi$$

$$= \int_{\frac{\pi}{2}+\delta}^{\pi} e^{i\alpha\sqrt{t^2-y^2}\sin\varphi} \frac{1 + \theta_2(i\cos\varphi - \sin\varphi)}{\cos^2\varphi + (\sin\varphi - \theta_2)^2} d\varphi.$$

Since φ is separated from $\pi/2$ and $1 \leqslant \theta_2 \leqslant 3/2$, we have

$$\left| \int_{\frac{\pi}{2}+\delta}^{\pi} e^{i\alpha\sqrt{t^2-y^2}\sin\varphi} \frac{1 + \theta_2(i\cos\varphi - \sin\varphi)}{\cos^2\varphi + (\sin\varphi - \theta_2)^2} d\varphi \right| \leqslant \frac{\text{const}}{\sin^2\delta} \leqslant \text{const}. \tag{55}$$

The first integral is estimated analogously:

$$\left(\int_{-\pi}^{0} + \int_{0}^{\frac{\pi}{2}-\delta} \right) \left| e^{i\alpha\sqrt{t^2-y^2}\sin\varphi} \frac{[1 + \theta_2(i\cos\varphi - \sin\varphi)]}{\cos^2\varphi + (\sin\varphi - \theta_2)^2} d\varphi \right|$$

$$\leqslant \frac{\text{const}}{\theta_2^2} + \frac{\text{const}}{(\cos\delta - \theta_2)^2} \leqslant \text{const}. \tag{56}$$

The greatest difficulty is encountered in estimating the middle integral. Making the substitution $\varphi = \psi + \pi/2$, we get

$$\int_{\frac{\pi}{2}-\delta}^{\frac{\pi}{2}+\delta} e^{i\alpha\sqrt{t^2-y^2}\sin\varphi} \frac{\cos\varphi + i\sin\varphi}{\cos\varphi + i\sin\varphi - i\theta_2} d\varphi = \int_{-\delta}^{\delta} e^{i\alpha\sqrt{t^2-y^2}\cos\psi} \frac{i\cos\psi - \sin\psi}{i\cos\psi - \sin\psi - i\theta_2} d\psi$$

$$= \int_{-\delta}^{\delta} e^{i\alpha\sqrt{t^2-y^2}\cos\psi} \frac{1 - \theta_2\cos\psi - i\theta_2\sin\psi}{1 - 2\theta_2\cos\psi + \theta_2^2} d\psi = \int_{-\delta}^{\delta} e^{i\alpha\sqrt{t^2-y^2}\cos\psi} \frac{1 - \theta_2\cos\psi - i\theta_2\sin\psi}{\sin^2\psi + (\cos\psi - \theta_2)^2} d\psi$$

$$= 2\int_{0}^{\delta} e^{i\alpha\sqrt{t^2-y^2}\cos\psi} \frac{1 - \theta_2\cos\psi}{1 - 2\theta_2\cos\psi + \theta_2^2} d\psi. \tag{57}$$

Consider the sign of the rational part of the integrand in (57) for small but fixed $\delta > 0$. The denominator is clearly nonnegative for any θ_2, since

$$1 - 2\theta_2 \cos\psi + \theta_2^2 \geqslant (1 - \theta_2)^2 \geqslant 0.$$

On the other hand, the numerator $1 - \theta_2 \cos\psi$ has the single root $\psi_0 = \arccos 1/\theta_2$, since we are considering the case $1 \leqslant \theta_2 \leqslant 3/2$, i.e. it changes sign exactly once. We divide the interval of integration $(0, \delta)$ into the two intervals $(0, \psi_0)$ and (ψ_0, δ) and calculate the corresponding integrals separately.

Since on the interval $(0, \psi_0)$ the function

$$\frac{1 - \theta_2 \cos\psi}{1 - 2\theta_2 \cos\psi + \theta_2^2}$$

does not change sign while the function

$$e^{i\alpha\sqrt{t^2-y^2}\cos\psi} = \cos(\alpha\sqrt{t^2 - y^2}\cos\psi) + i\sin(\alpha\sqrt{t^2 - y^2}\cos\psi)$$

is bounded in modulus, the second mean value theorem implies

$$\int_0^{\psi_0} \cos(\alpha\sqrt{t^2 - y^2}\cos\psi)\frac{1 - \theta_2\cos\psi}{1 - 2\theta_2\cos\psi + \theta_2^2}\,d\psi = \cos(\alpha\sqrt{t^2 - y^2}\cos\psi_1)\int_0^{\psi_0}\frac{1 - \theta_2\cos\psi}{1 - 2\theta_2\cos\psi + \theta_2^2}\,d\psi.$$

With the use of the substitution $\gamma = \tan\psi/2$, the latter integral can be calculated as an indefinite integral in explicit form:

$$\int \frac{1 - \theta_2\cos\psi}{1 - 2\theta_2\cos\psi + \theta_2^2}\,d\psi = \frac{\psi}{2} + \arctan\left\{\frac{1 + \theta_2}{1 - \theta_2}\tan\frac{\psi}{2}\right\} \tag{58}$$

(see [16], 2nd ed., §276, Example 13; 4th ed., §288, Example 13).

Substituting the limits of integration, we get (since $\psi_0 \leqslant \delta$ and δ is a small fixed number)

$$\left|\cos(\alpha\sqrt{t^2 - y^2}\cos\psi_1)\int_0^{\psi_0}\frac{1 - \theta_2\cos\psi}{1 - 2\theta_2\cos\psi + \theta_2^2}\,d\psi\right|$$
$$= \left|\cos\left(\alpha\sqrt{t^2 - y^2}\cos\psi_1\right)\left(\frac{\psi_0}{2} + \arctan\left\{\frac{1 + \theta_2}{1 - \theta_2}\tan\frac{\psi_0}{2}\right\}\right)\right| \leqslant \frac{\psi_0}{2} + \frac{\pi}{2} \leqslant \text{const} \tag{59}$$

for any $\theta_2 \in [1, 3/2]$ inasmuch as the right side of the equality in (59) is an analytic function giving the value of the integral for all $\theta_2 \in [1, 3/2]$. Analogously

$$\left|\sin(\alpha\sqrt{t^2 - y^2}\cos\psi_1)\int_{\psi_0}^{\delta}\frac{1 - \theta_2\cos\psi}{1 - 2\theta_2\cos\psi + \theta_2^2}\,d\psi\right| \leqslant \text{const}. \tag{60}$$

Thus, if $\theta_2 \geqslant 1$,

$$\left| \int\limits_{|z|=1} \exp\left\{\frac{\alpha}{2}\sqrt{t^2-y^2}\left(z-\frac{1}{z}\right)\right\}\left(\frac{1}{z-i\theta_2}-\frac{1}{z+i\theta_2}\right)dz \right| \leqslant \text{const},$$

where the constant does not depend on α, t, y or θ_2.

2) Suppose $\theta_2 \leqslant 1$. In this case Cauchy's integral theorem implies that the integral around the circle $|z| = 1/\sigma$ in (49) can be replaced by an integral around the unit circle. (The poles now lie inside the unit circle.) We consider the two cases: a) $\theta_2 \leqslant 1/2$ and b) $1/2 \leqslant \theta_2 \leqslant 1$.

a) $\theta_2 \leqslant 1/2$. We have

$$\left|\frac{1}{z-i\theta_2}\right| + \left|\frac{1}{z+i\theta_2}\right| \leqslant \frac{2}{|1-\theta_2|} \leqslant 4,$$

which implies $|Q'_{11}| \leqslant 4$.

b) $1/2 \leqslant \theta_2 \leqslant 1$. In this case the estimate is carried out in basically the same way as in the case $1 \leqslant \theta_2 \leqslant 3/2$, with the exception of the last step,when an integral over the neighborhood $(\pi/2-\delta, \pi/2+\delta)$ is estimated. A formula of the form (57) is obtained as before. But now the numerator $1-\theta_2 \cos\psi$ does not change sign on the interval $(0, \delta)$, since $\theta_2 \leqslant 1$. It is therefore not necessary to divide the integral into two integrals, and we at once obtain with the use of the second mean value theorem the inequalities

$$\left| \int\limits_0^\delta \cos\left(\alpha\sqrt{t^2-y^2}\cos\psi\right) \frac{1-\theta_2\cos\psi}{1-2\theta_2\cos\psi+\theta_2^2}\, d\psi \right|$$
$$\leqslant \left|\cos\left(\alpha\sqrt{t^2-y^2}\cos\psi_1\right)\right| \cdot \left|\frac{\delta}{2} + \operatorname{arc\,tan}\left\{\frac{1+\theta_2}{1-\theta_2}\tan\frac{\delta}{2}\right\}\right| \leqslant \text{const.}$$

The modulus of the integral with $\sin\left(\alpha\sqrt{t^2-y^2}\cos\psi\right)$ is estimated analogously. We have thus proved that

$$|Q''_{11}| \leqslant \text{const}, \tag{61}$$

where the constant does not depend on α, β, t or y. It follows according to (47), (48) and (61) that $|Q_{11}| \leqslant \text{const}$. The estimate of $|Q_{12}|$ is carried out in exactly the same way as the estimate of $|Q_{11}|$, since, upon making the substitution $\zeta = z\sigma$ and expanding the rational part of the integrand in partial fractions, we get

$$Q_{12} = \frac{\sin\beta y}{2\pi i} \int\limits_{|z|=\frac{1}{\sigma}} \exp\left\{\frac{\alpha}{2}\sqrt{t^2-y^2}\left(z-\frac{1}{z}\right)\right\}\left(\frac{z}{z^2+\theta_2^2}-\frac{z}{z^2+\theta_1^2}\right)dz = \frac{\sin\beta y}{4\pi i}$$
$$\times \int\limits_{|z|=\frac{1}{\sigma}} \exp\left\{\frac{\alpha}{2}\sqrt{t^2-y^2}\left(z-\frac{1}{z}\right)\right\}\left(\frac{1}{z-i\theta_2}+\frac{1}{z+i\theta_2}-\frac{1}{z-i\theta_1}-\frac{1}{z+i\theta_1}\right)dz, \tag{62}$$

i.e. a linear combination of the same integrals as in the corresponding expression for Q_{11}, if we take into account the fact that the sum of the residues at the poles $z = \pm i\theta_2$ is now equal to

$$\sum \text{res} = 2\cos\left[\frac{\alpha}{2}\sqrt{t^2 - y^2}\left(\theta_2 + \frac{1}{\theta_2}\right)\right] = 2\cos(t - \beta y)$$

(cf. (51)) and hence is also bounded in modulus.

And since in (44), as we have already noted, Q_{13} is $-Q_{11}$ when $y = 0$ while the estimate of $|Q_{11}|$ does not depend on y, the lemma is proved.

LEMMA 2. *If* $0 \leqslant y_1 \leqslant y \leqslant t$ *and* $\alpha \leqslant \beta$ $(\alpha = \rho/r,\ \beta = x_3/r,\ \rho^2 = x_1^2 + x_2^2$ *and* $r^2 = \rho^2 + x_3^2)$, *then the integral*

$$Q_2(x, t, y, y_1) = \int_{y_1}^{y} J_0\left(\alpha\sqrt{t^2 - z^2}\right)\sin(\beta z)\,dz \tag{63}$$

is uniformly bounded in modulus:

$$|Q_2(x, t, y, y_1)| \leqslant C_2, \tag{64}$$

by an absolute constant C_2.

PROOF. As in the proof of Lemma 1, we substitute the representation (43) for the Bessel function $J_0(\alpha\sqrt{t^2 - z^2})$ in (63) and integrate with respect to z. This gives us

$$Q_2 = \sum_{i=1}^{4} Q_{2i}, \tag{65}$$

where

$$Q_{21} = \frac{\sin(\beta y)}{2\pi i}\int_{|\zeta|=1}\exp\left\{\frac{\alpha t}{2}\left(\zeta - \frac{1}{\zeta}\right) + \frac{\alpha y}{2}\left(\zeta + \frac{1}{\zeta}\right)\right\}\frac{\alpha/2\,(\zeta^2 + 1)\,d\zeta}{\alpha^2/4\,(\zeta^2 + 1)^2 + \zeta^2\beta^2},$$

$$Q_{22} = -\frac{\cos(\beta y)}{2\pi i}\int_{|\zeta|=1}\exp\left\{\frac{\alpha t}{2}\left(\zeta - \frac{1}{\zeta}\right) + \frac{\alpha y}{2}\left(\zeta + \frac{1}{\zeta}\right)\right\}\frac{\beta\zeta\,d\zeta}{\alpha^2/4\,(\zeta^2 + 1)^2 + \zeta^2\beta^2}, \tag{66}$$

$$Q_{23} = -Q_{21} \quad \text{when} \quad y = y_1,$$

$$Q_{24} = -Q_{22} \quad \text{when} \quad y = y_1,$$

i.e. we obtain the same expressions as in the estimate of $|Q_{11}|$ upon letting

$$\cos(\beta y) \to \sin(\beta y), \qquad \sin(\beta y) \to -\cos(\beta y).$$

The lemma is proved.

§3. Asymptotic behavior of the solution of the Cauchy problem as $t \longrightarrow \infty$. Influence of rotation

We thus consider the homogeneous Cauchy problem (1*), (2). Its solution has the form (31), in which the kernels have the two alternative representations (32), (32′), (32″) and (35), (37)–(39). Both of these representations will be used in the present section to prove the basic Theorem 1 formulated in the Introduction.

PROOF OF THEOREM 1. We first prove the theorem for the vector $\mathbf{v}(x, t)$, which was represented by us in the form

$$\mathbf{v}(x, t) = \int_{E_3} \{\mathbf{v}^0(y) K_1(x-y, t) + [\mathbf{v}^0(y), \mathbf{K}_2(x-y, t)]\} \, dy, \tag{67}$$

in which

$$K_1(x, t) = \frac{1}{2(2\pi)^{3/2}(\nu t)^{3/2}\sqrt{\omega t}} \int_{-\infty}^{+\infty} \lambda^2 e^{-\lambda^2} \frac{J_{1/2}[\omega t \sqrt{a^2(\lambda+\beta/a)^2+\alpha^2}]}{(\sqrt{a^2(\lambda+\beta/a)^2+\alpha^2})^{1/2}} \, d\lambda$$

$$= \frac{1}{4\sqrt{2}\,(2\pi\nu t)^{3/2}\omega t} \int_0^{\omega t} J_0\left(\alpha\sqrt{(\omega t)^2-y^2}\right) \exp\left\{-\frac{a^2y^2}{4}\right\} \cos(\beta y)(2-a^2y^2)\, dy,$$

$$\mathbf{K}_2(x, t) = \mathbf{K}_2' + \mathbf{K}_2'', \tag{68}$$

$$K_{2i}'(x, t) = \frac{x_i}{2(2\pi\nu t)^{3/2}(\omega t)^{3/2}} \int_{-\infty}^{+\infty} \lambda^3 e^{-\lambda^2} \frac{J_{3/2}[\omega t \sqrt{a^2(\lambda+\beta/a)^2+\alpha^2}]}{(\sqrt{a^2(\lambda+\beta/a)^2+\alpha^2})^{3/2}} \, d\lambda$$

$$= \frac{x_i}{\rho} \frac{a}{4\sqrt{2}\,(2\pi\nu t)^{3/2}(\omega t)^2} \int_0^{\omega t} J_1\left(\alpha\sqrt{(\omega t)^2-y^2}\right)$$

$$\times \sqrt{(\omega t)^2-y^2} \exp\left\{-\frac{a^2y^2}{4}\right\} \sin(\beta y)\left(\frac{a^3y^3}{2} - 3ay\right) dy, \quad i = 1, 2, 3. \tag{69}$$

$$K_{23}''(x, t) = \frac{1}{2(2\pi\nu t)^{3/2}\sqrt{\omega t}} \int_{-\infty}^{+\infty} \lambda^2 e^{-\lambda^2} \frac{J_{3/2}[\omega t \sqrt{a^2(\lambda+\beta/a)^2+\alpha^2}]}{(\sqrt{a^2(\lambda+\beta/a)^2+\alpha^2})^{3/2}} \, d\lambda. \tag{70}$$

$$= \frac{1}{4\sqrt{2}\,(2\pi\nu t)^{3/2}\omega t} \int_0^{\omega t} J_1\left(\alpha\sqrt{(\omega t)^2-y^2}\right) \frac{\sqrt{(\omega t)^2-y^2}}{\omega t a} \exp\left\{-\frac{a^2y^2}{4}\right\} \cos(\beta y)(2-a^2y^2)\, dy,$$

where

$$a^2 = \frac{r^2}{(\nu t)(\omega t)^2}, \qquad \beta = \frac{x_3}{r}, \qquad \alpha = \frac{\rho}{r}, \qquad \rho^2 = x_1^2 + x_2^2, \qquad r^2 = \rho^2 + x_3^2.$$

In view of the invariance of the L_1 norm with respect to a shift, it suffices to prove that each relevant kernel $K(x, t)$ is representable in the form

$$K(x, t) = \frac{G(x, t)}{(\nu t)^{3/2}(\omega t)},$$

where $|G(x, t)| \leqslant C$, the constant C not depending on x or t.

This boundedness is most easily proved for the kernels $K'_{2i}(x, t)$, $i = 1, 2$. To this end we consider $K'_{2i}(x, t)$ $(i = 1, 2)$ in the form

$$K'_{2i}(x, t) = \frac{x_i}{\rho} \frac{1}{4\sqrt{2}\,(2\pi\nu t)^{3/2}(\omega t)} G'_{2i}(x, t), \tag{71}$$

where

$$G'_{2i}(x, t) = \int_0^{\omega t} J_1\left(\alpha\sqrt{(\omega t)^2 - y^2}\right)\frac{\sqrt{(\omega t)^2 - y^2}}{\omega t} \exp\left\{-\frac{a^2y^2}{4}\right\} \sin(\beta y)\left(\frac{a^4y^3}{2} - 3a^2y\right)dy.$$

Since $|x_i/\rho| \leqslant 1$ for $i = 1, 2$, it suffices to prove that $|G'_{2i}(x, t)| < \text{const}$. Since the Bessel function $J_1(x)$ of a real argument x satisfies the estimate $|J_1(x)| \leqslant 1/\sqrt{2}$ while $|\sqrt{(\omega t)^2 - y^2}/\omega t| \leqslant 1$, we have

$$|G'_{2i}(x, t)| \leqslant C\left[\int_0^{\omega t} e^{-a^2y^2/4} \cdot \frac{a^4y^3}{2}\,dy + 3\int_0^{\omega t} e^{-a^2y^2/4} \cdot a^2y\,dy\right]. \tag{72}$$

(We note that $a \geqslant 0$.) Let $ay = z$ in (72). Then

$$3\int_0^{\omega t} e^{-a^2y^2/4}a^2y\,dy = 3\int_0^{a\omega t} e^{-z^2/4}z\,dz$$

$$= -6\int_0^{a\omega t} de^{-z^2/4} = 6\left(1 - \exp\left\{-\frac{r^2}{4\nu t}\right\}\right) \leqslant 6 \tag{73}$$

for any $x \in E_3$ and $t > 0$. Analogously

$$\left|\frac{1}{2}\int_0^{\omega t} e^{-a^2y^2/4}a^4y^3\,dy\right| = \left|\frac{1}{2}\int_0^{a\omega t} \exp\left\{-\frac{z^2}{4}\right\} z^3\,dz\right|$$

$$= \left|-\int_0^{a\omega t} z^2\,de^{-z^2/4}\right| = \left|-\frac{r^2}{\nu t}e^{-r^2/4\nu t} + 2\int_0^{a\omega t} e^{-z^2/4} \cdot z\,dz\right|$$

$$= \left|4\left(1 - e^{-r^2/4\nu t}\right) - \frac{r^2}{\nu t}e^{-r^2/4\nu t}\right| \leqslant \text{const} \tag{74}$$

for any $x \in E_3$ and $t > 0$.

We have thus proved that

$$|G'_{2i}(x, t)| \leqslant \text{const}, \qquad i = 1, 2, \tag{75}$$

and consequently that

$$|K'_{2i}(x, t)| \leqslant \frac{\text{const}}{(\nu t)^{3/2}(\omega t)}, \qquad i = 1, 2,$$

where the constant does not depend on x or t.

Lemmas 1 and 2 of §2 will be used in estimating the kernels $K_1(x, t)$ and $K_{23}(x, t)$.

Estimation of the kernel $K_1(x, t)$. We consider the two cases 1) $\rho \geqslant |x_3|$ and 2) $\rho \leqslant |x_3|$.

1) Suppose $\rho \geqslant |x_3|$. Then $r/\rho = \sqrt{1 + x_3^2/\rho^2} \leqslant \sqrt{2}$.

Replacing the Bessel function in formula (68) for the kernel $K_1(x, t)$ by a trigonometric function according to the formula

$$J_{1/2}(\Omega) = \sqrt{\frac{2}{\pi\Omega}} \sin \Omega, \tag{76}$$

we get

$$|K_1(x, t)| \leqslant \frac{1}{(2\pi)^2 (\nu t)^{3/2} \omega t} \int_{-\infty}^{+\infty} \lambda^2 e^{-\lambda^2} \frac{|\sin [\omega t \sqrt{a^2 (\lambda + \beta/a)^2 + \alpha^2}]|}{|\sqrt{a^2 (\lambda + \beta/a)^2 + \alpha^2}|} d\lambda$$

$$\leqslant \frac{1}{(2\pi)^2 (\nu t)^{3/2} \omega t \alpha} \int_{-\infty}^{+\infty} \lambda^2 e^{-\lambda^2} d\lambda = \frac{\sqrt{2}}{(2\pi)^2 (\nu t)^{3/2} \omega t} \frac{\sqrt{\pi}}{2} = \frac{C_3}{(\nu t)^{3/2} \omega t}.$$

2) Suppose $\rho \leqslant |x_3|$. Then $\alpha \leqslant \beta$ and we find ourselves in the range of application of Lemma 1.

Consider the second representation of $K_1(x, t)$ in (68). Integrating by parts and using the formula

$$\int_0^{\omega t} J_0(\alpha \sqrt{(\omega t)^2 - y^2}) \cos (\beta y)\, dy = \sin (\omega t), \tag{77}$$

which is a special case of the more general formula (34), we get

$$K_1(x, t) = \frac{1}{4\sqrt{2}\, (2\pi\nu t)^{3/2} \omega t}$$
$$\times \int_0^{\omega t} \exp\left\{-\frac{a^2 y^2}{4}\right\} (2 - a^2 y^2)\, d\left(\int_0^y J_0(\alpha \sqrt{(\omega t)^2 - z^2}) \cos (\beta z)\right)$$
$$= \frac{\sin (\omega t)}{4\sqrt{2}\, (2\pi\nu t)^{3/2} \omega t} \exp\left\{-\frac{r^2}{4\nu t}\right\} \left(2 - \frac{r^2}{\nu t}\right) - \frac{1}{4\sqrt{2}\, (2\pi\nu t)^{3/2} \omega t}$$
$$\times \int_0^{\omega t} \left(\int_0^y J_0(\alpha \sqrt{(\omega t)^2 - z^2}) \cos (\beta z)\, dz\right) \exp\left\{-\frac{a^2 y^2}{4}\right\} \left(\frac{a^4 y^3}{2} - 3a^2 y\right) dy. \tag{78}$$

Here Lemma 1, the fact that

$$\left|\sin (\omega t) \exp\left\{-\frac{r^2}{4\nu t}\right\} \left(2 - \frac{r^2}{\nu t}\right)\right| \leqslant \text{const}$$

and estimates (73) and (74) imply

$$|K_1(x,t)| \leqslant \frac{\text{const}}{(\nu t)^{3/2}\,\omega t} + \frac{\text{const}}{(\nu t)^{3/2}\,\omega t}\left|\int_0^{\omega t} e^{-a^2y^2/4}\left(\frac{a^4y^3}{2} - 3a^2y\right)dy\right| \leqslant \frac{C_4}{(\nu t)^{3/2}\,\omega t}. \tag{79}$$

Thus

$$|K_1(x,\ t)| \leqslant \frac{C_5}{(\nu t)^{3/2}\,\omega t}, \tag{80}$$

where $C_5 = \min(C_3, C_4)$.

Estimation of the kernel $K_{23}(x,t) = K'_{23}(x,t) + K''_{23}(x,t)$. We again consider the same two cases.

1) Suppose $|x_3| \leqslant \rho$. Replacing the Bessel function $J_{3/2}(\Omega)$ in formula (70) for the kernel $K''_{23}(x,t)$ by trigonometric functions according to the formula

$$J_{3/2}(\Omega) = \sqrt{\frac{2}{\pi\Omega}}\left(\frac{\sin\Omega}{\Omega} - \cos\Omega\right). \tag{81}$$

we get

$$K''_{23}(x,t) = \frac{1}{2(2\pi\nu t)^{3/2}\sqrt{\omega t}}\int_{-\infty}^{+\infty} \lambda^2 e^{-\lambda^2}\frac{J_{3/2}\left[\omega t\sqrt{a^2(\lambda+\beta/a)^2+\alpha^2}\right]}{\left[\sqrt{a^2(\lambda+\beta/a)^2+\alpha^2}\right]^{3/2}}\,d\lambda$$

$$= \frac{1}{4\pi^2(\nu t)^{3/2}\,\omega t}\int_{-\infty}^{+\infty}\lambda^2 e^{-\lambda^2}\left\{\frac{\sin\left[\omega t\sqrt{a^2(\lambda+\beta/a)^2+\alpha^2}\right]}{\omega t\,[a^2(\lambda+\beta/a)^2+\alpha^2]^{3/2}} - \frac{\cos\left[\omega t\sqrt{a^2(\lambda+\beta/a)^2+\alpha^2}\right]}{[a^2(\lambda+\beta/a)^2+\alpha^2]}\right\}d\lambda,$$

which implies, since $1/\alpha^2 = r^2/\rho^2 \leqslant 2$ and $\omega t \geqslant t_0 > 0$, the estimate

$$|K''_{23}(x,\ t)| \leqslant \frac{\text{const}}{(\nu t)^{3/2}\omega t}\int_{-\infty}^{+\infty}\lambda^2 e^{-\lambda^2}d\lambda = \frac{\sqrt{\pi}}{2}\,\frac{\text{const}}{(\nu t)^{3/2}\,\omega t}. \tag{82}$$

The other summand $K'_{23}(x,t)$ is estimated for $|x_3| \leqslant \rho$ from the second part of formula (69) in exactly the same way as the $K'_{2i}(x,t)$ for $i = 1, 2$ were estimated:

$$|K'_{23}(x,t)| = \left|\frac{x_3}{\rho}\,\frac{a}{4\sqrt{2}(2\pi\nu t)^{3/2}\omega t}\int_0^{\omega t} J_1\left(\alpha\sqrt{(\omega t)^2-y^2}\right)\frac{\sqrt{(\omega t)^2-y^2}}{\omega t}\exp\left\{-\frac{a^2y^2}{4}\right\}\right.$$

$$\left.\times\sin(\beta y)\left(\frac{a^3y^3}{2} - 3ay\right)dy\right| \leqslant \frac{\text{const}}{(\nu t)^{3/2}\,\omega t}.$$

Thus when $|x_3| \leqslant \rho$

$$|K_{23}(x,\ t)| \leqslant \frac{C_6}{(\nu t)^{3/2}\omega t}, \tag{83}$$

where C_6 does not depend on x or t.

2) Suppose $|x_3| \leqslant \rho$. We first modify the form of $K''_{23}(x,t)$. Integrating by parts in the second part of (70), we get

$$K''_{23}(x,\ t) = \frac{r}{x_3}\,\frac{1}{4\sqrt{2}\,(2\pi\nu t)^{3/2}\,\omega t}\int_0^{\omega t} J_1\left(\alpha\sqrt{(\omega t)^2 - y^2}\right)\frac{\sqrt{(\omega t)^2 - y^2}}{(\omega t)\,\alpha}$$

$$\times \exp\left\{-\frac{a^2y^2}{4}\right\}(2 - a^2y^2)\,d\sin(\beta y) = -\frac{r}{x_3}\,\frac{1}{4\sqrt{2}\,(2\pi\nu t)^{3/2}(\omega t)^2}\int_0^{\omega t}\sin(\beta y)\,e^{-a^2y^2/4}$$

$$\times\left\{\frac{J_1\left(\alpha\sqrt{(\omega t)^2 - y^2}\right)}{\alpha}\sqrt{(\omega t)^2 - y^2}\left(\frac{a^4y^3}{2} - 3a^2y\right) - J_1'\left(\alpha\sqrt{(\omega t)^2 - y^2}\right)y\,(2 - a^2y^2)\right.$$

$$\left.-\frac{J_1\left(\alpha\sqrt{(\omega t)^2 - y^2}\right)}{\alpha\sqrt{(\omega t)^2 - y^2}}\,y\,(2 - a^2y^2)\right\}dy.$$

Since $J_1'(x) + x^{-1}J_1(x) = J_0(x)$,* it follows that

$$K''_{23}(x,\ t) = \frac{r}{x_3}\,\frac{1}{4\sqrt{2}\,(2\pi\nu t)^{3/2}\,\omega t}\int_0^{\omega t}\sin(\beta y)\exp\left\{-\frac{a^2y^2}{4}\right\}\left[\frac{\sqrt{(\omega t)^2 - y^2}}{\omega t}\right.$$

$$\left.\times\frac{J_1\left(\alpha\sqrt{(\omega t)^2 - y^2}\right)}{\alpha}\left(3a^2y - \frac{a^4y^3}{2}\right) + J_0\left(\alpha\sqrt{(\omega t)^2 - y^2}\right)\frac{y}{\omega t}(2 - a^2y^2)\right]dy. \qquad (84)$$

Taking the sum now of $K'_{23}(x, t)$ and $K''_{23}(x, t)$ as given by the second part of formula (69) for $i = 3$ and by formula (84) respectively, we get

$$K_{23}(x,\ t) = \frac{r}{x_3}\,\frac{1}{4\sqrt{2}\,(2\pi\nu t)^{3/2}\,\omega t}\int_0^{\omega t}\sin(\beta y)\exp\left\{-\frac{a^2y^2}{4}\right\}J_0\left(\alpha\sqrt{(\omega t)^2 - y^2}\right)$$

$$\times\frac{y}{\omega t}(2 - a^2y^2)\,dy + \left(\frac{r}{x_3} - \frac{x_3}{r}\right)\frac{r}{\rho}\,\frac{1}{4\sqrt{2}(2\pi\nu t)^{3/2}(\omega t)}\int_0^{\omega t}\sin(\beta y)\exp\left\{-\frac{a^2y^2}{4}\right\}$$

$$\times J_1\left(\alpha\sqrt{(\omega t)^2 - y^2}\right)\frac{\sqrt{(\omega t)^2 - y^2}}{\omega t}\left(3a^2y - \frac{a^4y^3}{2}\right)dy. \qquad (85)$$

Since

$$\frac{r}{\rho}\left|\left(\frac{r}{x_3} - \frac{x_3}{r}\right)\right| = \left|\frac{\rho}{x_3}\right| \leqslant 1,$$

the modulus of the second summand in (85) is estimated with the use of (73) and (74) in exactly the same way as the $K_{2i}(x, t)$ for $i = 1, 2$ were estimated. It remains to estimate the first summand in (85).

To this end, using the formula

$$\frac{1}{\omega t}\int_0^{\omega t} z\sin(\beta z)\,J_0\left(\alpha\sqrt{(\omega t)^2 - z^2}\right)dz = \frac{x_3}{r}\left[\frac{\sin(\omega t)}{\omega t} - \cos(\omega t)\right], \qquad (86)$$

which is derived by differentiating (34) with respect to q, we integrate by parts as follows:

*Editor's note. See, for example, G. N. Watson, *A treatise on the theory of Bessel functions*, 2nd ed., Cambridge Univ. Press, 1945, p. 18, formula (3).

$$\frac{r}{x_3}\int_0^{\omega t} \exp\left\{-\frac{a^2y^2}{4}\right\}(2-a^2y^2)\, d\left(\int_0^y \frac{z}{\omega t}\sin(\beta z)\, J_0\left(\alpha\sqrt{(\omega t)^2-z^2}\right) dz\right)$$

$$=\left[\frac{\sin(\omega t)}{\omega t}-\cos(\omega t)\right]\exp\left\{-\frac{r^2}{4\nu t}\right\}\left(2-\frac{r^2}{\nu t}\right)$$

$$-\frac{r}{x_3}\int_0^{\omega t}\left(\int_0^y \frac{z}{\omega t}\sin(\beta z)\, J_0\left(\alpha\sqrt{(\omega t)^2-z^2}\right) dz\right)\exp\left\{-\frac{a^2y^2}{4}\right\}\left(\frac{a^4y^3}{2}-3a^2y\right) dy. \tag{87}$$

According to the second mean value theorem, the inner integral here can be represented in the form

$$\int_0^y \frac{z}{\omega t}\sin(\beta z)\, J_0\left(\alpha\sqrt{(\omega t)^2-z^2}\right) dz = \frac{y}{\omega t}\int_{y_1}^y \sin(\beta z)\, J_0\left(\alpha\sqrt{(\omega t)^2-z^2}\right) dz, \tag{88}$$

where $0\leqslant y_1\leqslant y\leqslant \omega t$.

The first summand in (87) is obviously bounded uniformly in x and t. The modulus of the second summand in (87) is estimated by using Lemma 2, formula (88), estimates (73) and (74) and the facts that $|r/x_3|\leqslant\sqrt{2}$ and $|y/\omega t|\leqslant 1$ for $0\leqslant y\leqslant\omega t$:

$$\left|\frac{r}{x_3}\int_0^{\omega t}\left(\frac{y}{\omega t}\int_{y_1}^y \sin(\beta z)\, J_0\left(\alpha\sqrt{(\omega t)^2-z^2}\right) dz\right)\exp\left\{-\frac{a^2y^2}{4}\right\}\left(3a^2y-\frac{a^4y^3}{2}\right) dy\right|$$

$$\leqslant\sqrt{2}\, C_7\int_0^{\omega t}\exp\left\{-\frac{a^2y^2}{4}\right\}\left(3a^2y+\frac{a^4y^3}{2}\right) dy\leqslant C_8,$$

where C_8 does not depend on x or t. We have thus proved that

$$|K_{23}(x,\ t)|\leqslant\frac{C_9}{(\nu t)^{3/2}\,\omega t}, \tag{89}$$

where C_9 is an absolute constant.

The proof of the theorem for the function $P(x, t)$ is carried out by the same method. We recall from (31) that

$$P(x,\ t)=\int_{E_3}\{v_1^0(y)\, K_{32}(x-y,t)-v_2^0(y)\, K_{31}(x-y,t)-v_3^0(y)\, K_4(x-y,t)\}\, dy. \tag{90}$$

Let us estimate the kernels K_{3j}, $j=1, 2$, and K_4. We recall from (39) that

$$K_{3j}(x,\ t)=-\frac{x_j}{\rho}\,\frac{\omega a}{4\sqrt{2}\,(2\pi)^{3/2}\,(\nu t)\,(\omega t)}$$

$$\times\int_0^{\omega t} J_1\left(\alpha\sqrt{(\omega t)^2-y^2}\right)\frac{\sqrt{(\omega t)^2-y^2}}{\omega t}\, e^{-a^2y^2/4}\cos(\beta y)\,(2-a^2y^2)\, dy.$$

Since $|x_j|/\rho \leqslant 1$ $(j=1,2)$, $|J_1(\cdot)|<1$ and $|\cos(\beta y)| \leqslant 1$, it follows upon making the substitution $ay = 2z$ and integrating by parts that

$$|K_{3j}(x, t)| \leqslant \frac{\omega}{\sqrt{2}\,(2\pi)^{3/2}(\nu t)(\omega t)} \int_0^{\frac{a\omega t}{2}} e^{-z^2}(1+2z^2)\,dz$$

$$\leqslant \frac{\text{const}}{(\nu t)(\omega t)}\left[\Phi\left(\frac{a\omega t}{2}\right)+\frac{r}{4(\nu t)^{1/2}}\,e^{-r^2/4\nu t}\right]. \quad (91)$$

But the error function $\Phi(x)$ satisfies the estimate $|\Phi(x)| \leqslant \sqrt{\pi/2}$ for all x, and hence

$$|K_{3j}(x, t)| \leqslant \frac{C_{11}}{(\nu t)(\omega t)}. \quad (92)$$

In estimating $K_4(x,t)$, we consider (as before in estimating $K_1(x,t)$) two cases.

1) Suppose $|x_3| \leqslant \rho$. From (32″) and (23) we have

$$|K_4(x, t)| = \left|\frac{\omega}{(2\pi)^2(\nu t)(\omega t)} \int_{-\infty}^{+\infty} \lambda \exp\{-\lambda^2\} \frac{\sin\left[\omega t\sqrt{a^2(\lambda+\beta/a)^2+\alpha^2}\right]}{\sqrt{a^2(\lambda+\beta/a)^2+\alpha^2}}\,d\lambda\right|$$

$$\leqslant \frac{r}{\rho}\,\frac{\omega}{2\pi^2(\nu t)(\omega t)} \int_0^{\infty} \lambda e^{-\lambda^2}d\lambda = \frac{r}{\rho}\,\frac{\omega}{(2\pi)^2(\nu t)(\omega t)} \leqslant \frac{\sqrt{2}\,\omega}{(2\pi)^2(\nu t)(\omega t)}. \quad (93)$$

2) Suppose $|x_3| \geqslant \rho$. Integrating by parts in the second of formulas (39), we get

$$K_4(x,t) = -\frac{\omega}{2(2\pi)^{3/2}(\nu t)(\omega t)} \int_0^{\omega t} ay \exp\left\{-\frac{a^2y^2}{4}\right\} d\left(\int_0^y J_0\left(\alpha\sqrt{(\omega t)^2-z^2}\right)\sin(\beta z)\,dz\right)$$

$$= -\frac{\omega}{2(2\pi)^{3/2}(\nu t)(\omega t)}\,\frac{r}{\sqrt{\nu t}} \exp\left\{-\frac{r^2}{4\nu t}\right\} \int_0^{\omega t} J_0\left(\alpha\sqrt{(\omega t)^2-z^2}\right)\sin(\beta z)\,dz$$

$$+\frac{\omega}{2(2\pi)^{3/2}(\nu t)(\omega t)} \int_0^{\omega t}\left(\int_0^y J_0\left(\alpha\sqrt{(\omega t)^2-z^2}\right)\sin(\beta z)\,dz\right)\exp\left\{-\frac{a^2y^2}{4}\right\} a\left(1-\frac{a^2y^2}{2}\right)dy.$$

It follows from Lemma 2 in the same way as in the estimation of K_{3j} $(j=1,2)$ that

$$|K_4(x, t)| \leqslant \frac{C_{12}}{(\nu t)(\omega t)}, \quad (94)$$

where the constant again does not depend on x or t.

We have thus obtained the following uniform estimates for all $x \in E_3$ and $t>0$:

$$|K_1(x, t)|,\ |\mathbf{K}_{23}(x, t)| \leqslant \frac{C}{(\nu t)^{3/2}\,\omega t},$$

$$|K_{3j}(x, t)|,\ |K_4(x, t)| \leqslant \frac{C}{(\nu t)(\omega t)},$$

where the constants C do not depend on x or t. Substituting these estimates into the solution of the Cauchy problem, we get

$$|\mathbf{v}(x, t)| \leqslant \frac{C}{(\nu t)^{3/2}\,\omega t}\|\mathbf{v}^0\|_{L_1(E_3)},$$

$$|P(x, t)| \leqslant \frac{C}{(\nu t)\,(\omega t)}\|\mathbf{v}^0\|_{L_1(E_3)}.$$

The theorem is proved.

REMARK 1. If $\omega = 0, \nu > 0$ and the initial function $\mathbf{v}^0(x)$ is solenoidal and belongs to L_1, the vector function $\mathbf{v}(x, t)$ will decrease like $(\nu t)^{-3/2}$ while the scalar function $P(x, t)$ will be identically equal to zero.

The proof follows from the following explicit representation of the solution of the Cauchy problem (1*), (2) in the absence of the terms $[\mathbf{v}, \omega]$:

$$\begin{cases} \mathbf{v}(x, t) = \dfrac{1}{8(\pi\nu t)^{3/2}} \displaystyle\int_{E_3} \exp\left\{-\frac{r^2}{4\nu t}\right\} \mathbf{v}^0(y)\,dy, \\ P(x, t) = 0. \end{cases}$$

REMARK 2. As was previously proved by us (see [4], §3), if $\nu = 0, \omega \neq 0$ and the initial function $\mathbf{v}^0(x)$ is solenoidal and decreases sufficiently rapidly as $|x| \to \infty$, the functions $\mathbf{v}(x, t)$ and $P(x, t)$ will decrease like $O(1/\omega t)$ as $t \to \infty$.

REMARK 3. If $\nu = 0$ and $\omega = 0$, i.e. system (1*) takes the form

$$\begin{cases} \dfrac{\partial \mathbf{v}}{\partial t} + \operatorname{grad} P = 0, \\ \operatorname{div} \mathbf{v} = 0 \end{cases}$$

and the initial data is solenoidal, then a solution that decreases as $|x| \to \infty$ (so that the Fourier transform can be applied, i.e. the preceding situation can be realized) will have the form

$$\mathbf{v}(x, t) = \frac{1}{4\pi} \int_{E_3} (-\Delta \mathbf{v}^0) \frac{1}{r}\,dy = \mathbf{v}^0(x), \quad P(x, t) = 0.$$

Thus the solution is trivial: it does not depend on t and is equal to the initial function. The solution will therefore not decrease with increasing t.

REMARK 4. As indicated by the theorem just proved, the power of t at which the solution decreases as $t \to \infty$ in the case of a rotating viscous fluid is the sum of the power of t at which the solution decreases in the case of a viscous fluid when rotation is not taken into account and the power of t at which the solution decreases in the case of a Sobolev system, i.e. when rotation but not viscosity is taken into account.

REMARK 5. V. P. Maslov has directed my attention to the fact that the degeneration of homogeneous isotropic turbulence in an incompressible viscous fluid, viz. the law of decay of M. D. Millionščikov [14], [15], is of the order of $O(t^{-5/2})$.

In deducing this asymptotic law in [14], Millionščikov considered the Cauchy problem for the nonlinear Navier-Stokes system

$$\begin{cases} \frac{\partial \mathbf{v}}{\partial t}+(\mathbf{v}, \nabla)\mathbf{v}-\nu\Delta\mathbf{v}+\frac{1}{\rho}\operatorname{grad} P=0, \\ \operatorname{div}\mathbf{v}=0 \end{cases}$$

with the initial condition

$$\mathbf{v}(\mathrm{x}, t)|_{t=0}=\mathbf{v}^0(x), \qquad x=(x_1, x_2, x_3).$$

In the study of turbulent motions the initial data are considered to be random variables. Therefore a general statement of the problem of turbulent motion is: given the probability distribution for the initial functions, find the probability distribution for the random functions $\mathbf{v}$ and P of a point Q at any subsequent moment of time $t > 0$.

In [14] Millionščikov studied not the probability distributions themselves but their second moments (characterizing the mean square of the pulsation rate). For these moments he obtained in five-dimensional $y = (y_1, \cdots, y_5)$ space the heat equation

$$\nu\sum_{i=1}^{5}\frac{\partial^2 L_{dd}}{\partial y_i^2}=\frac{\partial L_{dd}}{\partial t},$$

for which the solution of the Cauchy problem, as is well known, has the form

$$L_{dd}=\frac{1}{(2\nu\pi t)^{5/2}}\int_{E_5} L_{dd}^0(z)\exp\left\{-\frac{r^2}{4\nu t}\right\}dz,$$

which implies the law of decay $O((\nu t)^{-5/2})$. It is interesting to note that the vector $\mathbf{v}(x, t)$ in the case of a rotating viscous fluid considered by us, i.e. in the case of the existence of vortices in an infinite fluid stream, decays at the same rate.

BIBLIOGRAPHY

1. V. N. Maslennikova, *Explicit representations and a priori estimates for the solutions of boundary problems for Sobolev systems*, Sibirsk. Mat. Ž. **9** (1968), 1182–1198 = Siberian Math. J. **9** (1968), 883–897. MR **38** #3621.

2. C. W. Oseen, *Neuere Methoden und Ergebnisse in der Hydrodynamik*, Akademische Verlag, Leipzig, 1927.

3. J. Leray, *Essai sur les mouvements plans d'un liquide visqueux que limitent des parois*, J. Math. Pures Appl. (9) **13** (1934), 331–418.

4. V. N. Maslennikova, *L_p-estimates and the asymptotic behavior as $t \to \infty$ of a solution of the Cauchy problem for a Sobolev system*, Trudy Mat. Inst. Steklov. **103** (1968), 117–141 = Proc. Steklov Inst. Math. **103** (1968), 123–150. MR **41** #5758; #6643.

5. ———, *An explicit representation and asymptotic behavior for $t \to \infty$ of the solution of the Cauchy problem for a linearized system of a rotating compressible fluid*, Dokl. Akad. Nauk SSSR **187** (1969), 989–992 = Soviet Math. Dokl. **10** (1969), 978–982. MR **40** #3043.

6. V. A. Ditkin and A. P. Prudnikov, *Handbook of operational calculus*, "Vysš. Škola", Moscow, 1965; French transl., Masson, Paris, 1967. MR **36** #1931.

7. G. E. Šilov, *Mathematical analysis. Special course*, 2nd ed., Fizmatgiz, Moscow, 1961; English transl., Pergamon Press, New York, 1965. MR **24** #A1180; **32** #2519.

8. N. Ja. Sonin, *Recherches sur les fonctions cylindriques et le développement des fonctions continues en séries*, Math. Ann. **16** (1880), 1–80.

9. S. L. Sobolev, *On a new problem of mathematical physics*, Izv. Akad. Nauk SSSR Ser. Mat. **18** (1954), 3–50. (Russian) MR **16**, 1029.

10. H. Weyl, *The method of orthogonal projection in potential theory*, Duke Math. J. **7** (1940), 411–444. MR **2**, 202.

11. V. N. Maslennikova, *Solution of a mixed problem for nonstationary motion of a rotating viscous fluid and a study of the differential properties of the solution*, Sibirsk. Mat. Ž. **2** (1961), 708–718. (Russian) MR **25** #4712.

12. ———, *Mathematical problems of the hydrodynamics of a rotating fluid, and Sobolev systems*, Doctoral Dissertation, Moscow, 1971. (Russian)

13. M. A. Lavrent′ev and B. V. Šabat, *Methods of the theory of functions of a complex variable*, 3rd corrected ed., "Nauka", Moscow, 1965; German transl., Mathematik für Naturwiss. und Technik, Band 13, VEB Deutscher Verlag, Berlin, 1967. MR **35** #344, #345.

14. M. D. Millionščikov, *Decay of homogeneous isotropic turbulence in a viscous incompressible fluid*, C. R. (Dokl.) Acad. Sci. URSS **22** (1939), 231–235.

15. A. N. Kolmogorov, *On degeneration of isotropic turbulence in an incompressible viscous liquid*, C. R. (Dokl.) Acad. Sci. URSS **31** (1941), 538–540. MR **3**, 26.

16. G. M. Fihtengol′z, *A course in differential and integral calculus*. Vol. II, 2nd ed., GITTL, Moscow, 1948; 4th ed., Fizmatgiz, Moscow, 1959; German transl., VEB Deutscher Verlag, Berlin, 1966. MR **39** #1b.

Trudy Mat. Inst. Steklov.
126 (1973)

Proc. Steklov Inst. Math.
126 (1973)

ASYMPTOTIC BEHAVIOR FOR LARGE VALUES OF TIME OF THE SOLUTIONS OF THE SECOND AND THIRD EXTERIOR BOUNDARY VALUE PROBLEMS FOR THE WAVE EQUATION WITH TWO SPATIAL VARIABLES

UDC 517.947.4

L. A. MURAVEĬ

ABSTRACT. In this paper asymptotic expansions as $t \to \infty$ are obtained for solutions of the second and third exterior boundary value problems for the two-dimensional wave equation in the case of a convex boundary contour and initial data with compact support. The possibility of analytic continuation into part of the region $\operatorname{Im} k > 0$ is established for the Green's function of the corresponding boundary value problem for the operator $\Delta + k^2$. A representation of the Green's function for small k in the form of a convergent series is obtained as well as estimates as $k \to \infty$.

Bibliography: 21 items.

Introduction. Formulation of the basic results

Let $u(x, t)$ denote a function satisfying the following conditions in the exterior Ω of a closed convex contour Γ:

$$\partial^2 u(x, t)/\partial t^2 = \Delta u(x, t), \qquad x \in \Omega, \qquad t > 0; \tag{0.1}$$

$$\partial u(x, t)/\partial t|_{t=0} = f_0(x), \qquad u(x, t)|_{t=0} = f_1(x), \qquad x \in \bar{\Omega}; \tag{0.2}$$

$$\partial u(x, t)/\partial n + g(x) u(x, t)|_{x \in \Gamma} = 0, \qquad t \geqslant 0, \tag{0.3}$$

where $x = (x_1, x_2)$, $\Delta = \partial^2/\partial x_1^2 + \partial^2/\partial x_2^2$, and $n = (n_1, n_2)$ is the inward (outward with respect to Ω) unit normal to Γ.

It is presupposed that the contour Γ is twice smooth, having a positive curvature. We will assume that the function $g(x)$ is nonnegative and continuously differentiable on Γ and that the functions $f_\alpha(x)$ $(\alpha = 0, 1)$ are $\alpha + 1$ times continuously differentiable in $\bar{\Omega}$, vanish for sufficiently large values of $|x|$ and satisfy the consistency conditions

$$\partial f_\alpha(x)/\partial n + g(x) f_\alpha(x)|_{x \in \Gamma} = 0, \qquad \alpha = 0, 1.$$

The paper is devoted to an investigation of the asymptotic behavior of the solution $u(x, t)$ of problem (0.1)–(0.3) at an arbitrary point $x \in \bar{\Omega}$ for $t \to \infty$.

AMS (MOS) subject classifications (1970). Primary 35L05; Secondary 35B40.

Before formulating the results, we note that it has been established in the case of three spatial variables that the solutions of all three exterior boundary value problems for the wave equation decrease exponentially on any compactum with respect to x as $t \to \infty$. This was done for the first boundary value problem $(u|_\Gamma = 0)$ by Lax, Morawetz and Phillips [1] under the assumption that the surface Γ is star-shaped, and for the second and third boundary value problems by Mihaĭlov [2] under the assumption that the surface Γ has positive Gaussian curvature. The method employed in [1] made use of the weaker estimate $O(1/t)$ of the solution of the first boundary value problem obtained earlier by Morawetz [3], whereas the method employed in [2] was based on estimates obtained in [2] of the analytic continuations of the Green functions of the corresponding boundary value problems for the Helmholtz equation.

In the present paper, using the same method as in [2], we investigate the asymptotic behavior of the solutions of the second and third boundary value problems for the wave equation with two spatial variables. Our basic result is

THEOREM 1. *Let* $u_\alpha(x, t)$ $(\alpha = 0, 1)$ *denote the solution of problem* (0.1)–(0.3) *when* $f_{1-\alpha} \equiv 0$, *and suppose* $x \in \bar{\Omega}$ *and* $t \to \infty$. *Then when* $g(x) = 0$ (*the second boundary value problem*)

$$u_\alpha(x, t) = \frac{(-1)^\alpha}{2\pi t^{1+\alpha}} \iint_\Omega f_\alpha(y)\, dy + (-1)^{1+\alpha}(2+\alpha)!\, \frac{\operatorname{meas}(E_2 \,|\, \Omega)}{\pi} \frac{\ln t}{t^{3+\alpha}} \iint_\Omega f_\alpha(y)\, dy$$
$$+ \sum_{n=1}^{N} \sum_{m=0}^{n} c_{m,n}^{(\alpha)}(x) \frac{\ln^m t}{t^{2n+1+\alpha}} + \tilde{u}_N^{(\alpha)}(x, t), \qquad \alpha = 0, 1, \tag{0.4}$$

and when $g(x) \not\equiv 0$ (*the third boundary value problem*)

$$u_\alpha(x, t) = \frac{(-1)^\alpha \varphi_\alpha(t)}{\pi t^{1+\alpha} \ln^2 t} \nu(x) \iint_\Omega \mu(y) f_\alpha(y)\, dy + \sum_{n=1}^{N} \sum_{m=0}^{n-1} c_{m,n}^{(\alpha)}(x) \frac{\ln^m t}{t^{2n+1+\alpha}}$$
$$+ \sum_{n=1}^{N} \sum_{m=2}^{n+2} d_{m,n}^{(\alpha)}(x) \frac{\varphi_{m,n}^{(\alpha)}(t)}{t^{2n+1+\alpha} \ln^m t} + \tilde{\tilde{u}}_N^{(\alpha)}(x, t). \tag{0.5}$$

In (0.4) *the function* $c_{1,1}^{(\alpha)}(x) \equiv 0$, *the functions* $c_{m,n}^{(\alpha)}(x)$ *are continuous in* $\bar{\Omega}$, *and the functions* $\tilde{u}_N^{(\alpha)}(x, t)$ *satisfy the estimates*

$$|\tilde{u}_N^{(\alpha)}(x, t)| \leqslant C_\alpha \frac{\ln^{N+1} t}{t^{2N+3+\alpha}}, \qquad \alpha = 0, 1, \tag{0.6}$$

for $x \in \bar{\Omega}_R$ *and* $t \geqslant T_\alpha$, *where* N *is any positive integer*, $\Omega_R = \Omega \cap \{|x| < R\}$, *and the positive numbers* T_α *and* C_α *depend on* N, Γ, R *and* f_α. *In* (0.5) *the functions* $\nu(x)$ *and* $\mu(x)$ *are harmonic in* Ω *and continuous in* $\bar{\Omega}$, *the functions* $c_{m,n}^{(\alpha)}(x)$ *and* $d_{m,n}^{(\alpha)}(x)$ *are continuous in* $\bar{\Omega}$, *the functions* $\phi_\alpha(t)$ *and* $\phi_{m,n}^{(\alpha)}(t)$ *depend continuously on* t *and have finite nonzero limits as* $t \to \infty$, *and the functions* $\tilde{\tilde{u}}_N^{(\alpha)}(x, t)$ *satisfy the estimates*

$$|\tilde{u}_N^{(\alpha)}(x, t)| \leqslant C_\alpha \frac{\ln^N t}{t^{2N+3+\alpha}}, \qquad \alpha = 0, 1, \tag{0.7}$$

for $x \in \bar{\Omega}_R$ *and* $t \geqslant T_\alpha$, *where the positive numbers* T_α *and* C_α *depend on* N, Γ, g, R *and* f_α *(the functions* $\nu(x)$, $\mu(x)$ *and* $\phi_\alpha(t)$ *will be indicated below).*

Let us compare the asymptotic behavior of the solutions of the boundary value problems with the behavior as $t \to \infty$ of the solution of the Cauchy problem for the wave equation. Suppose each function $f_\alpha(x)$ $(\alpha = 0, 1)$ has a compact support in E_2 and let $w_\alpha(x, t)$ denote the solution when $f_{1-\alpha} \equiv 0$ of the Cauchy problem for equation (0.1) with the initial data of (0.2) in E_2. As is well known, the functions $w_\alpha(x, t)$ can be written in the form

$$w_\alpha(x, t) = \frac{\partial^\alpha}{\partial t^\alpha}\left\{\frac{1}{2\pi}\iint\limits_{|y|\leqslant t} \frac{f_\alpha(x+y)}{\sqrt{t^2 - |y|^2}}\,dy\right\}, \qquad \alpha = 0, 1.$$

It readily follows that at any point $x \in E_2$

$$w_\alpha(x, t) = \frac{(-1)^\alpha}{2\pi t^{1+\alpha}}\iint\limits_{E_2} f_\alpha(y)\,dy + \sum_{n=1}^{N}\frac{c_n^{(\alpha)}(x)}{t^{2n+1+\alpha}} + \tilde{w}_N^{(\alpha)}(x, t) \tag{0.8}$$

as $t \to \infty$, the functions $c_n^{(\alpha)}(x)$ being continuous in E_2 and the functions $\widetilde{w}_N^{(\alpha)}(x, t)$ satisfying the estimates

$$|\widetilde{w}_N^{(\alpha)}(x, t)| \leqslant C_\alpha / t^{2N+\alpha+3}$$

for $x \in \bar{\Omega}_R$ and $t \geqslant T_\alpha$, where $Q_R = \{|x| < R\}$ and the numbers T_α and C_α depend on N, R and f_α.

Comparing (0.4) with (0.8), we conclude that the leading term of the asymptotic expansion as $t \to \infty$ of the solution of the second boundary value problem does not depend on x and coincides with the leading term of the asymptotic expansion of the solution of the corresponding Cauchy problem when the initial functions f_α are extended onto the complement of Ω by setting them equal to zero there. The second term of (0.4) characterizes the scattering of waves (the difference between the solutions of the Cauchy problem and a boundary value problem) by an obstacle $E_2 \backslash \Omega$ with the second boundary condition on its boundary Γ, has the form $\ln t / t^{3+\alpha}$ as a function of t, does not depend on x and is proportional to the area of the obstacle. This result was obtained in our papers [4]–[6] under stronger restrictions on the smoothness of the contour Γ.

An asymptotic expansion analogous to (0.4) of the Green function of the second boundary value problem for the wave equation that was derived under the same assumptions concerning the obstacle $E_2 \backslash \Omega$ as were used by us in [4]–[6] has since been announced by V. M. Babič in the note [7] (without elucidating the dependence of the coefficients on the characteristics of the obstacle).

Comparing (0.5) with (0.8), we see that the solution $u(x, t)$ of the third boundary value problem generally tends to zero more rapidly than the solution of the Cauchy problem as $t \to \infty$ and has the following finite norm, for example, at any point $x \in \bar{\Omega}$:

$$\|u(x, t)\|_{L_1(0, \infty)} = \int_0^\infty |u(x, t)|\, dt.$$

The proof of Theorem 1 is based on the following properties of the Green function $G(x, y, k)$ of the corresponding boundary value problem for the equation $\Delta v(x') + k^2 v(x') = f(x')$, $x' \in \Omega$, where $x, y \in \Omega$ and $\operatorname{Im} k \geqslant 0$ $(k \neq 0)$. Let K_β denote a domain of the complex k plane of the form

$$K_\beta = \{k\colon |\operatorname{Im} k| < \beta(1 + |\operatorname{Re} k|^{1/3});\ \operatorname{Re} k \neq 0 \text{ for } -\beta < \operatorname{Im} k \leqslant 0\}, \quad (0.9)$$

where β is an arbitrary positive number.

For continuations made relative to the contour Γ and the function $g(x)$ we have

THEOREM 2. *There exists a positive number* $\beta = \beta(\Gamma, g)$ *such that for any* $x, y \in \bar{\Omega}$ $(x \neq y)$ *the function* $G(x, y, k)$ *admits an analytic continuation into the domain* K_β.

When $k \in \bar{K}_\beta \cap \{|\operatorname{Re} k| \geqslant \kappa\}$ *the function* $G(x, y, k)$ *satisfies the estimates*

$$|G(x, y, k)| \leqslant C|k|^{1/6}\left[\int_{|k|^{-1/3}}^{1} \tau^{1/2} \exp\{-B|k||x-y|\tau^2\}\, d\tau + |\ln(|k||x-y|)| \exp\{-B|k||x-y|\}\right] + C'|k|^{-1/3} \exp\{-B|k|^{1/3}|x-y|\}, \quad x, y \in \Gamma, \quad (0.10)$$

$$|G(x, y, k)| \leqslant C'|k|^{-1/6}|x-y|^{-1/2} \exp\{-\operatorname{Im} k|x-y| + D|\operatorname{Im} k|\},$$
$$x \in \Gamma,\ y \in \bar{\Omega}; \quad y \in \Gamma,\ x \in \bar{\Omega}; \quad (0.11)$$

$$|G(x, y, k)| \leqslant C'|k|^{1/3}|x-y|^{-1/2} \exp\{-\operatorname{Im} k|x-y| + D|\operatorname{Im} k|\} \times [1 + \exp\{-\operatorname{Im} k \sup_{x' \in \Gamma}(|x'-x| + |x'-y|)\}], \quad x, y \in \bar{\Omega}, \quad (0.12)$$

where κ *is a positive number depending on* Γ *and* g, *the constants* B, C *and* D *depend on* Γ *and the constant* C' *depends on* Γ *and* g.

When $k \in \bar{K}_\beta \cap \{|k| \geqslant \beta\} \cap \{|\operatorname{Re} k| \leqslant \kappa\}$, *the function* $G(x, y, k)$ *satisfies the estimate*

$$|G(x, y, k)| \leqslant C' \exp\{-\operatorname{Im} k|x-y|\}|x-y|^{-1/2}, \quad x, y \in \bar{\Omega}. \quad (0.13)$$

Suppose $k \in K_\beta \cap \{|k| < \beta\}$, $y \in \bar{\Omega}$ *and* $x \in \Omega$. *Then the function* $G(x, y, k)$ *has the following representations: when* $g(x) = 0$ (*the second boundary value problem*)

$$G(x, y, k) = G_0^{(0)}(x, y) - \frac{1}{2}\left(\frac{1}{\pi}\ln k + \gamma\right) + \frac{\operatorname{meas}(E_2 \mid \Omega)}{4\pi^2} k^2 \ln^2 k + \sum_{n=1}^{\infty}\sum_{m=0}^{n+1} G_{m,n}(x, y) k^{2n} \ln^m k \qquad (0.14)$$

and when $g(x) \not\equiv 0$ (*the third boundary value problem*) *either*

$$G(x, y, k) = G_0^{(1)}(x, y) + \frac{\mu^{(1)}(x)\nu^{(1)}(y)}{\delta\left(\frac{1}{\pi}\ln k + \gamma\right) - 1} + \sum_{n=1}^{\infty}\sum_{m=0}^{n} G_{m,n}^{+}(x, y) k^{2n} \ln^m k + \sum_{n=1}^{\infty}\sum_{m=1}^{n+1} G_{m,n}^{-}(x, y) k^{2n}\left[\delta\left(\frac{1}{\pi}\ln k + \gamma\right) - 1\right]^{-m}, \qquad (0.15)$$

or

$$G(x, y, k) = G_0^{(2)}(x, y) + \frac{\mu^{(2)}(x)\nu^{(2)}(y)}{\frac{1}{\pi}\ln k + \gamma} + \sum_{n=1}^{\infty}\sum_{m=0}^{n} G_{m,n}^{+}(x, y) k^{2n} \ln^m k + \sum_{n=1}^{\infty}\sum_{m=1}^{n+1} G_{m,n}^{-}(x, y) k^{2n}\left(\frac{1}{\pi}\ln k + \gamma\right)^{-m}. \qquad (0.16)$$

In (0.14) *the function* $G_{1,2}(x, y) \equiv 0$ *and*

$$G_0^{(0)}(x, y) = -\frac{1}{2\pi}\ln|x - y| - \frac{1}{2\pi}\int_\Gamma \frac{(n', x - x')}{|x - x'|^2}\mu_0^{(0)}(x', y)\,dx', \qquad x \in \Omega, \quad y \in \bar{\Omega},$$

where n' *is the unit normal to* Γ *that is outward relative to* Ω *and the function* $\mu_0^{(0)}(x, y)$ *is a solution of the equation*

$$\mu_0^{(0)}(x, y) + \frac{1}{\pi}\int_\Gamma \frac{(n', x - x')}{|x - x'|^2}\mu_0^{(0)}(x', y)\,dx' = -\frac{1}{\pi}\ln|x - y|, \qquad x \in \Gamma, \quad y \in \bar{\Omega}.$$

The representation (0.15) *is valid when there are no nonzero solutions* $\mu_0(x)$ *of the equation*

$$\mu_0(x) + \frac{1}{\pi}\int_\Gamma \left[\frac{(n', x - x')}{|x - x'|^2} - g(x')\ln|x - x'|\right]\mu_0(x')\,dx' = 0, \qquad x \in \Gamma. \qquad (0.17)$$

In this case

$$G_0^{(1)}(x, y) = -\frac{1}{2\pi}\ln|x-y| - \frac{1}{2\pi}\int_\Gamma\left[\frac{(n', x-x')}{|x-x'|^2} - g(x')\ln|x-x'|\right]$$

$$\times\left[\mu_0^{(1)}(x', y) + \frac{1}{\delta}\mu_1(x')\right]dx', \qquad x\in\Omega, \qquad y\in\bar\Omega,$$

$$\mu^{(1)}(x) = \frac{1}{2\pi}\int_\Gamma\left[\frac{(n', x-x')}{|x-x'|^2} - g(x')\ln|x-x'|\right]\mu_1(x')\,dx' - \frac{1}{2}, \qquad x\in\Omega, \tag{0.18}$$

$$\nu^{(1)}(y) = \frac{1}{\delta}\int_\Gamma g(x')\mu_0^{(1)}(x', y)\,dx', \qquad y\in\bar\Omega, \tag{0.19}$$

where $\mu_0^{(1)}(x, y)$ *is a solution of the equation*

$$\mu_0^{(1)}(x, y) + \frac{1}{\pi}\int_\Gamma\left[\frac{(n', x-x')}{|x-x'|^2} - g(x')\ln|x-x'|\right]\mu_0^{(1)}(x', y)\,dx'$$

$$= -\frac{1}{\pi}\ln|x-y| - 1/\delta, \qquad x\in\Gamma, \qquad y\in\bar\Omega,$$

and $\mu_1(x)$ *is a solution of the equation*

$$\mu_1(x) + \frac{1}{\pi}\int_\Gamma\left[\frac{(n', x-x')}{|x-x'|^2} - g(x')\ln|x-x'|\right]\mu_1(x')\,dx' = 1, \quad x\in\Gamma,$$

$$\delta = \int_\Gamma g(x')\mu_1(x')\,dx' \neq 0.$$

The representation (0.16) *is valid when there exists a nonzero solution* $\mu_0(x)$ *of equation* (0.17). *It is then possible to choose a solution* $\mu_0(x)$ *of this equation and a solution* $\nu_0(x)$ *of the adjoint equation of* (0.17) *such that*

$$\int_\Gamma \mu_0(x')g(x')\,dx' = 1, \qquad \int_\Gamma \nu_0(x')\,dx' = 1.$$

In this case

$$G_0^{(2)}(x, y) = -\frac{1}{2\pi}\ln|x-y| - \frac{1}{2\pi}\int_\Gamma\left[\frac{(n', x-x')}{|x-x'|^2} - g(x')\ln|x-x'|\right]$$

$$\times\left[\mu_0^{(2)}(x', y) + \mu_0(x')\right]dx' + \frac{1}{2\pi}\int_\Gamma \nu_0(x')\ln|x'-y|\,dx', \qquad x\in\Omega, \qquad y\in\bar\Omega,$$

$$\mu^{(2)}(x) = -\frac{1}{2\pi}\int_\Gamma\left[\frac{(n', x-x')}{|x-x'|^2} - g(x')\ln|x-x'|\right]\mu_0(x')\,dx' + \frac{1}{2}, \qquad x\in\Omega, \tag{0.21}$$

$$\nu_2(y) = \frac{1}{\pi}\int_\Gamma \nu_0(x')\ln|x'-y|\,dx', \qquad y\in\bar\Omega, \tag{0.22}$$

where $\mu_0^{(2)}(x, y)$ *is a solution of the equation*

$$\mu_0^{(2)}(x, y) + \frac{1}{\pi}\int_\Gamma\left[\frac{(n', x-x')}{|x-x'|^2} - g(x')\ln|x-x'|\right]\mu_0^{(2)}(x', y)\,dx'$$

$$= -\frac{1}{\pi}\ln|x-y| + \frac{1}{\pi}\int_\Gamma \nu_0(x')\ln|x'-y|\,dx', \qquad x\in\Gamma, \qquad y\in\bar\Omega,$$

satisfying the condition

$$\int_{\Gamma} \mu_0^{(2)}(x', y)\, g(x')\, dx' = 0, \qquad y \in \bar{\Omega}.$$

The functions $\mu^{(1)}, \mu^{(2)}, \nu^{(1)}$ *and* $\nu^{(2)}$ *are harmonic in* Ω *and continuous in* $\bar{\Omega}$; *the functions* $G_{m,n}$ *and* $G^{\pm}_{m,n}$ *are continuous with respect to* $x, y \in \bar{\Omega}$ *and satisfy the inequalities*

$$|G_{m,n}| + |G^{\pm}_{m,n}| \leqslant D(R)\, B^n(\Gamma, g) \tag{0.23}$$

for $x, y \in \bar{\Omega}_R$; *and the complex number* γ *is equal to* $-i/2 + (\mathbf{C} - \ln 2)/\pi$, *where* $\mathbf{C}$ *is Euler's constant.*

The representations (0.14)–(0.16) *minus some distinguished terms converge absolutely and uniformly in* $k \in \{|k| \leqslant \beta\}$ *for* $x, y \in \bar{\Omega}_R$.

In (0.5) *the functions* $\mu(x)$ *and* $\nu(x)$ *are respectively given by either* (0.18)–(0.19) *or* (0.21)–(0.22), *depending on whether or not* (0.17) *has a nontrivial solution, while the function* $\phi_\alpha(t)$ $(\alpha = 0, 1)$ *has the form*

$$\varphi_\alpha(t) = b_\alpha \ln^2 t \int_0^\infty \frac{\tau^\alpha e^{-a\tau}\, d\tau}{\ln^2(\tau/t) + \pi^2},$$

where the numbers a *and* b_α *are respectively given by* $a^{(1)} = 2\exp\{-\pi/\delta - \mathbf{C}\}$, $b_\alpha^{(1)} = \pi^2 [a^{(1)}]^{1+\alpha}/\delta$, $a^{(2)} = 2e^{-\mathbf{C}}$, $b_\alpha^{(2)} = \pi^2 [a^{(2)}]^{1+\alpha}$.

We note that in a series of papers beginning with a paper of Ursell [8] (e.g. [9], [10]) estimates of the Green function $G(x, y, k)$ of the second boundary value problem $(g = 0)$ for the Helmholtz equation were established for various reciprocal locations of the points x and y in the case of real $k \to \infty$ (the short wave diffraction problem). It is asserted in the note [11] that the analytic continuation of the Green function of the second boundary value problem does not have any poles in a domain of type (0.9) for $\operatorname{Im} k < 0$. In all of these papers the restrictions on the smoothness of the contour Γ are greater than in our Theorem 2.

We note further that the solutions of the equation $\Delta v(x) + k^2 v(x) = 0, x \in \Omega$, satisfying the radiation condition at infinity and one of the boundary conditions $v|_\Gamma = f(x, k)$ or $\partial v/\partial n|_\Gamma = f(x, k)$ were studied under the assumption that the functions $f(x, k)$ have a special form for $k \to 0$ by MacCamy [12].

Our paper consists of four sections.

In §1 we prove Theorem 1.

In §2 we prove the assertions of Theorem 2 concerning the properties of an analytic continuation of the Green function in a neighborhood of the point $k = 0$.

In §3 we establish estimates (from Theorem 2) of the analytic continuation of the Green function for large values of $|k|$.

In §4 we establish estimates of the analytic continuation of the Green function of the second boundary value problem for the Helmholtz equation in the exterior of a disk. These estimates are used in §3.

The author wishes to express his gratitude to V. P. Mihaĭlov for his constant interest in the present work.

§1. On the asymptotic behavior of the solutions of the second and third boundary value problems for the wave equation

By using the methods of [13], [14], it is not difficult to show that there exists under our assumptions a unique function $u(x, t)$ belonging to the Sobolev space $W_2^2(\Omega)$ for any $t \geqslant 0$ and satisfying the wave equation (for almost all $(x, t) \in \Omega \times (0 < t)$) together with the conditions

$$\int_\Gamma (\partial u(x, t)/\partial n + g(x) u(x, t))^2 dx = 0, \qquad t \geqslant 0, \tag{1.1}$$

$$\iint_\Omega [(\partial u(x, 0)/\partial t - f_0(x))^2 + (u(x, 0) - f_1(x))^2]\, dx = 0. \tag{1.2}$$

We then conclude on the basis of an imbedding theorem of Sobolev [13] that $u(x, t)$ is continuous in $\bar{\Omega} \times (0 \leqslant t)$.

In this section we investigate the behavior of $u(x, t)$ at an arbitrary point $x \in \bar{\Omega}$ as $t \to \infty$.

In the subsequent considerations we will make use of the inequality

$$\|u(x, t)\|_{W_2^2(\Omega)} \leqslant C(\Gamma, g, f_0, f_1)(t+1), \quad t \geqslant 0, \tag{1.3}$$

which will be proved at the end of this section.

It follows from (1.3) on the basis of Sobolev's imbedding theorem that

$$\max_{x \in \bar{\Omega}} |u(x, t)| \leqslant C'(\Gamma, g, f_0, f_1)(t+1), \qquad t \geqslant 0, \tag{1.4}$$

which implies the existence in the halfplane $\operatorname{Im} k > 0$ of the analytic function

$$U(x, k) = \int_0^\infty e^{ikt} u(x, t)\, dt. \tag{1.5}$$

Consider for $\operatorname{Im} k > 0$ the sequence of functions

$$U^{(m)}(x, k) = \int_0^m e^{ikt} u(x, t)\, dt, \qquad m = 1, 2, \dots.$$

Using (1.1), (1.4) and the Cauchy-Bunjakovskiĭ inequality, we deduce that when $\operatorname{Im} k > 0$ the functions $U^{(m)}$ satisfy the conditions

$$\|U^{(m)}(x, k)\|_{W_2^2(\Omega)} \leqslant C(\Gamma, g, f_0, f_1, \operatorname{Im} k), \tag{1.6}$$

$$\int_{\Gamma} |\partial U^{(m)}(x, k)/\partial n + g(x) U^{(m)}(x, k)|^2 \, dx = 0, \tag{1.7}$$

and, since when $\operatorname{Im} k > 0$

$$\|U - U^{(m)}\|^2_{W_2^2(\Omega)} \leqslant \int_m^\infty e^{-\operatorname{Im} kt} dt \int_m^\infty e^{-\operatorname{Im} kt} \|u(x, t)\|^2_{W_2^2(\Omega)} dt$$

$$\leqslant C^2(\Gamma, g, f_0, f_1) \left\{ \int^\infty e^{-\operatorname{Im} kt}(1+t)\, dt \right\}^2 \to 0$$

as $m \to \infty$, the function $U(x, t)$ also satisfies conditions (1.6) and (1.7).

We will show that $U(x, k) \equiv \widetilde{U}(x, k)$ for $\operatorname{Im} k > 0$, where $\widetilde{U}(x. k)$ is the classical solution of the problem

$$\left.\begin{array}{ll} \Delta \widetilde{U}(x, k) + k^2 \widetilde{U}(x, k) = f_0(x) + ikf_1(x) - f_0(x), & x \in \Omega \\ \partial \widetilde{U}(x, k)/\partial n + g(x) \widetilde{U}(x, k) = 0, & x \in \Gamma \\ \partial \widetilde{U}(x, k)/\partial |x| - ik\widetilde{U}(x, k) = e^{-\operatorname{Im} k |x|} o(|x|^{-1/2}), & |x| \to \infty \end{array}\right\} \tag{1.8}$$

(f_0 and f_1 are the functions in (0.2)).

Let $\{\psi(x)\}$ denote the set of functions that are twice continuously differentiable in $\bar{\Omega}$, vanish for sufficiently large values of $|x|$ and satisfy the condition

$$\partial \psi(x)/\partial n + g(x) \psi(x) = 0, \qquad x \in \Gamma. \tag{1.9}$$

Integrating by parts and taking into account conditions (0.2), (0.3) and (1.9), we get for any $T \geqslant 0$ the equalities

$$\int_0^T (T-t)\, dt \iint_{\Omega} (\partial^2 u/\partial t^2 - \Delta u) \psi dx = -T \iint_{\Omega} f_0(x) \psi(x)\, dx$$

$$-\iint_{\Omega} f_1(x) \psi(x)\, dx + \iint_{\Omega} u(x, T) \psi(x)\, dx$$

$$-\int_0^T (T-t)\, dt \iint_{\Omega} u(x. t) \Delta \psi(x)\, dx = 0.$$

Since $\psi(x) \equiv 0$ for sufficiently large values of $|x|$ and estimate (1.4) holds, the latter equality can be multiplied by e^{ikT} and termwise integrated with respect to T for $\operatorname{Im} k > 0$. As a result we get the equality

$$\iint_{\Omega} (k^2 \psi(x) + \Delta \psi(x)) U(x, k)\, dx = \iint_{\Omega} (ikf_1(x) - f_0(x)) \psi(x)\, dx.$$

Since $U(x, k)$ satisfies (1.6) and (1.7) and $\psi(x)$ satisfies (1.9), this equality can be rewritten in the form

$$\iint_{\Omega} [k^2 U \psi - (\nabla U \cdot \nabla \psi)]\, dx = \int_{\Gamma} g U \psi dx + \iint_{\Omega} (ikf_1 - f_0) \psi dx. \tag{1.10}$$

By using the methods of [15], it is not difficult to show that there exists for $\operatorname{Im} k > 0$ and under our restrictions on the contour Γ and the functions g, f_0 and f_1 a unique solution $\widetilde{U}(x, k)$ of problem (1.8), which obviously belongs to the space $W_2^1(\Omega)$. Therefore, if we multiply the equation of (1.8) by $\psi(x)$ and integrate by parts, taking into account the boundary condition of (1.8), we can conclude that (1.10) also holds for $\widetilde{U}(x, k)$; hence the difference $U_0(x, k) = U(x, k) - \widetilde{U}(x, k)$ satisfies the equality

$$\iint_\Omega [(\nabla U_0 \cdot \nabla\psi) - k^2 U_0\psi]\, dx + \int_\Gamma g U_0 \psi\, dx = 0 \tag{1.11}$$

for $\operatorname{Im} k > 0$ and $\psi \in \{\psi(x)\}$. Since the set $\{\psi(x)\}$ is everywhere dense in $W_2^1(\Omega)$, equality (1.11) holds for any function $\psi \in W_2^1(\Omega)$ and, in particular, for the function $\overline{U}_0$ (the complex conjugate of U_0), which, as the difference of two functions from $W_2^1(\Omega)$, also belongs to this space. Thus the following equalities hold for $\operatorname{Im} k > 0$:

$$\iint_\Omega \{|\nabla U_0|^2 + [(\operatorname{Im} k)^2 - (\operatorname{Re} k)^2]\,|U_0|^2\}\, dx + \int_\Gamma g\,|U_0|^2\, dx = 0,$$

$$\operatorname{Im} k \operatorname{Re} k \iint_\Omega |U_0|^2\, dx = 0.$$

When $\operatorname{Re} k \neq 0$ and $\operatorname{Im} k > 0$ the second equality, and when $\operatorname{Re} k = 0$, $\operatorname{Im} k > 0$ and $g \geqslant 0$ the first equality, implies that $U_0(x, k) \equiv 0$; hence the equality $U(x, k) = \widetilde{U}(x, k)$ is established for $\operatorname{Im} k > 0$. It follows that $\widetilde{U}(x, k)$ is analytic in the domain $\operatorname{Im} k > 0$ and that the inversion formula for the Laplace transform has the form

$$u(x, t) = \frac{1}{2\pi} \int_{\operatorname{Im} k = \varepsilon} e^{-ikt}\widetilde{U}(x, k)\, dk, \tag{1.12}$$

where ϵ is an arbitrary positive number.

Equality (1.12) permits one to reduce the study of the behavior of $u(x, t)$ as $t \to \infty$ to an investigation of the analytic continuation of the function $\widetilde{U}(x, k)$ into the domain $\operatorname{Im} k > 0$.

Let $\widetilde{U}_\alpha(x, k)$ $(\alpha = 0, 1)$ denote the solution of the problem (1.8) when $f_{1-\alpha} = 0$. When $\operatorname{Im} k > 0$ the functions $\widetilde{U}_\alpha(x, k)$ can be represented in the form

$$\widetilde{U}_\alpha(x, k) = (ik)^\alpha \left[v_\alpha(x, k) - \int_\Gamma G(x', x, k) \left(\frac{\partial v_\alpha(x', k)}{\partial n'} + g(x') v_\alpha(x', k) \right) dx' \right], \tag{1.13}$$

where

$$v_\alpha(x, k) = \frac{i}{4} \iint_{E_2} H_0^{(1)}(k\,|x - y|) f_\alpha(y)\, dy, \qquad \alpha = 0, 1, \tag{1.14}$$

$(i/4)H_0^{(1)}(k|x - y|)$ is a fundamental solution of the Helmholtz equation, $G(y, x, k)$

is the Green function of problem (1.8) and the functions $f_\alpha(x)$ $(\alpha = 0, 1)$ in (1.14) are assumed to be extended onto $E_2\backslash\bar{\Omega}$ with preservation of smoothness $(f_\alpha \in C^{1+\alpha}(E_2))$.

By making use of Theorem 2 and the representation (1.13), it is not difficult to show that the functions $\widetilde{U}_\alpha(x, k)$ admit analytic continuations into the domain

$$K_{\beta,\sigma} = \{k\colon |\operatorname{Im} k| < \beta(1 + |\operatorname{Re} k|^{1/3-\sigma});$$
$$\operatorname{Re} k \neq 0 \quad \text{for} \quad -\beta < \operatorname{Im} k \leqslant 0\} \quad \left(0 \leqslant \sigma \leqslant \frac{1}{3}\right)$$

satisfying the estimates

$$|\widetilde{U}_\alpha(x, k)| \leqslant C_\alpha |k|^{-(\sigma+1/6)} e^{|\operatorname{Im} k|(R+D_\alpha)},$$
$$x \in \bar{\Omega}_R, \qquad k \in \bar{K}_{\beta,\sigma} \cap \{|k| \geqslant \beta\}, \tag{1.15}$$

where the number β depends on Γ and g, the number R is arbitrary, the constants D_α depend on Γ and $\operatorname{diam}(\operatorname{supp} f_\alpha)$ and the constants C_α depend on Γ, g, f_α and R.

In fact, using the estimate

$$|H_0^{(1)}(k|y|)| \leqslant 3(|k||y|)^{-1/2} e^{-\operatorname{Im} k|y|}, \qquad k \in \bar{K}_\beta, \tag{1.16}$$

which directly follows from the integral representation 8.421,9 in [16], and the fact that the functions f_α have compact supports, we deduce from (1.14) that the functions $v_\alpha(x, k)$ are analytic in the domain K_β for any $x \in \bar{\Omega}$ and satisfy the inequalities

$$|v_\alpha(x, k)| \leqslant C_\alpha |k|^{-1/2} e^{|\operatorname{Im} k|(R+D_\alpha)}, \qquad \alpha = 0, 1, \tag{1.17}$$

for $x \in \bar{\Omega}_R$ and $k \in \bar{K}_\beta \cap \{|k| \geqslant \beta\}$, where C_α depends on f_α and $D_\alpha = \operatorname{diam}(\operatorname{supp} f_\alpha)$. By going over to polar coordinates ρ and ϕ, we can rewrite the functions $v_\alpha(x, k)$ in the form

$$v_\alpha(x, k) = \frac{i}{4}\int_0^\infty\int_0^{2\pi} \rho H_0^{(1)}(k\rho) f_\alpha(x_1 + \rho\cos\varphi,\ x_2 + \rho\sin\varphi)\, d\varphi\, d\rho, \qquad \alpha = 0, 1.$$

Since the functions f_α in this integral are continuously differentiable and have compact supports, we can integrate by parts and transform the resultant expression into the form

$$v_\alpha(x, k) = \int_0^\infty\int_0^{2\pi} g_\alpha(x, \rho, \varphi)\, d\varphi \left(\int_0^\rho \frac{i}{4} H_0^{(1)}(k\rho')\, d\rho'\right) d\rho, \tag{1.18}$$

where

$$g_\alpha(x, \rho, \varphi) = -\frac{\partial}{\partial\rho}(\rho f_\alpha(x_1 + \rho\cos\varphi,\ x_2 + \rho\sin\varphi))$$

$(\alpha = 0, 1)$ is a continuous function of x, ρ and ϕ that vanishes for $\rho \geqslant r = r(x)$.

Using the integral representation 8.421,5 in [16] for the function $H_0^{(1)}(\xi)$ $(\xi > 0)$, we get

$$\int_0^{\rho} \frac{i}{4} H_0^{(1)}(k\rho')\,d\rho' = \frac{1}{k}\int_0^{k\rho} \frac{i}{4} H_0^{(1)}(z)\,dz = \frac{1}{k}\int_0^{\operatorname{Re} k\rho} \frac{i}{4} H_0^{(1)}(\xi)\,d\xi$$

$$-\frac{1}{4k}\int_0^{\operatorname{Im} k\rho} H_0^{(1)}(\operatorname{Re} k\rho + i\zeta)\,d\zeta = \frac{1}{2\pi k}\int_1^{\infty} \frac{e^{i\operatorname{Re} k\rho\tau} - 1}{\tau\sqrt{\tau^2-1}}$$

$$-\frac{1}{4k}\int_0^{\operatorname{Im} k\rho} H_0^{(1)}(\operatorname{Re} k\rho + i\zeta)\,d\zeta$$

for $k \in K_\beta \cap \{|k| > \beta\} \cap \{\operatorname{Re} k > 0\}$. By taking into account the form of the functions $g_\alpha(x, \rho, \phi)$, we can rewrite the functions $v_\alpha(x, k)$ in the following way:

$$v_\alpha(x, k) = \frac{1}{k}\int_0^{\infty}\int_0^{2\pi} g_\alpha(x, \rho, \varphi)\left[\frac{1}{2\pi}\int_1^{\infty}\frac{e^{i\operatorname{Re} k\rho\tau}}{\tau\sqrt{\tau^2-1}}\,d\tau - \frac{1}{4}\int_0^{\operatorname{Im} k\rho} H_0^{(1)}(\operatorname{Re} k\rho + i\zeta)\,d\zeta\right] d\rho\, d\varphi. \tag{1.19}$$

Let us estimate the innermost integrals in (1.19). We have

$$\left|\frac{1}{2\pi}\int_1^{\infty}\frac{e^{i\operatorname{Re} k\rho\tau}\,d\tau}{\tau\sqrt{\tau^2-1}}\right| \leqslant \frac{1}{2\pi}\int_1^{\infty}\frac{d\tau}{\tau\sqrt{\tau^2-1}} \leqslant 1.$$

A more exact estimate holds for $\operatorname{Re} k\rho \geqslant 1$:

$$\frac{1}{2\pi}\int_1^{\infty}\frac{e^{i\operatorname{Re} k\rho\tau}\,d\tau}{\tau\sqrt{\tau^2-1}} = -\frac{1}{2\pi}e^{-i\operatorname{Re} k\rho}\int_1^{\infty}\left[\int_0^{\tau}\frac{e^{i\operatorname{Re} k\rho\tau_1}\,d\tau_1}{\sqrt{\tau_1}}\right] d\left(\frac{1}{(\tau+1)\sqrt{\tau+2}}\right),$$

$$\left|\int_0^{\tau}\frac{e^{i\operatorname{Re} k\rho\tau_1}\,d\tau_1}{\sqrt{\tau_1}}\right| \leqslant \frac{1}{\sqrt{\operatorname{Re} k\rho}}\left|\int_0^{\operatorname{Re} k\rho\tau}\frac{e^{i\tau_1}\,d\tau_1}{\sqrt{\tau_1}}\right| \leqslant \frac{\pi}{\sqrt{\operatorname{Re} k\rho}},$$

and hence

$$\left|\frac{1}{2\pi}\int_1^{\infty}\frac{e^{i\operatorname{Re} k\rho\tau}\,d\tau}{\tau\sqrt{\tau^2-1}}\right| \leqslant \frac{1}{1+\sqrt{\operatorname{Re} k\rho}}. \tag{1.20}$$

Using (1.16), we get the estimate

$$\left|\frac{1}{4}\int_0^{\operatorname{Im} k\rho} H_0^{(1)}(\operatorname{Re} k\rho + i\zeta)\,d\zeta\right| \leqslant e^{|\operatorname{Im} k|\rho}\int_0^{|\operatorname{Im} k|\rho}\frac{d\zeta}{[(\operatorname{Re} k\rho)^2+\zeta^2]^{1/4}}. \tag{1.21}$$

The inequality $|\operatorname{Im} k| \leqslant 2\beta$ holds for $k \in \bar{K}_\beta \cap \{|k| \geqslant \beta\} \cap \{0 \leqslant \operatorname{Re} k \leqslant 1\}$ (see (0.9)), so that the integral on the right in (1.21) can be estimated in the following way:

$$\int_0^{|\mathrm{Im}\, k|\rho} \frac{d\zeta}{[(\mathrm{Re}\, k)^2 + \zeta^2]^{1/4}} \leqslant \int_0^{|\mathrm{Im}\, k|\rho} \frac{d\zeta}{\sqrt{\zeta}} \leqslant 4\sqrt{\beta\rho}.$$

When $k \in \overline{K}_{\beta,\sigma} \cap \{|k| \geqslant \beta\} \cap \{\mathrm{Re}\, k \geqslant 1\}$ we estimate the integral on the right in (1.21) as follows:

$$\int_0^{|\mathrm{Im}\, k|\rho} \frac{d\zeta}{[(\mathrm{Re}\, k\rho)^2 + \zeta^2]^{1/4}} \leqslant \frac{|\mathrm{Im}\, k|\rho}{[\mathrm{Re}\, k\rho]^{1/2}} \leqslant 2\rho^{1/2}\beta\, (\mathrm{Re}\, k)^{-1/6-\sigma}.$$

It follows from (1.19) upon taking into account (1.20), (1.21) and the latter two inequalities that when $x \in \overline{\Omega}_R$ and $k \in \overline{K}_{\beta,\sigma} \cap \{|k| \geqslant \beta\} \cap \{\mathrm{Re}\, k \geqslant 0\}$ the functions $v_\alpha(x, k)$ satisfy the estimates

$$|v_\alpha(x, k)| \leqslant C_\alpha |k|^{-7/6-\sigma} \exp\{|\mathrm{Im}\, k|(R + D_\alpha)\}, \tag{1.22}$$

where D_α $(\alpha = 0, 1)$ depends on $\mathrm{diam}(\mathrm{supp}\, f_\alpha)$ and C_α $(\alpha = 0, 1)$ depends on f_α, Γ, g and R (since $\beta = \beta(\Gamma, g)$). Bearing in mind the well-known relation $H_0^{(1)}(k\rho) = \overline{H}_0^{(1)}(\bar{k}e^{i\pi}\rho)$, $k \in K_\beta$, we conclude that estimates (1.22) are valid for all $k \in \overline{K}_{\beta,\sigma} \cap \{|k| \geqslant \beta\}$.

Estimates (1.15) follow directly from the representations (1.13), inequalities (1.16) and (1.22) and estimate (0.11) in Theorem 2 if one takes into account the fact that under our assumptions the functions $\partial v_\alpha(x, k)/\partial n|_{x \in \Gamma}$ can be written in the form

$$\partial v_\alpha(x, k)/\partial n = \frac{i}{4} \iint_{E_2} H_0^{(1)}(k|y|)\, f'_\alpha(x, y)\, dy,$$

where

$$f'_\alpha(x, y) = (n \cdot \nabla f_\alpha(x + y)) \in C^\alpha(x \in \Gamma) \times C^\alpha(y \in E_2).$$

Therefore (1.16) holds for $\partial v_0(x, k)/\partial n$ while (1.22) holds for $\partial v_1(x, k)/\partial n$.

Estimates (1.15) are not sharp, but they nevertheless permit us to verify the natural assumption that in our case the asymptotic behavior of the functions $u_\alpha(x, t)$ as $t \to \infty$ are determined by the behavior of the functions $\widetilde{U}_\alpha(x, k)$ in a neighborhood of the singular point $k = 0$.

For suppose $1/12 < \sigma < 1/3$. We have established that the functions $\widetilde{U}_\alpha(x, k)$ are analytic in $K_{\beta,\sigma}$ for any $x \in \overline{\Omega}$ and satisfy estimates (1.15) for $x \in \overline{\Omega}_R$ and $k \in \overline{K}_{\beta,\sigma} \cap \{|k| \geqslant \beta\}$; hence for any $x \in \overline{\Omega}_R$ and $t \geqslant T_\alpha - 3(R + D_\alpha)$ we have

$$\left| \int_{-\beta(1+|\mathrm{Re}\, k|^{1/3-\sigma})}^{\beta} e^{-ikt}\widetilde{U}_\alpha(x, \mathrm{Re}\, k + i\zeta)\, d\zeta \right| \leqslant C_\alpha(\Gamma, g, f_\alpha, R)\, e^{\beta t}\, |\mathrm{Re}\, k|^{-(2\sigma - 1/6)} \to 0, \quad |\mathrm{Re}\, k| \to \infty,$$

so that when $x \in \overline{\Omega}_R$ and $t \geqslant T_\alpha$ (R is arbitrary) the functions

$$u_\alpha(x,\ t)=\frac{1}{2\pi}\int\limits_{\operatorname{Im}k=\beta} e^{-ikt}\tilde{U}_\alpha(x,\ k)\,dk$$

can be rewritten in the form

$$u_\alpha(x,\ t)=\frac{1}{2\pi}\int\limits_{|k|=\beta} e^{-ikt}\tilde{U}_\alpha(x,\ k)\,dk + \frac{1}{2\pi}\int\limits_{\operatorname{Im}k=-\beta(1+|\operatorname{Re}k|^{1/3-\sigma})} e^{-ikt}\tilde{U}_\alpha(x,\ k)\,dk = u_\alpha^{(1)}(x,\ t)+u_\alpha^{(2)}(x,\ t). \tag{1.23}$$

On the basis of (1.15) we have for any $x\in\bar{\Omega}_R$ and $t\geqslant T_\alpha$ the estimates

$$|u_\alpha^{(2)}(x,\ t)|\leqslant C_\alpha^{(1)}e^{-\beta t/3}\int\limits_0^\infty \xi^{-\sigma-1/6}e^{-T_\alpha\xi^{(1/3-\sigma)}}\,d\xi\leqslant C_\alpha^{(2)}e^{-\beta t/3}, \tag{1.24}$$

where the constant $C_\alpha^{(2)}$ $(\alpha=0,\ 1)$ depends on Γ, g, f_α and R.

Let us consider the behavior of the functions $u_\alpha^{(1)}(x,\ t)$ as $t\to\infty$. To this end we represent the functions $\tilde{U}_\alpha(x,\ k)$ for $\operatorname{Im}k\geqslant 0$ $(k\neq 0)$ in the form

$$\tilde{U}_\alpha(x,\ k)=(ik)^\alpha\iint\limits_\Omega G(x',\ x,\ k)f_\alpha(x')\,dx',\qquad \alpha=0,\ 1,\qquad x\in\Omega,$$

where $G(x',x,k)$ is the Green function of problem (1.8). From Theorem 2 we get that for any $x\in\bar{\Omega}$ the function $\tilde{U}_\alpha(x,\ k)$ $(\alpha=0,\ 1)$ can be analytically continued into the domain $K_\beta\cap\{|k|<\beta\}$ $(\beta=\beta(\Gamma,\ g))$ in the form of a series: when $g(x)=0$ (the second boundary value problem)

$$\tilde{U}_\alpha(x,\ k)=(ik)^\alpha\Bigg[\iint\limits_\Omega G_0^{(0)}(x',\ x)f_\alpha(x')\,dx'-\frac{1}{2\pi}(\ln k+\gamma)\iint\limits_\Omega f_\alpha(x')\,dx' + \frac{\operatorname{meas}(E_2\,|\,\Omega)}{4\pi^2}k^2\ln^2 k\iint\limits_\Omega f_\alpha(x')\,dx' + \sum_{n=1}^{\infty}\sum_{m=0}^{n+1}U_{(m,\ n)}^{(\alpha,\ 0)}(x)k^{2n}\ln^m k\Bigg], \tag{1.25}$$

and when $g(x)\not\equiv 0$ (the third boundary value problem) either

$$\tilde{U}_\alpha(x,\ k)=(ik)^\alpha\Bigg\{\iint\limits_\Omega G_0^{(1)}(x',\ x)f_\alpha(x')\,dx' + \frac{\nu^{(1)}(x)}{\delta\left(\frac{1}{\pi}\ln k+\gamma\right)-1}\iint\limits_\Omega \mu^{(1)}(x')f_\alpha(x')\,dx' + \sum_{n=1}^{\infty}\sum_{m=0}^{n}U_{m,\ n}^{(\alpha,\ 1)+}(x)k^{2n}\ln^m k + \sum_{n=1}^{\infty}\sum_{m=1}^{n+1}U_{m,\ n}^{(\alpha,\ 1)-}(x)k^{2n}\left[\delta\left(\frac{1}{\pi}\ln k+\gamma\right)-1\right]^{-m}\Bigg\}, \tag{1.26}$$

or

$$\tilde{U}_\alpha(x, k) = (ik)^\alpha \Bigg\{ \iint_\Omega G_0^{(2)}(x', x) f_\alpha(x')\, dx'$$

$$+ \frac{\nu^{(2)}(x)}{\frac{1}{\pi}\ln k + \gamma} \iint_\Omega \mu^{(2)}(x') f_\alpha(x')\, dx' + \sum_{n=1}^{\infty} \sum_{m=0}^{n} U_{m,n}^{(\alpha,2)+}(x) k^{2n} \ln^m k$$

$$+ \sum_{n=1}^{\infty} \sum_{m=1}^{n+1} U_{m,n}^{(\alpha,2)-}(x) k^{2n} \Big(\frac{1}{\pi}\ln k + \gamma\Big)^{-m} \Bigg\}. \tag{1.27}$$

Here the functions $U_{m,n}^{(\alpha,0)}(x)$ and $U_{m,n}^{(\alpha,j)\pm}(x)$ $(\alpha = 0, 1; j = 1, 2)$ are continuous in $x \in \bar{\Omega}$ and satisfy the inequalities

$$|U_{m,n}^{(\alpha,0)}| + |U_{m,n}^{(\alpha,j)\pm}| \leqslant C(R, f_\alpha) B^n(\Gamma, g), \quad x \in \bar{\Omega}_R, \tag{1.28}$$

and the series (1.25)–(1.27) minus some distinguished terms converge absolutely and uniformly in $k \in \{|k| \leqslant \beta\}$ and $x \in \bar{\Omega}_R$. Therefore in any of the cases being considered we have for $|k| \leqslant \beta$ and $x \in \bar{\Omega}_R$ the estimates

$$|\tilde{U}_\alpha(x, k)| \leqslant C(\Gamma, g, f_\alpha, R) |k^\alpha \ln k|.$$

It follows from these estimates that for any $x \in \bar{\Omega}$ and $t > 0$

$$\Bigg| \int_{|k|=\delta} e^{-ikt} \tilde{U}_\alpha(x, k)\, dk \Bigg| \to 0 \qquad \text{as} \qquad \delta \to 0.$$

Therefore the function $u_\alpha^{(1)}(x, t)$ $(\alpha = 0, 1)$ in (1.23) can be rewritten in the form

$$u_\alpha^{(1)}(x, t) = \frac{i}{2\pi} \int_0^\beta (-\tau)^\alpha e^{-\tau t} \Big[\Big(\tilde{U}_\alpha\Big(x, \tau e^{-i\frac{\pi}{2}}\Big) - \tilde{U}_\alpha\Big(x, \tau, e^{i\frac{3\pi}{2}}\Big)\Big] d\tau. \tag{1.29}$$

Let N be an arbitrary positive integer and let $x \in \bar{\Omega}$. Taking into account (1.25), we transform expression (1.29) when $g = 0$:

$$u_\alpha^{(1)}(x, t) = \frac{(-1)^\alpha}{2\pi} \iint_\Omega f_\alpha(x')\, dx' \int_0^\beta \tau^\alpha e^{-\tau t} d\tau$$

$$+ \frac{(-1)^\alpha}{2\pi} \operatorname{meas}(E_2 | \Omega) \iint_\Omega f_\alpha(x')\, dx' \int_0^\beta \tau^{2+\alpha} \ln \tau e^{-\tau t} d\tau$$

$$+ \sum_{n=1}^{N} \sum_{m=0}^{n} v_{m,n}^{\alpha}(x) \int_0^\beta \tau^{2n+\alpha} \ln^m \tau e^{-\tau t} d\tau + \int_0^\beta v_N^{(\alpha)}(x, \tau) e^{-\tau t} d\tau, \tag{1.30}$$

where the functions $v_{m,n}^{(\alpha)}(x)$ are continuous in $\bar{\Omega}$ and, on the basis of (1.28), the functions $v_N^{(\alpha)}(x, \tau)$ satisfy for $x \in \bar{\Omega}_R$ the estimate

$$|v_N^{(\alpha)}(x, \tau)| \leqslant C(\Gamma, g, f_\alpha, R, N) \tau^{2N+2+\alpha} |\ln^N \tau|$$

and hence the estimate

$$\left|\int_0^\beta v_N^{(\alpha)}(x,\ \tau)\, e^{-\tau t} d\tau\right| \leqslant C_1 \frac{\ln^N t}{t^{2N+3+\alpha}}. \qquad (1.31)$$

We transform the integrals in (1.30) in the following manner:

$$\int_0^\beta \tau^\alpha e^{-\tau t} d\tau = \frac{1}{t^{1+\alpha}}\left[\int_0^\infty \tau_1^\alpha e^{-\tau_1} d\tau_1 - \int_{\beta t}^\infty \tau_1^\alpha e^{-\tau_1} d\tau_1\right] = \frac{1}{t^{1+\alpha}} + O(e^{-\beta t/2}), \quad (1.32)$$

$$\int_0^\beta \tau^{2+\alpha} \ln \tau e^{-t\tau} d\tau = \frac{1}{t^{3+\alpha}} \int_0^{\beta t} \tau_1^{2+\alpha} \ln \tau_1 e^{-\tau_1} d\tau_1 - \frac{\ln t}{t^{3+\alpha}} \int_0^{\beta t} \tau_1^{2+\alpha} e^{-\tau_1} d\tau_1$$

$$= -(2+\alpha)!\, \frac{\ln t}{t^{3+\alpha}} + \frac{1}{t^{3+\alpha}} \int_0^\infty \tau_1^{2+\alpha} \ln \tau_1 e^{-\tau_1} d\tau_1 + O(e^{-\beta t/2}), \text{ etc.} \qquad (1.33)$$

Equality (1.30) can be transformed into the form (0.4) with the functions $\widetilde{u}_N^{(\alpha)}(x,\ t)$ satisyfing estimates (0.6) on the basis of (1.24) and (1.31)–(1.33).

When $g(x) \not\equiv 0$ (the third boundary value problem) both cases are considered in the same way; we therefore confine ourselves to the case when the functions $\widetilde{U}_\alpha(x,\ k)$ can be represented in the form (1.26). Analogously to the case $g = 0$, we transform expression (1.29) into the form (taking into account the fact that $\gamma = -i/2 + (\mathbf{C} - \ln 2)/\pi$, where $\mathbf{C}$ is Euler's constant)

$$u_\alpha^{(1)}(x,\ t) = \frac{(-1)^\alpha}{\pi t^{1+\alpha}} \nu^{(1)}(x) \iint_\Omega \mu^{(1)}(x') f_\alpha(x')\, dx' \frac{i}{2} \int_0^\infty \tau^\alpha e^{-\tau}\left[\frac{1}{z(\tau/t)} - \frac{1}{\bar{z}(\tau/t)}\right] d\tau$$

$$+ \sum_{n=1}^\infty \sum_{m=0}^{n-1} c_{m,\,n}^{(\alpha)}(x) \frac{\ln^m t}{t^{2n+1+\alpha}} + \sum_{n=1}^\infty \sum_{m=1}^{n+1} d_{m,\,n}^{(\alpha)}(x)$$

$$\times \frac{1}{t^{2n+1+\alpha}} \int_0^\infty \tau^{2n+\alpha} e^{-\tau}\left[\left(\frac{1}{z(\tau/t)}\right)^m - \left(\frac{1}{\bar{z}(\tau/t)}\right)^m\right] d\tau + \tilde{\tilde{u}}_N^{(\alpha)}(x,\ t), \qquad (1.34)$$

where $z(\tau/t) = \delta(\ln(\tau/2t) + \mathbf{C})/\pi - 1 + \delta i$, $\bar{z}(\tau/t)$ is the complex conjugate of $z(\tau/t)$ and the functions $\tilde{\tilde{u}}_N^{(\alpha)}(x,\ t)$ satisfy (0.7).

Consider the functions

$$\Phi_\alpha(t) = \frac{i}{2\pi\delta} \int_0^\infty \tau^\alpha e^{-\tau}\left[\frac{1}{z(\tau/t)} - \frac{1}{\bar{z}(\tau/t)}\right] d\tau$$

$$= \frac{1}{\pi} \int_0^\infty \frac{\tau^\alpha e^{-\tau} d\tau}{\left[\frac{\delta}{\pi}(\ln(\tau/2t) + \mathbf{C}) - 1\right]^2 + \delta^2}$$

$$= \frac{\pi}{\delta^2} a^{1+\alpha} \int_0^\infty \frac{\tau^\alpha e^{-a\tau} d\tau}{\ln^2(\tau/t) + \pi^2}, \qquad \alpha = 0,\ 1, \qquad (1.35)$$

where $a = 2\exp\{\pi/\delta - \mathbf{C}\}$. The function $\ln^2(\tau/t) + \pi^2$ monotonely decreases for $0 \leqslant \tau \leqslant t$; hence $\ln^2(\tau/t) + \pi^2 \geqslant (1-\epsilon)^2 \ln^2 t$ for $0 \leqslant \tau \leqslant t^\epsilon$, where $\epsilon\,(t)$ is a sufficiently small (large) positive number, and therefore (1.35) implies that as $t \to \infty$

$$0 \leqslant \Phi_\alpha(t) = \frac{\pi}{\delta^2} a^{1+\alpha} \left[\int_0^{t^\varepsilon} \frac{\tau e^{-a\tau} d\tau}{\ln^2(\tau/t) + \pi^2} + \int_{t^\varepsilon}^{\infty} \frac{\tau e^{-a\tau} d\tau}{\ln^2(\tau/t) + \pi^2} \right]$$

$$\leqslant A_1(\varepsilon, \delta) \ln^{-2} t + A_2(\varepsilon, \delta) e^{-1/2 t^\varepsilon} \leqslant A(\varepsilon, \delta)/\ln^2 t.$$

Thus the functions $\Phi_\alpha(t) \ln^2 t$ are bounded as $t \to \infty$. Moreover,

$$\frac{d}{dt}(\Phi_\alpha(t) \ln^2 t) = \frac{\pi a^{1+\alpha}}{\delta^2 t} \int_0^\infty \frac{2 \ln t\, |\ln^2(\tau/t) + \pi^2| + 2 \ln(\tau/t)}{|\ln^2(\tau/t) + \pi^2|^2} d\tau \geqslant 0$$

as $t \to \infty$. Consequently the functions $\Phi_\alpha(t) \ln^2 t$ have finite (and nonzero) limits as $t \to \infty$ and the functions $\delta\Phi_\alpha(t)$ can be represented in the form

$$\delta\Phi_\alpha(t) = \frac{\varphi_\alpha(t)}{\ln^2 t},$$

where the functions $\phi_\alpha(t)$ are continuous in t and have finite limits as $t \to \infty$, as also asserted in Theorem 1. It is proved in exactly the same way that

$$\frac{i}{2\pi} \int_0^\infty \tau^{2n+\alpha} e^{-\tau} \left[\left(\frac{1}{z(\tau/t)} \right)^m - \left(\frac{1}{\bar{z}(\tau/t)} \right)^m \right] d\tau = \frac{\varphi_{m,n}^{(\alpha)}(t)}{\ln^{m+1} t}$$

as $t \to \infty$, where the functions $\phi_{m,n}^{(\alpha)}(t)$ have finite and nonzero limits as $t \to \infty$.

To complete the proof of Theorem 1 it remains to establish inequality (1.3).

Under our assumptions the solution $u(x, t)$ of problem (0.1)–(0.3) satisfies the "conservation of energy principles" [13]

$$\iint_\Omega [(\partial u(x, t)/\partial t)^2 + |\nabla u(x, t)|^2]\, dx + \int_\Gamma g(x) u^2(x, t)\, dx$$

$$= \iint_\Omega [f_0^2(x) + |\nabla f_1(x)|^2]\, dx + \int_\Gamma g(x) f_1^2(x)\, dx,$$

$$\iint_\Omega [(\Delta u(x, t))^2 + |\nabla \partial u(x, t)/\partial t|^2]\, dx + \int_\Gamma g(x) (\partial u(x, t)/\partial t)^2\, dx$$

$$= \iint_\Omega [(\Delta f_1(x))^2 + |\nabla f_0(x)|^2]\, dx + \int_\Gamma g(x) f_0^2(x)\, dx,$$

from which (taking into account the fact that $g(x) \geqslant 0$) we deduce the inequality

$$\iint_\Omega [(\partial u(x, t)/\partial t)^2 + |\nabla u(x, t)|^2 + (\Delta u(x, t))^2]\, dx \leqslant C(f_0, f_1, \Gamma, g). \quad (1.36)$$

By virtue of the finiteness of "the speed of propogation of disturbances for the wave equation," we have $u(x, t) = 0$ for $|x| \geqslant t + \max_{\alpha=0,1}(\text{diam supp } f_\alpha)$. It therefore follows from the representation

$$u(x, t) = f_1(x) + \int_0^t \frac{\partial u(x, \tau)}{\partial \tau}\, d\tau$$

taking into account (1.36) and the Cauchy-Bunjakovskiĭ inequality, that

$$\iint_\Omega u^2(x, t)\, dx \leqslant 2 \iint_\Omega f_1^2(x)\, dx + 2\iint_\Omega \left[\int_0^t d\tau \int_0^t \left(\frac{\partial u(x, \tau)}{\partial \tau}\right)^2 d\tau\right] dx \leqslant C(1+t)^2. \tag{1.37}$$

On the basis of (1.36) and (1.37) we conclude that inequality (1.3) holds if

$$\|u(x, t)\|_{W_2^2(\Omega)} \leqslant C(\Gamma, g)\,(\|\Delta u(x, t)\|_{L_2(\Omega)} + \|\,|\nabla u(x, t)|\,\|_{L_2(\Omega)} + \|u(x, t)\|_{L_2(\Omega)}).$$

This inequality can be proved by establishing that any function $v(x) \in C^2(\bar{\Omega}_R)$ (R is an arbitrary positive number) satisfying the conditions

$$\partial v/\partial n + gv\,|_{x\in\Gamma} = 0, \quad v\,|_{|x|=R} = 0 \tag{1.38}$$

satisfies the inequality

$$\|v\|_{W_2^2(\Omega)} \leqslant C(\|\Delta v\|_{L_2(\Omega)} + \|\,|\nabla v|\,\|_{L_2(\Omega)} + \|v\|_{L_2(\Omega)}) \tag{1.39}$$

with a constant $C = C(\Gamma, g)$ not depending on R. We note that inequalities of type (1.39) for a function $v(x)$ satisfying the condition $v|_\Gamma = 0$ are well known [17], [18]. Inequality (1.39) is established under conditions (1.38) by similar considerations:

$$\begin{aligned} 2\iint_\Omega \left[\frac{\partial^2 v}{\partial x_1^2}\frac{\partial^2 v}{\partial x_2^2} - \left(\frac{\partial^2 v}{\partial x_1 \partial x_2}\right)^2\right] dx &= \left\{\int_\Gamma + \int_{|x|=R}\right\}\left[\frac{\partial^2 v}{\partial x_1^2}\frac{\partial v}{\partial x_2}\cos(n, x_2) - \frac{\partial^2 v}{\partial x_1 \partial x_2}\frac{\partial v}{\partial x_2}\cos(n, x_1)\right. \\ &\quad \left. + \frac{\partial^2 v}{\partial x_2^2}\frac{\partial v}{\partial x_1}\cos(n, x_1) - \frac{\partial^2 v}{\partial x_1 \partial x_2}\frac{\partial v}{\partial x_1}\cos(n, x_2)\right] ds = I_1 + I_2, \end{aligned} \tag{1.40}$$

where s is the arc length along the curve Γ (or the curve $\Gamma_R = \{|x| = R\}$) and the normal in the integrals I_1 and I_2 is assumed to be outward relative to Ω_R.

By virtue of the second of the conditions in (1.38) we have (see [17])

$$I_2 = \frac{1}{R}\int_{\Gamma_R} \left(\frac{\partial v}{\partial n}\right)^2 ds. \tag{1.41}$$

Introducing an orthogonal coordinate system (ξ, η) with origin at an arbitrary point

$x_0 = x(s_0) \in \Gamma$ and the ξ axis directed along the outward (relative to Ω) normal and taking into account the condition

$$\frac{\partial v}{\partial \xi}\bigg|_{\xi=\eta=0} = \frac{\partial v}{\partial n}\bigg|_{x=x_0} = -g(s_0)\,v(s_0), \tag{1.42}$$

we transform the integrand in I_1 into the form

$$-\frac{\partial^2 v}{\partial \xi \partial \eta}\frac{\partial v}{\partial \eta} - gv\frac{\partial^2 v}{\partial \eta^2}. \tag{1.43}$$

There exists a neighborhood of the point $\xi = \eta = 0$ in which the equation of the contour Γ ($\in C^2$) has the form $\xi = \omega(\eta)$, with

$$\frac{d\omega}{d\eta}\bigg|_{\eta=0} = 0, \quad \frac{d^2\omega}{d\eta^2}\bigg|_{\eta=0} = a(s_0), \quad s = s_0 + \int_0^\eta \sqrt{1 + \left(\frac{d\omega}{d\eta}\right)^2}\,d\eta, \tag{1.44}$$

where $a(s_0)$ is the curvature of Γ at the point $s = s_0$. In this neighborhood condition (1.38) takes the form

$$\frac{\partial v(s(\eta))}{\partial \xi} - \frac{\partial v(s(\eta))}{\partial \eta}\frac{d\omega}{d\eta} + g(s(\eta))\,v(\eta)\sqrt{1 + \left(\frac{d\omega}{d\eta}\right)^2} = 0.$$

Differentiating the latter equality with respect to η and taking (1.42) and (1.44) into account, we get

$$\frac{\partial v}{\partial \eta}\frac{\partial^2 v}{\partial \xi \partial \eta}\bigg|_{\xi=\eta=0} = a\left(\frac{\partial v}{\partial s}\right)^2 - \frac{\partial v}{\partial s}\frac{\partial}{\partial s}(gv)\bigg|_{s=s_0}. \tag{1.45}$$

Calculating the derivative $\partial^2 v/\partial \eta^2$ with the use of conditions (1.42) and (1.44), we also get

$$\frac{\partial^2 v}{\partial \eta^2}gv\bigg|_{\xi=\eta=0} = gv\frac{\partial^2 v}{\partial s^2} + a(gv)^2\bigg|_{s=s_0}. \tag{1.46}$$

On the basis of (1.43), (1.45) and (1.46), the integral I_1 can be transformed into the form

$$I_1 = -\int_\Gamma \left[a\left(\frac{\partial v}{\partial s}\right)^2 + a(gv)^2 + gv\frac{\partial^2 v}{\partial s^2} - \frac{\partial v}{\partial s}\frac{\partial}{\partial s}(gv)\right]ds$$
$$= -\int_\Gamma \left[a|\nabla v|^2 - 2\frac{\partial v}{\partial s}\frac{\partial}{\partial s}(gv)\right]ds. \tag{1.47}$$

Using (1.41)–(1.47) and taking into account the fact that $a(s)$ ($g(s)$) is a continuous (continuously differentiable) function on Γ, we deduce from (1.40) that

$$2\iint_\Omega \left(\frac{\partial^2 v}{\partial x_1 \partial x_2}\right)^2 dx \leqslant 2\iint_\Omega \frac{\partial^2 v}{\partial x_1^2}\frac{\partial^2 v}{\partial x_2^2}\,dx + B(\Gamma, g)\int_\Gamma (|\nabla v|^2 + v^2)\,ds. \tag{1.48}$$

Let $\Omega' = \Omega \cap \{|x| < R'\}$ denote the domain with boundaries Γ and $\Gamma' = \{|x| = R'\}$, where $R' = R'(\Gamma)$ is a certain fixed number. We apply Young's inequality to the function v:

$$\int_\Gamma |\nabla v|^2 ds \leqslant \varepsilon \iint_{\Omega'} \left[\left(\frac{\partial^2 v}{\partial x_1^2}\right)^2 + 2\left(\frac{\partial^2 v}{\partial x_1 \partial x_2}\right)^2 + \left(\frac{\partial^2 v}{\partial x_2^2}\right)^2\right] dx + C(\varepsilon, \Omega') \iint_{\Omega'} |\nabla v|^2 dx.$$

Here ϵ is an arbitrary positive number. Setting $\epsilon = 1/2B(\Gamma, g)$, we obtain the inequality

$$B(\Gamma, g) \int_\Gamma |\nabla v|^2 ds \leqslant \frac{1}{2} \iint_{\Omega'} \left[\left(\frac{\partial^2 v}{\partial x_1^2}\right)^2 + 2\left(\frac{\partial^2 v}{\partial x_1 \partial x_2}\right)^2 + \left(\frac{\partial^2 v}{\partial x_2^2}\right)^2\right] dx + C(\Gamma, g) \iint_{\Omega'} |\nabla v|^2 dx,$$

which, when applied to (1.48), leads to the estimate

$$\frac{1}{2} \iint_{\Omega} \left[\left(\frac{\partial^2 v}{\partial x_1^2}\right)^2 + 2\left(\frac{\partial^2 v}{\partial x_1 \partial x_2}\right)^2 + \left(\frac{\partial^2 v}{\partial x_2^2}\right)^2\right] dx \leqslant \iint_{\Omega} [(\Delta v)^2 + C(\Gamma, g) |\nabla v|^2] dx + B(\Gamma, g) \int_\Gamma v^2 ds.$$

This estimate clearly implies (1.39).

§2. On the properties of the analytic continuation of the Green function for the Helmholtz equation in a neighborhood of the point $k = 0$

Let $v(x, k)$ denote the classical solution of the following problem:

$$\Delta v + k^2 v = f(x), \quad x \in \Omega; \tag{2.1}$$

$$\partial v/\partial n + g(x) v = 0, \quad x \in \Gamma; \tag{2.2}$$

$$\partial v/\partial |x| - ikv = e^{-\operatorname{Im} k |x|} o(|x|^{-1/2}), \quad |x| \to \infty, \quad \operatorname{Im} k \geqslant 0, \quad k \neq 0, \tag{2.3}$$

where $f(x)$ is a continuous function in $\bar{\Omega}$ that vanishes for sufficiently large values of $|x|$ and $g(x)$ is a continuous nonnegative function on Γ.

As is well known, the solution of problem (2.1)–(2.3) can be represented in the form

$$v(x, k) = \iint_{\Omega} f(x') G(x', x, k) dx', \quad x \in \bar{\Omega}, \quad \operatorname{Im} k \geqslant 0, \quad k \neq 0, \tag{2.4}$$

where $G(x', x, k)$ is the Green function of problem (2.1)–(2.3), which on the basis of the Green formulas proved in [15] satisfies the relations

$$G(x, y, k) = \frac{i}{2} H_0^{(1)}(k\,|x - y|)$$
$$- \frac{i}{2} \int_\Gamma \left[\frac{\partial H_0^{(1)}(k|x - x'|)}{\partial n'} + g(x') H_0^{(1)}(k|x - x'|) \right] G(x', y, k)\, dx',$$
$$x \in \Gamma, \quad y \in \bar{\Omega}, \tag{2.5}$$

$$G(x, y, k) = \frac{i}{4} H_0^{(1)}(k\,|x - y|) - \frac{i}{4} \int_\Gamma \left[\frac{\partial H_0^{(1)}(k|x - x'|)}{\partial n'} \right.$$
$$\left. + g(x') H_0^{(1)}(k\,|x - x'|) \right] G(x', y, k)\, dx', \quad x \in \Omega, \quad y \in \bar{\Omega}. \tag{2.6}$$

Relation (2.5) is an integral equation for $G(x, y, k)$ the unique solvability of which for $\operatorname{Im} k \geqslant 0$ $(k \neq 0)$ is easily established by the methods of [15]. In this section we will obtain for sufficiently small $\beta = \beta(\Gamma, g) > 0$ a representation of $G(x, y, k)$ in the form of a series whose terms are analytic with respect to k in the domain $K_\beta \cap \{|k| < \beta\}$ and which (minus a finite number of terms) converges uniformly in $k \in \bar{K}_\beta \cap \{|k| \leqslant \beta\}, x, y \in \bar{\Omega}_R$, and thus determines the analytic continuation of $G(x, y, k)$ in the neighborhood $K_\beta \cap \{|k| < \beta\}$ of the point $k = 0$ (see the formulation of Theorem 2 in the Introduction).

Using the well-known expansions 8.441 and 8.444 in [16] of cylindrical functions, we get (i)

$$\frac{i}{2} H_0^{(1)}(k\,|x - x'|) = -\frac{1}{\pi} \ln|x - x'| - \left(\frac{1}{\pi} \ln k + \gamma\right)$$
$$+ \sum_{m=1}^{\infty} \left[a_m(x, x') + b_m(x, x') \left(\frac{1}{\pi} \ln k + \gamma\right) \right] k^{2m}, \tag{2.7}$$

where $\gamma = -i/2 + (\mathrm{C} - \ln 2)/\pi$, C being Euler's constant, and the functions $a_m(x, x')$ and $b_m(x, x')$ are continuous and symmetric with respect to x and x', vanish when $x = x'$ and are such that, in particular,

$$b_1(x, x') = \frac{1}{2} |x - x'|, \tag{2.8}$$

and, in general, the estimates

$$|a_m(x, x')| + |b_m(x, x')| \leqslant A \frac{R^{2m+1}}{m!\,(m-1)!}, \quad m = 1, 2, \ldots, \tag{2.9}$$

in which A is a certain positive number, hold for $x' \in \Gamma$ and $x \in \bar{\Omega}_R$; and (ii)

$$\frac{ik}{2} H_1^{(1)}(k\,|x - x'|) = \frac{1}{\pi\,|x - x'|} + \sum_{m=1}^{\infty} c_m(x, x') k^{2m}$$
$$+ \sum_{m=1}^{\infty} d_m(x, x') k^{2m} \left(\frac{1}{\pi} \ln k + \gamma\right), \tag{2.10}$$

where the functions $c_m(x, x')$ and $d_m(x, x')$ are continuous and symmetric with respect to x and x', vanish when $x = x'$ and are such that, in particular,

$$d_1(x, x') = -\frac{1}{2}|x - x'|, \tag{2.11}$$

and, in general, the estimates

$$|c_m(x, x')| + |d_m(x, x')| \leqslant A\frac{R^{2m}}{(m-1)!^2}, \qquad m = 1, 2, \ldots, \tag{2.12}$$

hold for $x' \in \Gamma$ and $x \in \bar{\Omega}_R$. It follows from (2.7), (2.10) and the equality

$$\frac{i}{2}\frac{\partial H_0^{(1)}(k|x-x'|)}{\partial n'} = \frac{ik}{2}H_1^{(1)}(k|x-x'|)\frac{(n', x-x')}{|x-x'|}$$

that the integral equation (2.5) can be written in the form

$$\begin{aligned} G(x, y, k) + \int_\Gamma M_0(x' x)G(x', y, k)\,dx' &= \left(\frac{1}{\pi}\ln k + \gamma\right)\int_\Gamma g(x')G(x', y, k)\,dx' \\ &- \int_\Gamma M(x', x, k)G(x', y, k)\,dx' - \frac{1}{\pi}\ln|x-y| \\ &- \left(\frac{1}{\pi}\ln k + \gamma\right) + H(x, y, k), \end{aligned} \tag{2.13}$$

where

$$M_0(x' x) = \frac{1}{\pi}\left[\frac{(n', x-x')}{|x-x'|^2} - g(x')\ln|x-x'|\right], \tag{2.14}$$

$$M(x', x, k) = \sum_{m=1}^{\infty}\left[A_m(x', x) + B_m(x,' x)\left(\frac{1}{\pi}\ln k + \gamma\right)\right]k^{2m}, \tag{2.15}$$

$$H(x, y, k) = \sum_{m=1}^{\infty}\left[a_m(x, y) + b_m(x, y)\left(\frac{1}{\pi}\ln k + \gamma\right)\right]k^{2m}, \tag{2.16}$$

$$A_m(x', x) = a_m(x', x)g(x') + \frac{(n', x-x')}{|x-x'|}c_m(x', x), \quad m = 1, 2, \ldots, \tag{2.17}$$

$$B_m(x', x) = b_m(x', x)g(x') + \frac{(n', x-x')}{|x-x'|}d_m(x', x), \quad m = 1, 2, \ldots. \tag{2.18}$$

In particular, from (2.8) and (2.11) we have

$$B_1(x', x) = \frac{1}{2}[(n', x-x') - |x-x'|g(x')]. \tag{2.19}$$

Since $\Gamma \in C^2$ and $g \in C$, the functions $A_m(x', x)$ and $B_m(x', x)$ are continuous in $\Gamma \times \bar{\Omega}$, vanish when $x = x'$ and, according to (2.9) and (2.12), satisfy the estimates

$$|A_m(x', x)| + |B_m(x', x)| \leqslant A(\Gamma, g)\frac{R^{2m+1}}{(m-1)!^2} \tag{2.20}$$

for $x' \in \Gamma$ and $x \in \bar{\Omega}_R$.

For the sequel we will need a number of properties of the solutions $\mu(x)$ of the equation

$$(T_0\mu)(x) = \mu(x) + \int_\Gamma M_0(x', x)\mu(x')\,dx' = h(x), \qquad x \in \Gamma, \tag{2.21}$$

and of the solutions $\nu(x)$ of the homogeneous adjoint equation

$$(T_0^*\nu)(x) = \nu(x) + \int_\Gamma M_0(x, x')\nu(x')\,dx' = 0, \qquad x \in \Gamma, \tag{2.22}$$

where the operators T_0 and T_0^* act from $L_2(\Gamma)$ into $L_2(\Gamma)$.

It is well known (see, for example, [19]) that equations (2.21) and (2.22) have only the zero solutions when $g = h = 0$. When $g(x) \not\equiv 0$ the homogeneous equations (2.21) and (2.22) may or may not have nonzero solutions. Let us show that equation (2.22) can have only one linearly independent solution when $g(x) \geqslant 0$. For this purpose it suffices to establish that every solution $\nu(x)$ of (2.22) satisfying the condition

$$\int_\Gamma \nu(x)\,dx = 0 \tag{2.23}$$

is identically equal to zero. We introduce the function

$$v(x) = -\frac{1}{\pi}\int_\Gamma \nu(x')\ln|x - x'|\,dx', \tag{2.24}$$

which is harmonic in Ω, continuous in $\bar{\Omega}$ and, by virtue of condition (2.23), tends to 0 as $|x| \to \infty$; it consequently attains its maximum (as well as its minimum) value on Γ. Since $\nu(x)$ satisfies equation (2.22) on Γ, we have the equality

$$\partial v/\partial n + gv = 0, \tag{2.25}$$

which implies on the basis of the principle of the maximum ($\partial v/\partial n > 0$ at a point where v attains its maximum value) and the condition $g \geqslant 0$ that $v(x) = 0$ on Γ. But then (see [19]) $\nu(x) = 0$. Consequently, if ν_1 and ν_2 are nonzero solutions of equation (2.22), there always exists a constant λ such that the solution $\nu = \lambda\nu_1 - \nu_2$ satisfies condition (2.23) and hence $\nu_2 = \lambda\nu_1$. By virtue of the Fredholm alternative the homogeneous equation (2.21) also can have only one linearly independent solution $\mu_0(x)$ when $g \geqslant 0$.

Let us show that if $g(x) \geqslant 0$ and $\mu_0(x)$ is a nonzero solution of the homogeneous equation (2.21), then

$$\int_\Gamma g(x)\mu_0(x)\,dx \neq 0. \tag{2.26}$$

Suppose (2.26) is not satisfied and $\mu_0 \not\equiv 0$ Then the equation

$$T_0^*\nu = g \tag{2.27}$$

is solvable, and since $\int_\Gamma \nu_0(x)\,dx \neq 0$, where ν_0 is a solution of equation (2.22) (this follows from the condition $\mu_0 \not\equiv 0$ by virtue of the Fredholm alternative), there exists a solution of (2.27) satisfying the condition $\int_\Gamma \nu(x)\,dx = 0$. Defining the function $v(x)$ by (2.24) and repeating the above arguments, we get that $\nu(x) = 0$, and this contradicts (2.27).

We will assume below that if $\mu_0 \not\equiv 0$ and $\nu_0 \not\equiv 0$, they have been chosen so as to satisfy the conditions

$$\int_\Gamma g(x)\mu_0(x)\,dx = 1, \quad \int_\Gamma \nu_0(x)\,dx = 1. \tag{2.28}$$

Let $\widetilde{L}_2(\Gamma)$ ($\widetilde{\widetilde{L}}_2(\Gamma)$) denote the subspace of functions from $L_2(\Gamma)$ that are orthogonal to $\nu_0(x)$ (to $g(x)$). The above remarks and the Fredholm alternative imply that the operator T_0 acting from $\widetilde{\widetilde{L}}_2(\Gamma)$ into $\widetilde{L}_2(\Gamma)$ has a bounded inverse, and from the form of the kernel $M_0(x, x')$, the properties of the contour Γ ($\in C^2$) and the function g ($\in C$) it follows that the operator T_0^{-1} is bounded from $C(\Gamma) \cap \widetilde{L}_2(\Gamma)$ into $C(\Gamma) \cap \widetilde{\widetilde{L}}_2(\Gamma)$.

Suppose the homogeneous equation (2.21) does not have any nonzero solutions and $g(x) \geqslant 0$ is not identically equal to zero. We show that the solution $\mu_1(x)$ of the equation

$$T_0\mu_1 = 1 \tag{2.29}$$

satisfies the condition

$$\int_\Gamma \mu_1(x)\,g(x)\,dx = \delta \neq 0. \tag{2.30}$$

For suppose condition (2.30) is not satisfied. Then the solution $\nu(x)$ of equation (2.27) satisfies the condition

$$\int_\Gamma \nu(x)\,dx = (\nu,\ 1) = (\nu,\ T_0\mu_1) = (T_0^*\nu,\ \mu_1) = (g,\ \mu_1) = 0,$$

which implies on the basis of the above remarks that $\nu(x) = 0$, contradicting (2.27).

Let us return to the investigation of the integral equation (2.13) for $k \to 0$. We first consider the case when there exist nonzero solutions μ_0 and ν_0 of the homogeneous equations (2.21) and (2.22). We show that in this case a solution $G(x, y, k)$ of (2.13) can be represented in the form

$$G(x,\ y,\ k) = \sum_{n=0}^{\infty} \sum_{m=-n-1}^{n} G_{m,n}(x,\ y)\,k^{2n}\left(\frac{1}{\pi}\ln k + \gamma\right)^m, \tag{2.31}$$

in which

$$G_{m,n}(x,\ y) = \mu_{m,n}(x,\ y) + \lambda_{m,n}(y)\,\mu_0(x),$$

where the functions $\mu_{m,n}(x, y)$ satisfy the conditions

$$\int_\Gamma \mu_{m,n}(x', y)\, g(x')\, dx' = 0, \qquad y \in \bar{\Omega}. \tag{2.32}$$

Substituting (2.31) into (2.13) and taking (2.32) into account, we obtain the equality

$$\sum_{n=0}^{\infty}\sum_{m=-n-1}^{n} (T_0\mu_{m,n})(x, y)\, k^{2n}\left(\frac{1}{\pi}\ln k + \gamma\right)^m = -\frac{1}{\pi}\ln|x-y| - \left(\frac{1}{\pi}\ln k + \gamma\right)$$
$$+ \sum_{n=1}^{\infty} a_n(x, y)\, k^{2n} + \sum_{n=1}^{\infty} b_n(x, y)\, k^{2n}\left(\frac{1}{\pi}\ln k + \gamma\right)$$
$$+ \sum_{n=0}^{\infty}\sum_{m=-n-1}^{n} \lambda_{m,n}(y)\, k^{2n}\left(\frac{1}{\pi}\ln k + \gamma\right)^{m+1} - \int_\Gamma \left[\sum_{p=1}^{\infty} A_p(x', x)\, k^{2p}\right.$$
$$\left. + \sum_{p=1}^{\infty} B_p(x', x)\, k^{2p}\left(\frac{1}{\pi}\ln k + \gamma\right)\right]$$
$$\times \left[\sum_{r=0}^{\infty}\sum_{q=-r-1}^{r} (\mu_{q,r}(x', y) + \lambda_{q,r}(y)\,\mu_0(x'))\, k^{2r}\left(\frac{1}{\pi}\ln k + \gamma\right)^q\right] dx',$$

which clearly holds if the functions $\mu_{m,n}$ and $\lambda_{m,n}$ are connected by relations such that the coefficients of each term of the form

$$k^{2n}\left(\frac{1}{\pi}\ln k + \gamma\right)^m, \qquad n = 0, 1, \ldots, \qquad m = -n-1, -n, \ldots, n, n+1,$$

are equal.

We first write out the relations corresponding to $n = 0$:

$$(T_0\mu_{0,0})(x, y) = -\frac{1}{\pi}\ln|x-y| + \lambda_{-1,0}(y), \qquad m = 0, \tag{2.33}$$
$$(T_0\mu_{-1,0})(x, y) = 0, \qquad m = -1, \tag{2.34}$$
$$0 = -1 + \lambda_{0,0}(y), \qquad m = 1. \tag{2.35}$$

Equation (2.33) is solvable if for any $y \in \bar{\Omega}$ the function $\lambda_{-1,0}(y) - \ln|x-y|/\pi$ belongs to the subspace $\widetilde{L}_2(x \in \Gamma)$, i.e.

$$\lambda_{-1,0}(y) = \frac{1}{\pi}\int_\Gamma \nu_0(x')\ln|x'-y|\, dx'. \tag{2.36}$$

Then equation (2.33) has the unique solution $\mu_{0,0}(x, y)$ in the subspace $\widetilde{\widetilde{L}}_2(x \in \Gamma)$. The solution $\mu_{-1,0}(x, y)$ of equation (2.34) is identically equal to zero by virtue of condition (2.32). From (2.35) we find that $\lambda_{0,0}(y) = 1$. Thus the functions $G_{0,0}(x, y)$ and $G_{-1,0}(x, y)$ are defined.

We now write out the relations corresponding to $n = 1$:

$$(T_0\mu_{0,1})(x, y) = a_1(x, y) + \lambda_{-1,1}(y) - \int_\Gamma [A_1(x', x)G_{0,0}(x', y) + B_1(x', y)G_{-1,0}(x', y)]\,dx', \tag{2.37}$$

$$(T_0\mu_{-1,1})(x, y) = \lambda_{-2,1}(y) - \int_\Gamma A_1(x', x)G_{-1,0}(x', y)\,dx', \tag{2.38}$$

$$(T_0\mu_{-2,1})(x, y) = 0, \tag{2.39}$$

$$(T_0\mu_{1,1})(x, y) = b_1(x, y) + \lambda_{0,1}(y) - \int_\Gamma B_1(x', x)G_{0,0}(x', y)\,dx', \tag{2.40}$$

$$0 = \lambda_{1,1}(y). \tag{2.41}$$

Setting

$$\lambda_{-1,1}(y) = \int_\Gamma \nu_0(x)\left\{-a_1(x,y) + \int_\Gamma [A_1(x', x)G_{0,0}(x', y) + B_1(x', x)G_{-1,0}(x', y)]\,dx'\right\}dx,$$

we get that the right side of equation (2.37) belongs to $C(\Gamma) \cap \widetilde{L}_2(\Gamma) \times C(\bar{\Omega})$; hence there exists a unique solution $\mu_{0,1}(x, y)$ of (2.37) from $C(\Gamma) \cap \widetilde{L}_2(\Gamma) \times C(\bar{\Omega})$. The functions $\mu_{-1,1}(x, y)$, $\lambda_{-2,1}(y)$ and $\mu_{1,1}(x, y)$, $\lambda_{0,1}(y)$ are determined analogously from (2.38) and (2.40) respectively; hence the functions $G_{j,1}(x, y)$ $(j = -2, -1, 0, 1)$ are defined and there exist constants $D = D(R)$ and $B = B(\Gamma, g) > 1$ such that the estimates

$$|G_{j,1}(x, y)| \leqslant D(R)\,B(\Gamma, g) \tag{2.42}$$

hold for any $y \in \bar{\Omega}_R$ and any $x \in \Gamma$. Continuing this process, we obtain for arbitrary $n > 1$ the system of equations

$$(T_0\mu_{m,n})(x, y) = \begin{cases} a_n(x, y), & m = 0 \\ 0, & m \neq 0 \end{cases} + \lambda_{m-1,n}(y) + \begin{cases} b_n(x, y), & m = 1 \\ 0, & m \neq 1 \end{cases}$$
$$- \sum_{j=\max(0,\,m,\,-m-1)}^{n-1} \int_\Gamma A_{n-j}(x', x)G_{m,j}(x', y)\,dx'$$
$$- \sum_{j=\max(0,\,m-1,\,-m)}^{n} \int_\Gamma B_{n-j}(x', x)G_{m-1,j}(x', y)\,dx', \tag{2.43}$$

where

$$A_0 = B_0 = 0, \quad m = -n, -n+1, \ldots, n-1, n, \quad \mu_{-n-1,n} = \lambda_{n,n} = 0,$$

the $\lambda_{m-1,n}(y)$ $(\in C(\bar{\Omega}))$ being chosen in each equation of (2.43) so that this equation has a unique solution $\mu_{m,n}(x, y) \in C(\Gamma) \cap \widetilde{L}_2(\Gamma) \times C(\bar{\Omega})$. The estimates

$$|G_{m,n}(x, y)| \leqslant D(R)\,B^n(\Gamma, g)$$

are established for any $y \in \bar{\Omega}_R$ and any $x \in \Gamma$ by induction. Consequently, when $|k| \leqslant \beta$, where $\beta = \beta(\Gamma, g)$ is sufficiently small, the series (2.31) (minus the term $G_{0,0}$) converges uniformly in $x \in \Gamma, y \in \bar{\Omega}_R$ and determines a solution of (2.13).

Analogously to the derivation of (2.13), relation (2.6) can be transformed into the form

$$G(x, y, k) = -\frac{1}{2}\int_\Gamma M_0(x', x)G(x', y, k)\,dx' + \frac{1}{2}\left(\frac{1}{\pi}\ln k + \gamma\right)\int_\Gamma g(x')$$

$$\times G(x', y, k)\,dx' - \frac{1}{2}\int_\Gamma M(x', x, k)\,G(x', y, k)\,dx' - \frac{1}{2\pi}\ln|x-y|$$

$$-\frac{1}{2}\left(\frac{1}{\pi}\ln k + \gamma\right) + \frac{1}{2}H(x, y, k), \tag{2.44}$$

from which, taking into account (2.14)–(2.20) and (2.31), we deduce that when $x \in \Omega_R$ the function $G(x, y, k)$ can again be analytically continued into the domain $K_\beta \cap \{|k| < \beta\}$ in the form

$$G(x, y, k) = \sum_{n=0}^{\infty}\sum_{m=-n-1}^{n} G_{m,n}(x, y)k^{2n}\left(\frac{1}{\pi}\ln k + \gamma\right)^m, \tag{2.45}$$

in which the functions $G_{m,n}(x, y)$ are defined by the relations

$$G_{0,0}(x, y) = -\frac{1}{2}\int_\Gamma M_0(x', x)\,G_{0,0}(x', y)\,dx'$$

$$+\frac{1}{2}\int_\Gamma g(x')G_{0,0}(x', y)\,dx'\left(\frac{1}{\pi}\ln k + \gamma\right) - \frac{1}{2\pi}\ln|x-y| - \frac{1}{2}\left(\frac{1}{\pi}\ln k + \gamma\right)$$

$$+\frac{1}{2}\int_\Gamma g(x')\,G_{0,-1}(x', y)\,dx' = -\frac{1}{2\pi}\ln|x-y|$$

$$-\frac{1}{2\pi}\int_\Gamma\left[\frac{(n', x-x')}{|x-x'|^2} - g(x')\ln|x-x'|\right][\mu_{0,0}(x', y) + \mu_0(x')]\,dx'$$

$$+\frac{1}{2\pi}\int_\Gamma \nu_0(x')\ln|x'-y|\,dx', \tag{2.46}$$

$$G_{0,-1}(x, y) = -\frac{1}{2}\int_\Gamma M_0(x', x)\,G_{0,-1}(x', y)\,dx' = \mu^{(2)}(x)\,\nu^{(2)}(y), \tag{2.47}$$

where

$$\mu^{(2)}(x) = -\frac{1}{2\pi}\int_\Gamma\left[\frac{(n', x-x')}{|x-x'|^2} - g(x')\ln|x-x'|\right]\mu_0(x')\,dx', \tag{2.48}$$

$$\nu^{(2)}(y) = \frac{1}{\pi}\int_\Gamma \nu_0(x')\ln|x'\ \ y|\,dx', \tag{2.49}$$

and the relations

$$G_{m,n}(x, y) = -\frac{1}{2}\int_\Gamma M_0(x', x)\, G_{m,n}(x', y)\, dx' + \frac{1}{2}\int_\Gamma g(x')\, G_{m-1,n}(x', y)\, dx'$$

$$-\frac{1}{2}\sum_{\substack{j=\min[n-1,\\ \max(0,\, m,\, -m-1)]}}^{n-1} \int_\Gamma A_{n-j}(x', x)\, G_{m,j}(x', y)\, dx'$$

$$-\frac{1}{2}\sum_{\substack{j=\min[n-1,\\ \max(0,\, m-1,\, -m)]}}^{n-1} \int_\Gamma B_{n-j}(x' x)\, G_{m-1,j}(x', y)\, dx'$$

$$+\begin{cases} \frac{1}{2} a_n(x, y), & m = 0, \\ \frac{1}{2} b_n(x, y), & m = 1, \\ 0, & m \neq 0,\ 1. \end{cases}$$

Taking into account the well-known properties (see, for example, [15]) of the single and double layer potentials, we conclude that (i) the function $G_{0,0}(x, y)$ given by (2.46) is continuous in $(\Omega \times \bar{\Omega})\backslash(x = y)$ and tends to the function $G_{0,0}(x_0, y)$ in (2.31) as $x \to x_0 \in \Gamma$, (ii) the functions $\mu^{(2)}(x)$ and $\nu^{(2)}(y)$ are harmonic in Ω and continuous in $\bar{\Omega}$, and (iii) each function $G_{m,n}(x, y)$ $(n \geqslant 1)$ is continuous in $\Omega \times \bar{\Omega}$ and tends to the function $G_{m,n}(x_0, y)$ in (2.31) as $x \to x_0 \in \Gamma$, i.e. each function $G_{m,n}(x, y)$ is continuous in $\bar{\Omega} \times \bar{\Omega}$ and (as in the case when $x \in \Gamma$ and $y \in \bar{\Omega}_R$) satisfies the estimate

$$|G_{m,n}| \leqslant D_1(R)\, B^n(\Gamma, g)$$

for $x, y \in \bar{\Omega}_R$. Consequently in the case under consideration the series (2.45) determines for all $x, y \in \bar{\Omega}$ $(x \neq y)$ the analytic continuation of the Green function of the third boundary value problem into the domain $K_\beta \cap \{|k| < \beta\}$. Denoting the function $\mu_{0,0}(x, y)$ by $\mu_0^{(2)}(x, y)$, the function $G_{0,0}(x, y)$ by $G_0^{(2)}(x, y)$, each function $G_{m,n}(x, y)$ in (2.45) by $G^-_{-m,n}(x, y)$ for $m < 0$ and the coefficient of each term $k^{2n}\ln^m k$ in (2.45) by $G^+_{m,n}(x, y)$ for $m \geqslant 0$, we see that Theorem 2 is proved in the case when equation (0.17) has a nonzero solution.

We now consider the case when equation (0.17), i.e. the homogeneous equation (2.21), does not have any nonzero solutions. Equation (2.13) is rewritten in the form

$$G(x, y, k) + \int_\Gamma M_0(x', x)\, G(x', y, k)\, dx' = \frac{1}{\delta}\left[\delta\left(\frac{1}{\pi}\ln k + \gamma\right) - 1\right]$$

$$\times \int_\Gamma g(x')\, G(x' y, k)\, dx' + \frac{1}{\delta}\int_\Gamma g(x')\, G(x', y, k)\, dx'$$

$$-\int_\Gamma \sum_{n=1}^{\infty}\left\{\left[A_n(x', x) + \frac{1}{\delta} B_n(x', x)\right] k^{2n} + \frac{1}{\delta} B_n(x', x)\, k^{2n}\left[\delta\left(\frac{1}{\pi}\ln k + \gamma\right) - 1\right]\right\} \times$$

$$\times G(x', y, k)\,dx' - \frac{1}{\pi}\ln|x-y| - \frac{1}{\delta} - \frac{1}{\delta}\left[\delta\left(\frac{1}{\pi}\ln k + \gamma\right) - 1\right]$$
$$+ \sum_{n=1}^{\infty}\left[a_n(x, y) + \frac{1}{\delta}b_n(x, y)\right]k^{2n} + \sum_{n=1}^{\infty}\frac{1}{\delta}b_n(x, y)k^{2n}\left[\delta\left(\frac{1}{\pi}\ln k + \gamma\right) - 1\right], \quad (2.50)$$

where the nonzero number δ is given by (2.30). We show that a solution $G(x, y, k)$ of equation (2.50) for $k \in K_\beta \cap \{|k| < \beta = \beta(\Gamma, g)\}$ has the form

$$G(x, y, k) = \sum_{n=0}^{\infty}\sum_{m=-n-1}^{n} G_{m,n}(x, y)k^{2n}\left[\delta\left(\frac{1}{\pi}\ln k + \gamma\right) - 1\right]^m, \quad (2.51)$$

in which

$$G_{m,n}(x, y) = \mu_{m,n}(x, y) + \lambda_{m,n}(y)\mu_1(x), \quad (2.52)$$

where $\mu_1(x)$ is defined by (2.29) and the functions $\mu_{m,n}(x, y)$ and $\lambda_{m,n}(x, y)$ are determined from the system of equations obtained by substituting (2.51) into (2.50) and equating the coefficients of each term of the form

$$k^{2n}\left[\delta\left(\frac{1}{\pi}\ln k + \gamma\right) - 1\right]^m, \quad n = 0, 1, \ldots, \quad m = -n-1, \ldots, \quad n+1.$$

This system has the form

$$(T_0\mu_{m,n})(x, y) = \frac{1}{\delta}\int_\Gamma g(x')\mu_{m,n}(x', y)\,dx' + \lambda_{m-1,n}(y) + f_{m,n}(x, y),$$
$$n = 0, 1, \ldots, m = -n-1, -n, \ldots, n, n+1, \quad (2.53)$$

where

$$f_{0,0}(x, y) = -\frac{1}{\pi}\ln|x-y| - \frac{1}{\delta},$$

$$f_{m,n}(x, y) = -\sum_{\substack{j=\min[n-1,\\ \max(0,\, m,\, -m-1)]}}^{n-1}\int_\Gamma\left[A_{n-j}(x', x) + \frac{1}{\delta}B_{n-j}(x', x)\right]G_{m,j}(x', y)\,dx'$$
$$- \sum_{\substack{j=\min[n-1,\\ \max(0,\, m,\, -m-1)]}}^{n-1}\frac{1}{\delta}\int_\Gamma B_{n-j}(x', x)G_{m-1,j}(x', y)\,dx'$$
$$+\begin{cases} a_n(x, y) + \frac{1}{\delta}b_n(x, y), & m = 0, \\ \frac{1}{\delta}b_n(x, y), & m - 1, \\ 0, & m \neq 0, 1, \end{cases}$$

$$\lambda_{0,0}(y) = 1/\delta, \qquad \lambda_{n,n}(y) = 0, \qquad \mu_{-n-1,n}(x, y) = 0, \qquad n = 1, 2, \ldots.$$

We put

$$\mu_{m,n}(x, y) = (T_0^{-1}f_{m,n})(x, y),$$
$$\lambda_{m,n}(y) = -\frac{1}{\delta}\int_\Gamma g(x')\mu_{m,n}(x', y)\,dx'.$$

Then all of the functions $G_{m,n}(x, y)$ in (2.51) are defined and satisfy the estimates

$$|G_{m,n}(x, y)| \leqslant D(R) B^n(\Gamma, g), \qquad n = 1, 2, \ldots, x \in \Gamma, y \in \bar{\Omega}_R,$$

which are established by induction, and hence the series (2.51) (minus the term $G_{0,0}(x, y)$) converges uniformly in $k \in \{|k| \leqslant \beta = \beta(\Gamma, g)\}, x \in \Gamma, y \in \bar{\Omega}_R$ and determines a solution of (2.50).

Completely repeating further the arguments developed in the preceding case, we get that the series (0.15) determines the analytic continuation of $G(x, y, k)$ into the domain $K_\beta \cap \{|k| < \beta\}$ with properties (0.18), (0.19) and (0.23).

We consider, finally, the integral equation (2.13) when $g = 0$. This case differs from the preceding cases by the fact that when $k \to 0$ the bounded operator $T_0 + T_k$ (T_k is the integral operator with kernel $M(x, x', k)$) from $L_2(\Gamma)$ into $L_2(\Gamma)$ has a bounded inverse, while the norm of the right side of (2.13) is unbounded, and hence the norm of a solution $G(x, y, k)$ of this equation is also unbounded. It is immediately verified that in this case the series

$$G(x, y, k) = \sum_{n=0}^{\infty} \sum_{m=0}^{n+1} G_{m,n}(x, y) k^{2n} \left(\frac{1}{\pi} \ln k + \gamma\right)^m \tag{2.54}$$

determines a solution of equation (2.13) if

$$G_{m,n}(x, y) = (T_0^{-1} f_{m,n})(x, y),$$

where

$$f_{0,0}(x, y) = -\frac{1}{\pi} \ln|x - y|, \qquad f_{0,1}(x, y) = -1,$$

$$f_{m,n}(x, y) = -\sum_{j=\max(0, m-1)}^{n} \int_\Gamma A_{n-j}(x', x) G_{m,j}(x', y)\, dx'$$

$$\sum_{j=\max(0,m-2)}^{n-1} \int_\Gamma B_{n-j}(x', x) G_{m-1,j}(x', y)\, dx' + \begin{cases} a_n(x, y), & m = 0, \\ b_n(x, y), & m = 1, \\ 0, & m \neq 0, 1, \end{cases}$$

$$A_0 = 0, \qquad |k| < \beta = \beta(\Gamma, g).$$

We note that the well-known properties (see, for example, [19]) of the kernel $M_0(x', x)$ in (2.13) when $g = 0$

$$\left(\int_\Gamma M_0(x', x)\, dx' = 1, \; x \in \Gamma\right)$$

imply that $T_0^{-1} h = h/2$ for any constant h; hence $G_{0,1}(x, y) = -1/2$ and from (2.19) we have

$$f_{2,1}(x, y) = -\int_\Gamma B_1(x', x) G_{0,1}(x', y)\, dx = \frac{1}{4} \int_\Gamma (n', x - x')\, dx'$$

$$= -\frac{1}{4} \iint_{E_2 \setminus \Omega} \left[\frac{\partial}{\partial x_1'} (x_1 - x_1') + \frac{\partial}{\partial x_2'} (x_2 - x_2')\right] dx' = \frac{1}{2} \operatorname{meas}(E_2/\Omega),$$

which implies $G_{2,1}(x, y) = \text{meas}(E_2\backslash\Omega)/4$.

Using (2.6), (2.15)–(2.18) and (2.54) in the same way as in the cases considered earlier, we get that the series (0.14) determines the analytic continuation of the Green function of the second boundary value problem into the domain $K_\beta \cap \{|k| < \beta\}$ with the property (0.23).

From the definitions and properties (2.14)–(2.20) of the functions $M_0(x', x)$, $M(x', x, k)$ and $H(x, y, k)$ it follows that for any $k \in K_\beta$ (β is an arbitrary positive number) we have

$$\int_\Gamma\int_\Gamma |M_0 + M|^2 dx'dx < \infty, \qquad \int_\Gamma |H_0^{(1)}(k|x-y|)|^2 dx < \infty,$$

in which the functions $M_0 + M$ and $H_0^{(1)}$ are analytic with respect to $k \in K_\beta$, and hence equation (2.5) is uniquely solvable for $\text{Im}\, k \geqslant 0$ $(k \neq 0)$. Therefore, on the basis of a theorem of J. D. Tamarkin [20], the solution $G(x, y, k)$ of equation (2.5) belongs to $L_2(x \in \Gamma)$, is analytic with respect to $k \in K_\beta$ and has not more than a countable number of isolated poles of finite order. Taking into account the fact that $G(x, y, k)$ does not have any poles on the axis $\text{Im}\, k = 0$ as well as the established properties (0.14)–(0.16) of $G(x, y, k)$ for $k \in K_\beta \cap \{|k| < \beta = \beta(\Gamma, g)\}$ and the properties of the potentials with kernels $H_0^{(1)}(k|x - x'|)$ and $\partial H_0^{(1)}(k|x - x'|)/\partial n'$ (see [15]), we conclude that for any $\kappa > 0$ there exists a $\beta = \beta(\Gamma, g, \kappa) > 0$ such that for any $x, y \in \bar{\Omega}$ the function $G(x, y, k)$ admits an analytic continuation into the domain $K_\beta \cap \{|\text{Re}\, k| < \kappa\}$ with properties (0.13)–(0.16). Theorem 2 will consequently be completely proved if we show that there exist numbers $\kappa_1 = \kappa_1(\Gamma, g)$ and $\beta_1 = \beta_1(\Gamma, g)$ such that for any $x, y \in \bar{\Omega}$ the function $G(x, y, k)$ admits an analytic continuation into the domain $K_{\beta_1} \cap \{|\text{Re}\, k| > \kappa_1\}$ with properties (0.10)–(0.12).

§3. On estimates of the analytic continuation of the Green function for the Helmholtz equation for large values of $|k|$

In this section we first show that when $x, y \in \Gamma$ and $k \in K_{\beta_\epsilon} \cap \{|\text{Re}\, k| > \kappa_\epsilon\} \cap \{\text{Im}\, k \geqslant 0\}$ the function $G(x, y, k)$ satisfies the equation

$$G(x, y, k) - L_\epsilon(y, x, k) - \int_\Gamma [M_\epsilon(x', x, k) + g(x')L_\epsilon(x', x, k)]\, G(x', y, k)\, dx', \tag{3.1}$$

where ϵ is an arbitrary sufficiently small positive number and for any $x', x \in \Gamma$ $(x' \neq x)$ the functions L_ϵ and M_ϵ are analytic with respect to $k \in K(\beta_c, \kappa_c) = K_{\beta_\epsilon} \cap \{|\text{Re}\, k| > \kappa_\epsilon\}$ and satisfy the inequalities

$$|L_\varepsilon(x', x, k)| \leqslant C(\Gamma)|k|^{1/6}\left[\int_{|k|^{-1/3}}^{1} \tau^{1/2}\exp\{-B(\Gamma)|k||x'-x|\tau^2\}\,d\tau\right.$$
$$\left. + |\ln(|k||x-x'|)|\exp\{-B(\Gamma)|k||x-x'|\}\right]$$
$$+ C_1(\Gamma, \varepsilon)\exp\{-B_1(\Gamma, \varepsilon)|k|^{1/3}\},$$
$$|M_\varepsilon(x', x, k)| \leqslant C(\Gamma)\varepsilon|k|^{1/3}\exp\{-B(\Gamma)|k|^{1/3}|x-x'|\} \tag{3.2}$$
$$+ C_1(\Gamma, \varepsilon)\exp\{-B_1(\Gamma, \varepsilon)|k|^{1/3}\}. \tag{3.3}$$

Using estimates (3.2) and (3.3), we then show that for sufficiently small $\epsilon_1 = \epsilon_1(\Gamma, g)$, i.e. for sufficiently large κ_1, and sufficiently small β_1 equation (3.1) is solvable for all $k \in K(\beta_1, \kappa_1)$ by the iterative method and its solution $G(x, y, k)$ is the analytic continuation of the Green function of the exterior boundary value problem for the Helmholtz equation into the domain $K(\beta_1, \kappa_1)$ with estimates (0.10)–(0.12).

We proceed to construct the functions L_ϵ and M_ϵ. Let y be an arbitrary point on Γ and let R_y be the radius of curvature of Γ at y. Since $\Gamma \in C^2$ and $R_y \geqslant R_{\min} > 0$, for $\epsilon < R_{\min}$ we can consider the circle $\Gamma_{\epsilon y}$ of radius $R_{\epsilon y} = R_y - \epsilon$ that is tangent to Γ at y. Let $\Omega_{\epsilon y}$ denote the exterior of $\Gamma_{\epsilon y}$. Since $R_y \leqslant R_{\max} < \infty$, there exists a number $R^{(1)} = R^{(1)}(\Gamma) > R_{\max}$ such that if $\Gamma_y^{(1)}$ is the circle of radius $R^{(1)}$ that is tangent to Γ at y and $\Omega_y^{(1)}$ is its exterior, the curve Γ lies wholly in the complement of $\Omega_y^{(1)}$. Further, let $\Gamma_y^{(2)}$ denote the circle of radius $R^{(2)} = 2R^{(1)}$ that is tangent to Γ at y and let $\Omega_y^{(2)}$ denote its exterior (see Figure 1).

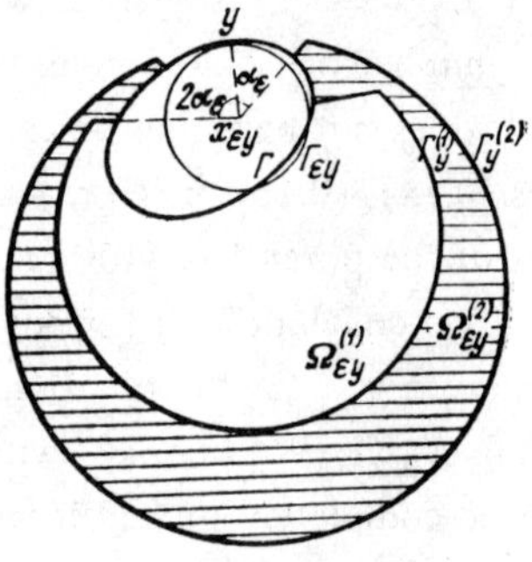

FIGURE 1

Let $\Phi_{\epsilon y}(x, y', k)$ denote the Green function of the second boundary value problem for the Helmholtz equation in the domain $\Omega_{\epsilon y}$. Then for any continuous function $f(y')$ on $\Gamma_{\epsilon y}$ the function $v(x) = \int_{\Gamma_{\epsilon y}} \Phi_{\epsilon y}(x, y', k) f(y')\,dy'$ is the classical solution of the problem

$$\Delta v + k^2 v = 0, \qquad x \in \Omega_{\varepsilon y},$$
$$\frac{\partial v}{\partial |x|} - ikv = e^{-\operatorname{Im} k|x|}o(|x|^{-1/2}), \qquad |x| \to \infty,$$
$$\frac{\partial v}{\partial n} = f(x), \qquad x \in \Gamma_{\varepsilon y}$$

where n is the outward (relative to $\Omega_{\epsilon y}$) unit normal to Γ and $\operatorname{Im} k \geqslant 0$ $(k \neq 0)$.

Let $\Phi_{\epsilon y}(x, k) = \Phi_{\epsilon y}(x, y, k)$. If we introduce the polar coordinate system

(ρ, ϕ) with origin at the center $x_{\epsilon y}$ of the circle $\Gamma_{\epsilon y}$ and coordinates $\rho = 1, \phi = 0$ for the point y (i.e. $\rho > 1, |\phi| \leqslant \pi$ for $x \in \Omega_{\epsilon y}$), then the function $\Phi_{\epsilon y}(x, k)$ can be written in the form

$$\Phi_{\varepsilon y}(x, k) = \Phi\left(\frac{|x - x_{\varepsilon y}|}{R_{\varepsilon y}}, \varphi, kR_{\varepsilon y}\right), \tag{3.4}$$

where $\Phi(\rho, \phi, k)$ is represented for $\rho > 1$ and $\operatorname{Im} k \geqslant 0$ $(k \neq 0)$ by the convergent series

$$\Phi(\rho, \varphi, k) = -\frac{1}{\pi k}\left[\frac{H_0(k\rho)}{H_0'(k)} + 2\sum_{m=0}^{\infty}\frac{H_m(k\rho)}{H_m'(k)}\cos m\varphi\right], \tag{3.5}$$

in which $H_m(z) = H_m^{(1)}(z)$, $H_m'(z) = dH_m(z)/dz$, $m = 0, 1, \cdots$, with $H_m^{(1)}(z)$ denoting, as usual, the Hankel function that tends to zero as $|z| \to \infty$ in the sector $0 < \arg z < \pi$ (equality (3.5) is easily established by the method of separation of variables). The properties of the function $\Phi(\rho, \phi, k)$, which are established in the lemma of §4, will be essentially used in the subsequent constructions. It follows from this lemma that there exist numbers $\beta = \beta(\Gamma)$ and $\kappa = \kappa(\Gamma)$ such that for any $x \in \Omega_{\epsilon y} \backslash \overline{\Omega}_y^{(2)}$ the function $\Phi_{\epsilon y}(x, k)$ and its first derivatives with respect to ρ and ϕ admit analytic continuations into the domain $K(\beta, \kappa)$ with estimates

$$\begin{aligned} |\Phi_{\varepsilon y}| \leqslant C(\Gamma)|k|^{1/6}\exp\{|\operatorname{Im} k|\sqrt{\rho^2 - 1}\} \\ \times\left[\int_{|k|^{-1/3}}^{1}\tau^{1/2}\exp\left\{-B(\Gamma)|k|\left(|\varphi| - \arccos\frac{1}{\rho}\right)\tau^2\right\}d\tau\right. \\ \left.+\left|\ln|k|\left(|\varphi| - \arccos\frac{1}{\rho}\right)\right|\exp\left\{-B(\Gamma)|k|\left(|\varphi| - \arccos\frac{1}{\rho}\right)\right\}\right], \end{aligned} \tag{3.6}$$

$$\begin{aligned} \left|\frac{\partial\Phi_{\varepsilon y}}{\partial\varphi}\right| \leqslant C(\Gamma)|k|^{1/3}\left(|\varphi| - \arccos\frac{1}{\rho}\right)^{-1} \\ \times\exp\left\{|\operatorname{Im} k|\sqrt{\rho^2 - 1} - B(\Gamma)|k|^{1/3}\left(|\varphi| - \arccos\frac{1}{\rho}\right)\right\}, \end{aligned} \tag{3.7}$$

$$\begin{aligned} \left|\frac{\partial\Phi_{\varepsilon y}}{\partial\rho}\right| \leqslant C(\Gamma)|k|^{1/3}(\rho - 1)\left(|\varphi| - \arccos\frac{1}{\rho}\right)^{-2} \\ \times\left[1 + (\rho - 1)\left(|\varphi| - \arccos\frac{1}{\rho}\right)^{-2}\right] \\ \times\exp\left\{|\operatorname{Im} k|\sqrt{\rho^2 - 1} - B(\Gamma)|k|^{1/3}\left(|\varphi| - \arccos\frac{1}{\rho}\right)\right\}, \end{aligned} \tag{3.8}$$

not depending on ϵ for sufficiently small ϵ (e.g. $\epsilon < R_{\min}/2$).

The function $\Phi_{\epsilon y}$ is generally not defined for all $x \in \Omega$, since the circle $\Gamma_{\epsilon y}$

can intersect the curve Γ. Inasmuch as $\Gamma \in C^2$ and $R_{\epsilon y} < R_y$, for any $\epsilon > 0$ there exists a number $\alpha_\epsilon = \alpha(\epsilon, \Gamma)$ such that $\rho(\phi) > 1$ for $0 < |\phi| \leqslant 2\alpha_\epsilon$, where $\rho(\phi)$ denotes the equation of the curve Γ (see Figure 1).

Let $\Omega_{\epsilon y}^{(1)}$ denote the domain with boundaries $\{\rho = \rho(\phi), |\phi| \geqslant 2\alpha_\epsilon\}$, $\{\rho = \rho^{(1)}(\phi), |\phi| \geqslant 2\alpha_\epsilon\}$, $\{\phi = -2\alpha_\epsilon, \rho(-2\alpha_\epsilon) \leqslant \rho \leqslant \rho^{(1)}(-2\alpha_\epsilon)\}$, $\{\phi = 2\alpha_\epsilon, \rho(2\alpha_\epsilon) \leqslant \rho \leqslant \rho^{(1)}(2\alpha_\epsilon)\}$, where $\rho^{(1)}(\phi)$ denotes the equation of the circle $\Gamma_y^{(1)}$, and let $\Omega_{\epsilon y}^{(2)}$ denote the domain with boundaries $\{\rho = \rho(\phi), \alpha_\epsilon \leqslant |\phi| \leqslant 2\alpha_\epsilon\}$, $\{\rho = \rho^{(1)}(\phi), 2\alpha_\epsilon \leqslant |\phi|\}$, $\{\rho = \rho^{(2)}(\phi), \alpha_\epsilon \leqslant |\phi|\}$, $\{\phi = -\alpha_a, \rho(-\alpha_\epsilon) \leqslant \rho \leqslant \rho^{(2)}(-\alpha_\epsilon)\}$, $\{\phi = \alpha_\epsilon, \rho(\alpha_\epsilon) \leqslant \rho \leqslant \rho^{(2)}(\alpha_\epsilon)\}$, $\{\phi = -2\alpha_\epsilon, \rho^{(1)}(-2\alpha_\epsilon) \leqslant \rho \leqslant \rho^{(2)}(-2\alpha_\epsilon)\}$, $\{\phi = 2\alpha_\epsilon, \rho^{(1)}(2\alpha_\epsilon) \leqslant \rho \leqslant \rho^{(2)}(2\alpha_\epsilon)\}$, where $\rho^{(2)}(\phi)$ denotes the equation of the circle $\Gamma_y^{(2)}$ (the domain $\Omega_{\epsilon y}^{(2)}$ is distinguished in Figure 1 by hatching). With the use of averaging functions, it is not difficult to construct a function $\chi_\epsilon(x, y) \in C^2(x \in \bar{\Omega}) \times C(y \in \Gamma)$, satisfying the conditions

$$\begin{aligned} &\chi_\varepsilon(x, y) = 0, \quad x \in \bar{\Omega}_{\varepsilon y}^{(1)};\\ &\chi_\varepsilon(x, y) = 1, \quad x \in \overline{\Omega \setminus (\Omega_{\varepsilon y}^{(1)} + \Omega_{\varepsilon y}^{(2)})};\\ &\left.\begin{aligned} &|\chi_\varepsilon(x, y)| \leqslant 1,\\ &|\nabla_x \chi_\varepsilon(x, y)| + |\Delta_x \chi_\varepsilon(x, y)| \leqslant C(\varepsilon, \Gamma) \end{aligned}\right\} x \in \bar{\Omega}. \end{aligned} \tag{3.9}$$

The function $\widetilde{\Phi}_\epsilon(x, y, k) = \Phi_{\epsilon y}(x, k)\chi_\epsilon(x, y)$ is then defined for all $x \in \bar{\Omega}$ and $y \in \Gamma$ $(x \neq y)$ and satisfies the conditions

$$\tilde{\Phi}_\varepsilon = 0, \quad x \in \bar{\Omega}_{\varepsilon y}^{(1)}; \tag{3.10}$$

$$\partial\tilde{\Phi}_\varepsilon / \partial |x| - ik\tilde{\Phi}_\varepsilon = e^{-\operatorname{Im} k |x|} o(|x|^{-1/2}), \quad |x| \to \infty, \tag{3.11}$$

$$\Delta_x \tilde{\Phi}_\varepsilon + k^2 \tilde{\Phi}_\varepsilon = f_\varepsilon(x, y, k) = \begin{cases} 0, & x \in \bar{\Omega} \setminus \Omega_{\varepsilon y}^{(2)},\\ (\nabla_x \Phi_{\varepsilon y} \cdot \nabla_x \chi_\varepsilon) + \Phi_\varepsilon \Delta_x \chi_\varepsilon, & x \in \Omega_{\varepsilon y}^{(2)}. \end{cases} \tag{3.12}$$

Consider the functions $\widetilde{\Phi}_\epsilon(x, y, k)$ and $\partial\widetilde{\Phi}_\epsilon(x, y, k)/\partial n_x$ for $x \in \Gamma$ $(x \neq y)$. The function $\rho(\phi) \in C^2(-\pi, \pi)$ giving the equation of the contour Γ has the properties $\rho(0) = 1, \rho'(0) = 0$ and $\rho''(0) = \epsilon/R_y$, and hence for $|\phi| \leqslant 2\alpha_\epsilon$, where $\alpha_\epsilon = \alpha(\epsilon, \Gamma)$ is sufficiently small, it satisfies the inequalities

$$\begin{aligned} &|\rho'(\varphi)| \leqslant A_1(\Gamma)\varepsilon|\varphi|, \quad 1 + A_2(\Gamma)\varphi^2 \leqslant \rho(\varphi) \leqslant 1 + A_3(\Gamma)\varphi^2,\\ &\arccos\frac{1}{\rho(\varphi)} \leqslant \sqrt{\varepsilon}\, A_4(\Gamma)|\varphi|, \quad |\varphi| - \arccos\frac{1}{\rho(\varphi)} \geqslant \frac{|\varphi|}{2} \end{aligned} \tag{3.13}$$

for $\epsilon \leqslant \epsilon_0 = \epsilon_0(\Gamma)$, from which for $x \in \Gamma \cap \{0 < |\phi| < 2\alpha_\epsilon\}$ we obtain the estimates

$$\begin{aligned} \left|\frac{\partial\Phi_{\varepsilon y}}{\partial n_x}\right| &\leqslant \frac{1}{R_{\varepsilon y}}\left[\frac{1}{\rho}\left|\frac{\partial\Phi_{\varepsilon y}}{\partial\varphi}\right| \frac{|\rho'(\varphi)|}{\rho\cos\varphi + \rho'\sin\varphi} + \left|\frac{\partial\Phi_{\varepsilon y}}{\partial\rho}\right| \frac{\rho(\varphi)}{\rho\cos\varphi + \rho'\sin\varphi}\right]\\ &\leqslant A_0(\Gamma)\left[\varepsilon|\varphi|\left|\frac{\partial\Phi_{\varepsilon y}}{\partial\varphi}\right| + \left|\frac{\partial\Phi_{\varepsilon y}}{\partial\rho}\right|\right]. \end{aligned} \tag{3.14}$$

On the basis of (3.6)–(3.10), (3.13) and (3.14) we conclude that when $x \in \Gamma$ $(x \neq y)$ the functions $\widetilde{\Phi}_\epsilon(x, y, k)$ and $\widetilde{M}_\epsilon(x, y, k) = \partial\widetilde{\Phi}_\epsilon(x, y, k)/\partial n_x$ admit analytic continuations into the domain $K(\beta(\Gamma), \kappa(\Gamma))$ which satisfy (i) for $x \in \Gamma \cap \{|x-y| \leqslant \delta_\epsilon = \delta(\epsilon, \Gamma)\}$ (in view of the obvious inequalities $A'(\Gamma)|x-y| \leqslant |\phi| \leqslant A''(\Gamma)|x-y|$ for $\Gamma \in C^2$) the estimates

$$|\tilde{\Phi}_\varepsilon(x, y, k)| \leqslant \tilde{C}(\Gamma)|k|^{1/6}\exp\{(\bar{D}(\Gamma)\varepsilon|\operatorname{Im} k| - 2\tilde{B}(\Gamma)|k|^{1/3})|x-y|\}$$
$$\times\Bigg[\int\limits_{|k|^{-1/3}}^{1} \tau^{1/2}\exp\{-2\tilde{B}(\Gamma)|k||x-y|\tau^2\}\,d\tau + |\ln(|k||x-y|)|\exp\{-2\tilde{B}(\Gamma)$$
$$\times|k||x-y|\}\Bigg] \leqslant \tilde{C}(\Gamma)|k|^{1/6}\int\limits_{|k|^{-1/3}}^{1}\tau^{1/2}\exp\{-\tilde{B}(\Gamma)|k||x-y|\tau^2\}\,d\tau$$
$$+\tilde{C}|k|^{1/6}|\ln(|k||x-y|)|\exp\{-\tilde{B}|k||x-y|\} \tag{3.15}$$

(if $\beta = \beta(\Gamma)$ is sufficiently small and $|x-y| \leqslant \delta_\epsilon$, i.e. $|\phi| \leqslant \alpha_\epsilon$),

$$|\tilde{M}_\varepsilon(x, y, k)| \leqslant \tilde{C}(\Gamma)\varepsilon|k|^{1/3}\exp\{-\tilde{B}(\Gamma)|k|^{1/3}|x-y|\}, \tag{3.16}$$

(ii) for $x \in \Gamma \cap \{|x - y| \geqslant \delta_\epsilon\}$, analogously, the estimates

$$|\tilde{\Phi}_\varepsilon(x, y, k)| \leqslant \tilde{C}(\Gamma)|k|^{1/6}\int\limits_{|k|^{-1/3}}^{1}\tau^{1/2}\exp\{-\tilde{B}(\Gamma)|k||x-y|\tau^2\}\,d\tau$$
$$+\tilde{C}|k|^{1/6}|\ln(|k||x-y|)|\exp\{-\tilde{B}|k||x-y|\}, \tag{3.17}$$
$$|\tilde{M}_\varepsilon(x, y, k)| \leqslant \tilde{C}(\Gamma, \varepsilon)\exp\{-\tilde{B}(\Gamma)|k|^{1/3}|x-y|\} \tag{3.18}$$

and (iii) for $x \in \bar{\Omega}^{(2)}_{\epsilon y}$ the estimate

$$|f_\varepsilon(x, y, k)| \leqslant \tilde{C}(\Gamma, \varepsilon)\exp\{-\tilde{B}(\Gamma)|k|^{1/3}|x-y|\}, \tag{3.19}$$

where the function f_ϵ is defined by (3.12).

The function $L_\epsilon(x, y, k)$ is defined for $x \in \Omega$ and $k \in K(\beta(\Gamma), \kappa(\Gamma))$ by the equality

$$L_\varepsilon(x, y, k) = \frac{i}{4}H_0^{(1)}(k|x-y|) - \frac{i}{4}\int\limits_\Gamma\Big[H_0^{(1)}(k|x-x'|)\frac{\partial\tilde{\Phi}_\varepsilon(x', y, k)}{\partial n'}$$
$$-\tilde{\Phi}_\varepsilon(x', y, k)\frac{\partial H_0^{(1)}(k|x-x'|)}{\partial n'}\Big]ds', \tag{3.20}$$

from which it follows that $L_\epsilon(x, y, k)$ satisfies the conditions

$$\Delta_x L_\varepsilon(x, y, k) + k^2 L_\varepsilon(x, y, k) = 0, \qquad x \in \Omega, \tag{3.21}$$
$$\frac{\partial L_\varepsilon(x, y, k)}{\partial|x|} - ikL_\varepsilon(x, y, k)$$
$$= \exp\{-\operatorname{Im} k|x|\}\,o(|x|^{-1/2}), \qquad |x| \to \infty, \qquad \operatorname{Im} k \geqslant 0. \tag{3.22}$$

On the basis of a Green's formula we have for $\operatorname{Im} k \geqslant 0$ and $x \in \Omega$ the relation

$$\tilde{\Phi}_\varepsilon(x, y, k) - \frac{i}{4} \int_{\Omega_{\varepsilon y}^{(2)}} H_0^{(1)}(k|x - x'|) f_\varepsilon(x', y, k)\, dx' = \frac{i}{4} H_0^{(1)}(k|x' - y|)$$

$$- \frac{i}{4} \int_\Gamma \left[H_0^{(1)}(k|x' - x|) \frac{\partial \Phi_\varepsilon(x' y, k)}{\partial n'} - \tilde{\Phi}_\varepsilon(x', y, k) \frac{\partial H_0^{(1)}(k|x - x'|)}{\partial n'} \right] ds',$$

which together with (3.20) implies that the representation

$$L_\varepsilon(x, y, k) = \tilde{\Phi}_\varepsilon(x, y, k) - \frac{i}{4} \iint_{\Omega_{\varepsilon y}^{(2)}} H_0^{(1)}(k|x' - x|) f_\varepsilon(x', y, k)\, dx' \tag{3.23}$$

is valid for $x \in \Omega$. This equality permits us to define the functions $L_\epsilon(x, y, k)$ and $M_\epsilon(x, y, k) = \partial L_\epsilon(x, y, k)/\partial n_x$ for $x, y \in \Gamma$ $(x \neq y)$. It follows on the basis of (3.17)–(3.19) and known properties of the Hankel functions that they admit for $x \neq y$ analytic continuations into the domain $K(\beta(\epsilon, \Gamma), \kappa(\Gamma))$ with estimates (3.2)–(3.3). Applying a Green's formula to the functions $G(x', y, k)$ and $L_\epsilon(x', x, k)$ for $k \in K(\beta, \kappa) \cap \{\operatorname{Im} k > 0\}$, we get that $G(x, y, k)$ satisfies the integral equation (3.1).

For $k \in K(\beta, \kappa) \cap \{\operatorname{Im} k \geqslant 0\}$ let

$$G_1(x, y, k) = G(x, y, k) \exp\{\sigma k^{1/3} |x - y|\}, \tag{3.24}$$

where

$$\sigma = \sigma(\Gamma, \varepsilon) = \min\left(\frac{1}{2} B(\Gamma),\ B_1(\Gamma, \varepsilon) \Big/ \int_\Gamma dx\right)$$

in which B and B_1 are the constants in (3.2)–(3.3).

Taking into account (3.1), we conclude that for $k \in K(\beta, \kappa) \cap \{\operatorname{Im} k \geqslant 0\}$ the function $G_1(x, y, k)$ satisfies the integral equation

$$G_1(x, y, k) = L_1(y, x, k) + \int_\Gamma M_1(x', x, y, k) G_1(x', y, k)\, dx', \tag{3.25}$$

where the functions

$$L_1(x, y, k) = L_\varepsilon(x, y, k) \exp\{\sigma k^{1/3} |x - y|\},$$

$$M_1(x', x, y, k) = -(M_\varepsilon(x', x, k) + g(x') L_\varepsilon(x', x, k)) \times \exp\{\sigma k^{1/3} (|x - y| - |x' - y|)\}$$

admit analytic continuations into $K(\beta, \kappa)$ with estimates

$$|L_1(x, y, k)| \leqslant C(\Gamma)|k|^{1/6}\left[\int_{|k|^{-1/3}}^{1} \tau^{1/2} \exp\left\{-\frac{B(\Gamma)}{2}|k||x-y|\tau^2\right\} d\tau + |\ln(|k||x-y|)| \exp\left\{-\frac{B(\Gamma)}{2}|k||x-y|\right\}\right] + C_1(\Gamma, \varepsilon) \exp\left\{-\frac{B_1(\Gamma, \varepsilon)}{2}|k|^{1/3}\right\}, \tag{3.26}$$

$$|M_1(x', x, y, k)| \leqslant C(\Gamma)\varepsilon|k|^{1/3} \exp\left\{-\frac{B(\Gamma)}{2}|k|^{1/3}|x-x'|\right\} + |\widetilde{M}(x', x, k)|, \tag{3.27}$$

$$|\widetilde{M}(x', x, k)| \leqslant C(\Gamma) \max_{x\in\Gamma}|g(x)||k|^{1/6}\left[\int_{|k|^{-1/3}}^{1} \tau^{1/2} \exp\left\{-\frac{B(\Gamma)}{2}|k||x-x'|\tau^2\right\} \times d\tau + |\ln(|k||x-y|)| \exp\left\{-\frac{B(\Gamma)}{2}|k||x-x'|\right\}\right] + C_1(\Gamma, \varepsilon, g) \exp\left\{-\frac{B_1(\Gamma, \varepsilon)}{2}|k|^{1/3}\right\}; \tag{3.28}$$

here we have taken into account the inequalities (3.2), (3.3) and

$$|\exp\{\sigma k^{1/3}(|x-y| - |x'-y|)\}| \leqslant \exp\{\sigma|k|^{1/3}|x-x'|\}).$$

Let us show that the solution $G_1(x, y, k)$ of equation (3.25) can be represented in $K(\beta, \kappa)$ in the form

$$G_1(x, y, k) = L_1(y, x, k) + N(x, y, k) + \int_\Gamma R(x', x, y, k) N(x', y, k)\, dx', \tag{3.29}$$

where

$$N(x, y, k) = \int_\Gamma M_1(x', x, y, k) L_1(y, x', k)\, dx', \tag{3.30}$$

$$R(x', x, y, k) = \sum_{m=1}^{\infty} M_m(x', x, y, k), \tag{3.31}$$

$$M_m(x', x, y, k) = \int_\Gamma M_{m-1}(x', x'', y, k) M_1(x'', x, y, k)\, dx'', \qquad m = 2, 3, \ldots.$$

On the basis of (3.26)–(3.27) the inequalities

$$\int_\Gamma |L_1(x, y, k)|\,dy + \int_\Gamma |L_1(y, x, k)|\,dy$$

$$\leqslant C|k|^{1/6} \int_{\Gamma\cap\{|x-y|<\delta(\varepsilon,\Gamma)\}} dy \int_{|k|^{-1/3}}^{1} \tau^{1/2} \exp\left\{-\frac{B}{2}|k||x-y|\tau^2\right\} d\tau$$

$$+ C|k|^{1/6} \int_{\Gamma\cap\{|x-y|<\delta\}} |\ln(|k||x-y|)| \exp\left\{-\frac{B}{2}|k||x-y|\right\} dy$$

$$+ C_3(\Gamma, \varepsilon)\exp\{-B_3(\Gamma, \varepsilon)|k|^{1/3}\}$$

$$\leqslant C_1(\Gamma)|k|^{1/6} \int_0^\delta d\xi \int_{|k|^{-1/3}}^{1} \tau^{1/2} \exp\{-B_4(\Gamma)|k|\tau^2\xi\}\,d\tau$$

$$+ C_1|k|^{-5/6} \int_0^\infty (1 + |\ln\xi|)\exp\{-B_4\xi\}\,d\xi + C_3\exp\{-B_3|k|^{1/3}\} \tag{3.32}$$

$$\leqslant C_4(\Gamma)|k|^{-5/6} \int_{|k|^{-1/3}}^{1} \tau^{-3/2} d\tau + C_5(\Gamma)|k|^{-5/6} + C_3\exp\{-B_3|k|^{1/3}\} \leqslant C_6(\Gamma)|k|^{-2/3},$$

$$\int_\Gamma |M_1(x', x, y, k)|\,dx' + \int_\Gamma |M_1(x, x', y, k)|\,dx' \leqslant C(\Gamma)\varepsilon|k|^{1/3}$$

$$\times \int_{\Gamma\cap\{|x-x'|<\delta\}} \exp\left\{-\frac{B}{2}|k|^{1/3}|x-x'|\right\} dx' + C'(\Gamma, g, \varepsilon)|k|^{-2/3}$$

$$\leqslant C''(\Gamma)\varepsilon + C'|k|^{-2/3} \leqslant \mu < 1$$

hold for sufficiently small $\epsilon = \epsilon_1(\Gamma)$ and sufficiently large $\kappa = \kappa_1(\Gamma, g)$. Consequently the series (3.31) converges in $L_1(\Gamma)$ for $k \in K(\beta_1, \kappa_1)$, and we have the estimate

$$\int_\Gamma |R(x', x, y, k)|\,dx' \leqslant C_3(\Gamma, g), \quad x, y \in \Gamma, \quad k \in K(\beta_1, \varkappa_1). \tag{3.33}$$

The properties of M_1 and L_1 imply that $N(x, y, k)$ is continuous with respect to $x, y \in \Gamma$, is analytic with respect to $k \in K(\beta_1, \kappa_1)$ and, analogously to (3.32), satisfies on the basis of (3.26)–(3.27) the estimate

$$|N(x, y, k)| \leqslant C(\Gamma)\varepsilon|k|^{1/3} \sup_{x, y\in\Gamma} \int_\Gamma |L_1(x', x, y, k)|\,dx'$$

$$+ \sup_{x, y\in\Gamma} \int_\Gamma |L_1(x', x, y, k)||\widetilde{M}(x', x, k)|\,dx' \leqslant C_4(\Gamma, g)|k|^{-1/3}.$$

From (3.29) we deduce that when $x, y \in \Gamma$ $(x \neq y)$ the function $G_1(x, y, k)$ admits an analytic continuation into $K(\beta_1, \kappa_1)$ with the estimate

$$|G_1(x, y, k)| \leqslant C_1(\Gamma)|k|^{1/6}\left[\int_{|k|^{-1/3}}^{1} \tau^{1/2} \exp\{-B_1(\Gamma)|k||x-y|\tau^2\}\, d\tau\right.$$
$$\left. + |\ln(|k||x-y|)| \exp\{-B_1(\Gamma)|k||x-y|\}\right] + C_2(\Gamma, g)|k|^{-1/3}.$$

Equality (3.24) determines the analytic continuation of $G(x, y, k)$ into $K(\beta_1, \kappa_1)$ with the estimate

$$|G(x, y, k)| \leqslant C_1(\Gamma)|k|^{1/6}\left[\int_{|k|^{-1/3}}^{1} \tau^{1/2} \exp\{-B_1(\Gamma)|k||x-y|\tau^2\}\, d\tau\right.$$
$$\left. + |\ln(|k||x-y|)| \exp\{-B_1(\Gamma)|k||x-y|\}\right]$$
$$+ C_2(\Gamma, g)|k|^{-1/3} \exp\{-B_2(\Gamma)|k|^{1/3}|x-y|\}. \tag{3.34}$$

Using a Green's formula, we obtain for $x \in \Omega, y \in \Gamma$ and $\operatorname{Im} k \geqslant 0$ $(k \neq 0)$ the equality

$$G(x, y, k) - \frac{i}{4} H_0^{(1)}(k|x-y|) - \frac{i}{4}\int_\Gamma \left[kH_1^{(1)}(k|x-x'|)\frac{(n', x-x')}{|x-x'|}\right.$$
$$\left. + g(x') H_0^{(1)}(k|x-x'|)\right] G(x', y, k)\, dx',$$

which determines the analytic continuation of $G(x, y, k)$ into $K(\beta_1, \kappa_1)$ for $x \in \Omega$, $y \in \Gamma$. From this result, (3.34) and the inequalities

$$|H_0^{(1)}(k|x-y|)| \leqslant 3(|k||x-y|)^{-1/2} \exp\{-\operatorname{Im} k|x-y|\},$$
$$|H_0^{(1)}(k|x-y|)| \leqslant [(|k||x-y|)^{-1/2} + (|k||x-y|)^{-1}] \exp\{-\operatorname{Im} k|x-y|\},$$

readily implied by the integral representations 8.421 in [16] for the Hankel functions, we deduce that $G(x, y, k)$ satisfies the following estimate for any $x \in \bar{\Omega}, y \in \Gamma$ and $k \in K(\beta_1, \kappa_1)$:

$$|G(x, y, k)| \leqslant \exp\{-\operatorname{Im} k|x-y| + |\operatorname{Im} k| D(\Gamma)\}\{(|k||x-y|)^{-1/2}$$
$$+ C(\Gamma, g)[|k|^{2/3}(I_1 + I_2) + |k|^{1/6}(I_3 + I_4 + I_5) + |k|^{-1/3} I_6]\}, \tag{3.35}$$

where

$$I_1 = \int_\Gamma \frac{|(n', x-x')|}{|x-x'|^{9/2}}\, dx' \int_{|k|^{-1/3}}^{1} \tau^{1/2} \exp\{-B|k||y-x'|\tau^2\}\, d\tau,$$
$$I_2 = \int_\Gamma \frac{|(n', x-x')|}{|x-x'|^{3/2}} |\ln(|k||x'-y|)| \exp\{-B|k||y-x'|\}\, dx',$$

$$I_3 = \int_\Gamma \frac{|(n', x-x')|}{|x-x'|^{3/2}} \exp\{-B|k|^{1/3}|y-x'|\}\,dx',$$

$$I_4 = \int_\Gamma \frac{|(n', x-x')|}{|x-x'|^2}\,dx' \int_{|k|^{-1/3}}^{1} \tau^{1/3}\exp\{-B|k||y-x'|\tau^2\}\,d\tau,$$

$$I_5 = \int_\Gamma \frac{|(n', x-x')|}{|x-x'|^2} |\ln(|k||x'-y|)|\exp\{-B|k||x'-y|\}\,dx',$$

$$I_6 = \int_\Gamma \frac{|(n', x-x')|}{|x-x'|^2} \exp\{-B|k|^{1/3}|x'-y|\}\,dx'.$$

Taking into account the inequalities

$$\int_{k|^{-1/3}}^{1} \tau^{1/2}\exp\{-B|k||y-x'|\tau^2\}\,d\tau$$

$$\leqslant \max_{0\leqslant\zeta} \zeta^{1/2}\exp\{-B\zeta\}(|x'-y||k|)^{-1/2}\int_{|k|^{-1/3}}^{1}\tau^{-1/2}d\tau$$

$$\leqslant C'(\Gamma)(|k||x'-y|)^{-1/2},\quad |\ln(|k||x'-y|)|\exp\{-B|k||x'-y\}$$

$$\leqslant C'(\Gamma)(|k||x'-y|)^{-1/2},\quad \exp\{-B|k|^{1/3}|x'-y|\}\leqslant C'(\Gamma)(|k|^{1/3}|x'-y|)^{-1/2},$$

we obtain the estimates

$$I_4 + I_5 \leqslant 2C'|k|^{-1/2}I_7,\qquad I_6 \leqslant C'|k|^{-1/6}I_7, \tag{3.36}$$

where

$$I_7 = \int_\Gamma \frac{|(n', x-x')|}{|x-x'|^2|x'-y|^{1/2}}\,dx'.$$

The integrals $I_1 - I_3$ are estimated analogously. We therefore estimate integrals I_1 and I_7. For $\min_{x'\in\Gamma}|x-x'| \geqslant \delta(\Gamma) > 0$ the following inequalities are obvious:

$$I_1 \leqslant C'(\Gamma)|x|^{-1/2}\int_\Gamma dx' \int_{|k|^{-1/3}}^{1} \tau^{1/2}\exp\{-B|k||y-x'|\tau^2\}\,d\tau$$

$$\leqslant C''(\Gamma)|x|^{-1/2}\,|k|^{-1}\int_{|k|^{-1/3}}^{1}\tau^{-3/2}d\tau \leqslant 3C''(\Gamma)|x-y|^{-1/2}|k|^{-5/6},$$

$$I_7 \leqslant C'(\Gamma)|x|^{-1}\int_\Gamma |x-'y|^{-1/2}dx' \leqslant C''(\Gamma)|x-y|^{-1/2}$$

(we assume that the point $x = (0, 0) \in E_2\backslash\bar{\Omega}$).

Suppose $\min|x - x'| = \rho < \delta$ $(x' \in \Gamma)$ and x_0 is a point on Γ such that $|x - x_0| = \rho$. We choose in a neighborhood of x_0 an orthogonal coordinate system

(ξ, η) so that the η axis is directed along the outward (relative to Ω) normal to Γ at x_0. If $\delta = \delta(\Gamma)$ is sufficiently small, the equation of Γ can be given in the form $\eta = \omega_0(\xi)$ for $|\xi| \leqslant 3\delta$, with the function $\omega_0(\xi)$ satisfying the inequalities

$$|\omega_0(\xi)| \leqslant A_1\xi^2, \qquad |d\omega_0(\xi)/d\xi| \leqslant A_2|\xi|, \qquad A_1\delta < 1, \tag{3.37}$$

in which the constants A_1 and A_2 depend only on Γ. Let $\Gamma_\sigma(x_0)$ denote the part of Γ having the form $\Gamma_\sigma(x_0) = \{x: x = (\xi, \omega_0(\xi)), |\xi| \leqslant \sigma\}$, where $0 < \sigma \leqslant 3\delta$. For $x' \in \Gamma_{3\delta}(x_0)$ we get

$$(n', x - x') = \frac{\rho - \omega_0(\xi) + \xi d\omega_0(\xi)/d\xi}{\sqrt{1 + (d\omega_0/d\xi)^2}}, \quad |x - x'|^2 = \xi^2 + (\rho - \omega_0(\xi))^2;$$

hence on the basis of (3.37) we have the inequalities

$$|x - x'|^{-1/2} \leqslant A(\Gamma)(\rho^2 + \xi^2)^{-1/4}, \quad |(n', x - x')| \leqslant A(\Gamma)(\rho + \xi^2). \tag{3.38}$$

It follows that for $\min|x' - x| = |x - x_0| = \rho < \delta$ and $|y - x_0| \geqslant 2\delta$

$$I_1 \leqslant C(\Gamma)\delta^{-1/2}(\Gamma) \int_{|k|^{-1/3}}^{1} \tau^{1/2} d\tau \int_{\Gamma_\delta(y)} \exp\{-B(\Gamma)|k|x' - y|\tau^2\} dx'$$

$$+ C_1(\Gamma)\exp\{-B(\Gamma)\delta(\Gamma)|k|^{1/3}\} \int_{\Gamma/\Gamma_\delta(y)} |x - x'|^{-1/2} dx'$$

$$\leqslant C_2(\Gamma) \int_{|k|^{-1/3}}^{1} \tau^{1/2} d\tau \int_0^\delta \exp\{-B|k|\xi\tau^2\} d\xi + C_1 \exp\{-B_1(\Gamma)|k|^{1/3}\}$$

$$\times \left[\int_{\Gamma_\delta(x_0)} |x - x'|^{-1/2} dx' + \delta^{-1/2}\int_\Gamma dx'\right] \leqslant C_3(\Gamma)|k|^{-1} \int_{|k|^{-1/3}}^{1} \tau^{-3/2} d\tau$$

$$+ C_1 \exp\{-B_1|k|^{1/3}\}\left[A'(\Gamma)\int_0^\delta \frac{d\xi}{(\rho^2 + \xi^2)^{1/4}} + A''(\Gamma)\right] \leqslant C_4(\Gamma)|k|^{-5/6},$$

$$I_7 \leqslant C_1(\Gamma)\delta^{-1}(\Gamma) \int_{\Gamma/\Gamma_\delta(x_0)} |x' - y|^{-1/2} dx' + C_2(\Gamma)\delta^{-1/2}(\Gamma) \int_{\Gamma_\delta(x_0)} \frac{|(n', x - x')|}{|x - x'|^2} dx'$$

$$\leqslant C_3(\Gamma) + C_4(\Gamma)\int_{-\delta}^{\delta} \frac{\rho d\xi}{\rho^2 + \xi^2} = C_3 + 2C_4 \arctan\frac{\delta}{\rho} \leqslant C_5(\Gamma).$$

Suppose $y \in \Gamma_{2\delta}(x_0)$. Then $y = (\xi_1, \omega_0(\xi_1))$ and for $|\xi_1| \geqslant 2\rho$

$$I_1 \leqslant C_1(\Gamma)|k|^{-5/6} + C_2(\Gamma)\int_{-3\delta}^{3\delta} \frac{\rho d\xi}{(\rho^2 + \xi^2)^{3/4}} \int_{|k|^{-1/3}}^{1} \tau^{1/2} \exp\{-B_1(\Gamma)|k||\xi - \xi_1|\tau^2\} d\tau$$

$$\leqslant C_1|k|^{-5/6} + C_2 \int_{-|\xi_1|/2}^{|\xi_1|/2} \frac{d\xi}{\sqrt{|\xi|}} (|k||\xi_1|)^{-1} \int_{|k|^{-1/3}}^{1} \tau^{-3/2} d\tau \times$$

$$\times \max_{0\leqslant\xi} \xi \exp\{-B_1\xi\} + C_3(\Gamma)|\xi_1|^{-1/2} \int_{|\xi|/2}^{3\delta} d\xi \int_{|k|^{-1/3}}^{1} \tau^{1/2}$$

$$\times \exp\{-B_1|k||\xi-|\xi_1||\tau^2\}\,d\tau \leqslant C_1(\Gamma)|k|^{-5/6} + C_2'(\Gamma)|\xi_1|^{-1/2}|k|^{-5/6}$$

$$+ C_3'(\Gamma)|\xi_1|^{-1/2}\int_0^{4\delta} d\xi' \int_{|k|^{-1/3}}^{1} \tau^{1/2}\exp\{-B_1|k|\xi'\tau^2\}\,d\tau \leqslant C_4(\Gamma)|k|^{-5/6}|\xi_1|^{-1/2}$$

$$\leqslant C_5(\Gamma)|k|^{-5/6}|x-y|^{-1/2},$$

$$I_7 \leqslant C_1 + C_2|\xi_1|^{-1/2}\int_{-|\xi_1|/2}^{|\xi_1|/2} \frac{\rho\,d\xi}{\rho^2+\xi^2} + C_3(\Gamma)\rho\int_{|\xi_1|/2}^{\infty} \frac{d\xi}{|\xi|^2|\xi-|\xi_1||^{1/2}}$$

$$\leqslant C_1 + 2C_2|\xi_1|^{-1/2}\arctan\left(\frac{|\xi_1|}{2\rho}\right) + C_3\rho|\xi_1|^{-3/2}\int_{1/2}^{\infty}\frac{d\xi'}{\xi'^2|1-\xi'|^{1/2}}$$

$$\leqslant C_4(\Gamma)|\xi_1|^{-1/2} \leqslant C_5(\Gamma)|x-y|^{-1/2}.$$

Finally, for $|\xi_1| \leqslant 2\rho$

$$I_1 \leqslant C_1|k|^{-5/6} + C_2\rho^{-1/2}\int_{-3\delta}^{3\delta} d\xi \int_{|k|^{-1/3}}^{1} \tau^{1/2}\exp\{-B_1|k||\xi-\xi_1|\tau^2\}\,d\tau$$

$$\leqslant C_3(\Gamma)|k|^{-5/6}\rho^{-1/2} \leqslant C_4'(\Gamma)|k|^{-5/6}|x-y|^{-1/2}$$

$$I_7 \leqslant C_1 + C_2\int_{-3\delta}^{3\delta}\frac{\rho\,d\xi}{(\rho^2+\xi^2)|\xi-\xi_1|^{1/2}} \leqslant C_1 + C_2\int_{-4\delta}^{4\delta}\frac{\rho\,d\xi'}{(\rho^2+(\xi_1+\xi')^2)|\xi'|^{1/2}}$$

$$\leqslant C_1 + C_2\rho^{-1}\int_{-\rho}^{\rho}\frac{d\xi'}{|\xi'|^{1/2}} + C_3\rho^{-1/2}\int_{-5\delta}^{5\delta}\frac{\rho\,d\xi}{\rho^2+\xi^2} \leqslant C_4(\Gamma)\rho^{-1/2} \leqslant C_5(\Gamma)|x-y|^{-1/2}.$$

We have thus obtained the inequalities

$$I_1 \leqslant C(\Gamma)|k|^{-5/6}|x-y|^{-1/2}, \quad I_7 \leqslant C(\Gamma)|x-y|^{-1/2}.$$

In exactly the same way we establish the estimates

$$I_2 \leqslant C(\Gamma)|k|^{-1}|x-y|^{-1/2}, \quad I_3 \leqslant C(\Gamma)|k|^{-1/3}|x-y|^{-1/2}.$$

Substituting these estimates and (3.36) into (3.35), we get

$$|G(x,y,k)| \leqslant C(\Gamma,g)|k|^{-1/6}|x-y|^{-1/2}\exp\{-\operatorname{Im}k|x-y| + |\operatorname{Im}k|D(\Gamma)\}. \tag{3.37}$$

From the well-known (for $\operatorname{Im} k \geqslant 0$) equality $G(x, y, k) = G(y, x, k)$ and the established analyticity of $G(x, y, k)$ with respect to $k \in K(\beta_1, \kappa_1)$ for $x \in \bar{\Omega}$ and $y \in \Gamma$ $(x \neq y)$ we conclude that $G(x, y, k)$ admits an analytic continuation into $K(\beta_1, \kappa_1)$ with estimate (3.39) for $x \in \Gamma$ and $y \in \bar{\Omega}$ $(x \neq y)$. Finally, from equality (2.2) it follows that $G(x, y, k)$ admits an analytic continuation into $K(\beta_1, \kappa_1)$ with estimate (0.12), which is established analogously to (3.39), for $x, y \in \bar{\Omega}$ $(x \neq y)$.

§4. On estimates of the analytic continuation into the domain $\operatorname{Im} k < 0$ of the Green function of the second boundary value problem for the equation $\Delta v + k^2 v = 0$ in the exterior of a disk

Let $\Phi(\rho, \phi - \psi, k)$ denote the Green function of the second boundary value problem for the Helmholtz equation in the exterior of the unit disk, i.e. let $\Phi(\rho, \phi, k)$ denote the solution for $\operatorname{Im} k \geqslant 0$ $(k \neq 0)$ of the problem

$$\Delta_{\rho,\varphi}\Phi + k^2\Phi = 0, \qquad \rho > 1,$$

$$\frac{\partial\Phi}{\partial\rho} - ik\Phi = e^{-\operatorname{Im} k\rho}\, o(\rho^{-1/2}), \qquad \rho \to \infty,$$

$$\lim_{\rho\to 1+0} \frac{\partial\Phi}{\partial\rho} = -\frac{1}{2\pi}\left(1 + 2\sum_{m=1}^{\infty} \cos m\varphi\right) = -\delta(\varphi),$$

where $\Delta_{\rho,\phi}$ is the Laplace operator in polar coordinates.

It can easily be shown with the use of the method of separation of variables that $\Phi(\rho, \phi, k)$ has the representation

$$\Phi(\rho, \varphi, k) = -\frac{1}{\pi k}\left[\frac{H_0(k\rho)}{H_0'(k)} + 2\sum_{m=1}^{\infty} \frac{H_m(k\rho)}{H_m'(k)} \cos m\varphi\right], \tag{4.1}$$

which converges for $\rho > 1$, $|\phi| \leqslant \pi$ and $\operatorname{Im} k \geqslant 0$ $(k \neq 0)$.

In the present section we obtain estimates of the analytic continuation of $\Phi(\rho, \phi, k)$ and its derivatives with respect to ρ and ϕ into the domain $K(\beta, \kappa)$ when β and κ are certain positive numbers and the points (ρ, ϕ) lie in the so-called "shadow zone," viz. the domain ω with boundaries $\rho = 1/\cos\phi$, $|\phi| \leqslant \pi/2$ and $\rho = 1$ (the unit circle and its tangent at the point $\rho = 1, \phi = 0$). Let $\omega_R = \omega \cap \{\rho < R\}$, where $R > 1$. We have the following

LEMMA. *For any* R *there exists a number* κ_R *such that when* $(\rho, \phi) \in \omega_R$ *the function* $\Phi(\rho, \phi, k)$ *and its derivatives with respect to* ρ *and* ϕ *admit analytic continuations into the domain* $K(\beta, \kappa_R)$ *with estimates*

$$\begin{aligned}\Phi| \leqslant C_R |k|^{1/6} \exp\{|\operatorname{Im} k|\sqrt{\rho^2 - 1}\}\Bigg[\int_{|k|^{-1/3}}^{1} \tau^{1/2} \\ \times \exp\left\{-B|k|\left(|\varphi| - \arccos\frac{1}{\rho}\right)\tau^2\right\} d\tau + \left|\ln|k|(|\varphi| - \arccos\frac{1}{\rho})\right| \\ \times \exp\left\{-B|k|\left(|\varphi| - \arccos\frac{1}{\rho}\right\}\right],\end{aligned} \tag{4.2}$$

$$\begin{aligned}\left|\frac{\partial\Phi}{\partial\varphi}\right| \leqslant C_R |k|^{1/3}\left(|\varphi| - \arccos\frac{1}{\rho}\right)^{-1} \exp\Big\{|\operatorname{Im} k|\sqrt{\rho^2 - 1} \\ - B|k|^{1/3}\left(|\varphi| - \arccos\frac{1}{\rho}\right)\Big\},\end{aligned} \tag{4.3}$$

$$\left|\frac{\partial\Phi}{\partial\rho}\right| \leqslant C_R|k|^{1/3}(\rho-1)\left(|\varphi|-\arccos\frac{1}{\rho}\right)^{-2}\Big[1+(\rho-1)$$
$$\times\left(|\varphi|-\arccos\frac{1}{\rho}\right)^{-2}\Big]\exp\Big\{|\operatorname{Im}k|\sqrt{\rho^2-1}-B|k|^{1/3}$$
$$\times\left(|\varphi|-\arccos\frac{1}{\rho}\right)\Big\}, \tag{4.4}$$

where β and B are certain positive numbers, while the dependence of the constants C_R and κ_R on R is such that $C_{R_2} \geqslant C_{R_1}$ and $\kappa_{R_2} \geqslant \kappa_{R_1}$ for $R_2 \geqslant R_1$.

REMARK. It can also be shown that when ϵ, δ and κ are respectively sufficiently small, sufficiently small and sufficiently large positive numbers the function Φ admits a continuously differentiable (with respect to ρ and ϕ) extension into the domain $1-\epsilon\phi^2<\rho<1$, $0<|\phi|<\delta\kappa^{-1/3}$ that satisfies the equation

$$\Delta_{\rho,\varphi}\Phi+k^2\Phi=0$$

for almost all (ρ, ϕ), is analytic with respect to $k \in K(\beta,\kappa)\cap\{\kappa<|\operatorname{Re}k|<\kappa+1\}$ and satisfies the inequalities

$$|\Phi|\leqslant C\left[\int_{\varkappa^{-1/3}}^{1}\exp\{-B\varkappa|\varphi|\tau^2\}\,d\tau+|\ln(\varkappa|\varphi|)|\exp\{-B\varkappa|\varphi|\}\right]$$
$$(1-\varepsilon\varphi^2\leqslant\rho\leqslant1+\varepsilon\varphi^2, \qquad |\varphi|\leqslant\delta\varkappa^{-1/3}),$$
$$\left|\frac{\partial\Phi}{\partial\varphi}\right|\leqslant C\varkappa^{1/3}|\varphi|^{-1}\exp\{-B\varkappa^{1/3}|\varphi|\},$$
$$\left|\frac{\partial\Phi}{\partial\rho}\right|\leqslant C\varkappa^{1/3}|\rho-1|\varphi^{-2}\exp\{-B\varkappa^{1/3}|\varphi|\},$$

where β, B and C are positive numbers.

The proof of the lemma makes use of a number of special properties of the Hankel functions of a complex argument and order, and is presented below in four subsections with their own enumerations of the formulas.

In §4.A we present preliminary information from the theory of cylindrical functions and establish the properties of these functions that are needed for the sequel.

In §4.B we study the distribution of the zeros of the function $H'_{k\xi}(k)$ in the halfplane $|\arg k\xi|\leqslant\pi/2$, where k is an arbitrary complex number from a certain sector $|\arg k|<\delta$, $|k|>\nu$.

In §4.C we prove that when $\rho>1$, $\phi\neq0$ and $k\in K\cap\{\operatorname{Im}k\geqslant0\}$

$$\Phi(\rho,\varphi,k)=2\sum_{n=1}^{\infty}\frac{H_{k\xi_n}(k\rho)}{\left[\frac{d}{d\xi}H'_{k\xi}(k)\right]_{\xi=\xi_n}}\frac{\cos k\xi_n(\pi-|\varphi|)}{\sin k\xi_n\pi}, \tag{4.5}$$

where the $\xi_n=\xi_n(k)$, $n=1,2,\cdots$, are the zeros of $H'_{k\xi}(k)$ in the halfplane $|\arg k\xi|\leqslant\pi/2$.

In §4.D we prove that when $(\rho, \phi) \in \omega_R$ the series (4.5) determines the analytic continuation of Φ into $K(\beta, \kappa_R)$ with estimates (4.2)–(4.4).

A. We will consider the Hankel function $H_{k\xi}(k\rho)$ and its derivative $H'_{k\xi}(k\rho)$ with respect to its argument for $k \in N(\delta, \nu)$, $\xi \in M(\theta, \mu)$ and $\rho \in (0, \infty)$, where $0 < \delta < \pi/2$, $\nu > 0$, $|\theta| < \delta$, $\mu > 0$ and

$$N(\delta, \nu) = \{k: |\arg k| = |\theta| < \delta, \ \nu < |k|\}, \tag{A.1}$$

$$M(\theta, \mu) = \{\xi: |\arg \xi + \theta| < \pi/2, \ \mu < |\xi|\}. \tag{A.2}$$

It follows from the work of Olver [21] that if we fix the numbers μ and δ, we can choose the number $\nu = \nu(\mu, \delta)$ so large that the following expansions converge uniformly in $k \in N(\delta, \nu)$, $\xi \in M(\theta, \mu)$ and $\rho \in (0, \infty)$:(1)

$$H_{k\xi}(k\rho) = \frac{2\sqrt{2}\, e^{-i\frac{\pi}{3}}}{(k\rho)^{1/3}} \left[\frac{w(\xi/\rho)}{\xi^2/\rho^2 - 1}\right]^{1/4} \Big\{ \mathrm{Ai}\left(e^{i\frac{2}{3}\pi} (k\rho)^{2/3} w(\xi/\rho)\right) + [1 + |(k\rho)^{2/3} w(\xi/\rho)|]^{-\frac{1}{4}} e^{\frac{2}{3} k\rho w^{3/2}(\xi/\rho)} O(|k\xi|^{-1}) \Big\}, \tag{A.3}$$

$$H'_{k\xi}(k\rho) = \frac{2\sqrt{2}\, e^{i\frac{\pi}{3}}}{(k\rho)^{1/3}} \left[\frac{\xi^2/\rho^2 - 1}{w(\xi/\rho)}\right]^{1/4} \Big\{ \mathrm{Ai}'\left(e^{i\frac{2\pi}{3}} (k\rho)^{2/3} w(\xi/\rho)\right) + [1 + |(k\rho)^{2/3} w(\xi/\rho)|]^{1/4} e^{\frac{2}{3} k\rho w^{3/2}(\xi/\rho)} O(|k\xi|^{-1}) \Big\}, \tag{A.4}$$

$$\frac{dH'_{k\xi}(k)}{d\xi} = -2\sqrt{2}\, \frac{dw(\xi)}{d\xi} \left[\frac{\xi^2 - 1}{w(\xi)}\right]^{1/4} \Big\{ \mathrm{Ai}''\left(e^{i\frac{2\pi}{3}} k^{2/3} w(\xi)\right) + [1 + |k^{2/3} w(\xi)|]^{3/4} e^{\frac{2}{3} k w^{3/2}(\xi)} O(|k\xi|^{-1}) \Big\}, \tag{A.5}$$

where the function $w(\zeta)$ is defined in the complex ζ plane with a cut along the real axis from $-\infty$ to -1 by the equality

$$\frac{2}{3} w^{3/2}(\zeta) = \zeta \ln(\zeta + \sqrt{\zeta^2 - 1}) - \sqrt{\zeta^2 - 1}, \tag{A.6}$$

in which the principal values of the root and logarithm are chosen, the function $w(\zeta)$ is positive for $\zeta > 1$, and $\mathrm{Ai}(z)$ is the solution of the Airy equation

---

(1) In expansions (A.3)–(A.5) the expression $e^{2k\rho w^{3/2}(\xi/\rho)/3}$ corresponds to ξ such that

$$|\arg e^{i\frac{2\pi}{3}} (k\rho)^{2/3} w(\xi/\rho)| \leqslant \pi - \delta' \quad (0 < \delta' < \pi);$$

for

$$\frac{\pi}{3} + \delta' \leqslant \arg e^{i\frac{2\pi}{3}} (k\rho)^{2/3} w(\xi/\rho) \leqslant \frac{5\pi}{3} - \delta' \quad \left(0 < \delta' < \frac{\pi}{3}\right)$$

this expression is replaced by $e^{|\mathrm{Re}\, 2k\rho w^{3/2}(\xi/\rho)/3|}$

$$\mathrm{Ai}''(z) = z\,\mathrm{Ai}(z) \tag{A.7}$$

that is real for real values of z (here and below $\mathrm{Ai}'(z) = d\,\mathrm{Ai}(z)/dz$ and $\mathrm{Ai}''(z) = d^2\mathrm{Ai}(z)/dz^2$).

We cite the properties (see, for example, [21]) of the extensively studied Airy function that will be needed for the sequel. The following asymptotic representations are valid for $|z| \to \infty$: when $|\arg z| \leqslant \pi - \delta'$ $(0 < \delta' < \pi)$

$$\mathrm{Ai}(z) = \frac{z^{-1/4}}{2\sqrt{\pi}} e^{-\frac{2}{3}z^{3/2}} \{1 + O_{\delta'}(|z|^{-3/2})\}, \quad \mathrm{Ai}'(z) = \frac{-z^{1/4}}{2\sqrt{\pi}} e^{-\frac{2}{3}z^{3/2}} \{1 + O_{\delta'}(|z|^{-3/2})\}; \tag{A.8}$$

when $\pi/3 + \delta' \leqslant \arg z \leqslant 5\pi/3 - \delta'$ $(0 < \delta' < \pi/3)$

$$\begin{aligned} \mathrm{Ai}(z) &= \frac{e^{i\frac{\pi}{4}} z^{-1/4}}{\sqrt{\pi}} \left\{ \cos\left(\frac{2}{3} i z^{3/2} - \frac{\pi}{4}\right) + e^{\left|\mathrm{Re}\frac{2}{3} z^{3/2}\right|} O_{\delta'}(|z|^{-3/2}) \right\}, \\ \mathrm{Ai}'(z) &= \frac{e^{-i\frac{\pi}{4}} z^{1/4}}{\sqrt{\pi}} \left\{ \sin\left(\frac{2}{3} i z^{3/2} - \frac{\pi}{4}\right) + e^{\left|\mathrm{Re}\frac{2}{3} z^{3/2}\right|} O_{\delta'}(|z|^{-3/2}) \right\}. \end{aligned} \tag{A.9}$$

The function $\mathrm{Ai}'(z)$ has a single sequence of zeros $z_n = e^{i\pi} a_n$, $n = 1, 2, \cdots$, where the a_n are positive numbers such that

$$a_n = \left(\frac{3\pi n}{2}\right)^{2/3} \left\{1 - \frac{1}{2n} + O\left(\frac{1}{n^2}\right)\right\}, \qquad n = 1, 2, \ldots. \tag{A.10}$$

The zeros z_n are simple and the values $\mathrm{Ai}(z_n)$ satisfy the asymptotic equalities

$$\mathrm{Ai}(z_n) = \frac{(-1)^{n-1}}{\sqrt{\pi}\left(\frac{3\pi n}{2}\right)^{1/6}} \left\{1 + \frac{1}{16n} + O\left(\frac{1}{n^2}\right)\right\}. \tag{A.11}$$

For $z = -re^{i\alpha}$, where $r > 0$ and $|\alpha| < \pi/3$, we have the estimates

$$\left| \frac{d^p\,\mathrm{Ai}(z)}{dz^p} \right| \leqslant C(1 + r)^{\frac{p}{2} - \frac{1}{4}} e^{r^{3/2}|\sin\alpha|}, \tag{A.12}$$

where $p = 0, 1, 2,$ and C is a positive constant.

It can easily be shown by representing the function $w(\zeta)$ from (A.6) as a composition of three functions:

$$\eta = \operatorname{arc\,cosh} \zeta, \quad \chi = \eta\cosh\eta - \sinh\eta,$$
$$w = \left(\frac{3}{2}\chi\right)^{2/3}, \tag{A.13}$$

and taking into account the equality $w(\overline{\zeta}) = \overline{w(\zeta)}$ that there exists a one-to-one mapping

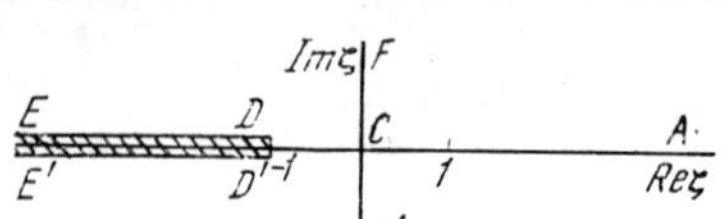

FIGURE 2

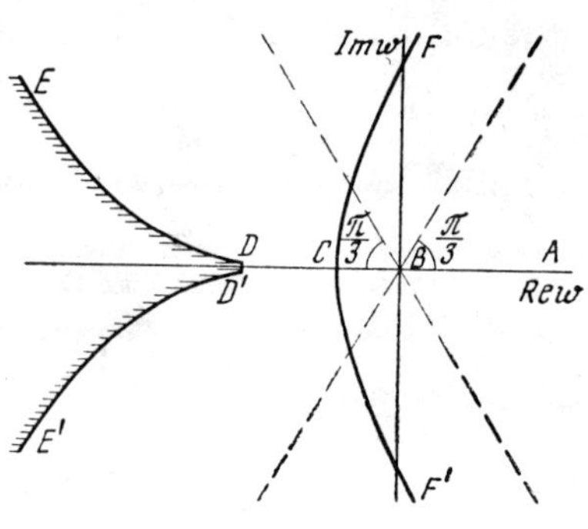

FIGURE 3

of the domain $\mathcal{D}_\zeta = \{\zeta: |\arg(1+\zeta)| < \pi\}$ (Figure 2) onto the domain $\mathcal{D}_w$ (Figure 3) of the complex w plane. The equation of the curve DE in Figure 3 has the form

$$w_{DE}(\tau) = \left(\frac{3}{2}\right)^{2/3}\left[e^{i\pi}\tau\cosh\tau + \sinh\tau + \pi e^{i\frac{3\pi}{2}}\cosh\tau\right]^{2/3}, \qquad \tau \geqslant 0,$$

$$w_{D'E'}(\tau) = \overline{w_{DE}(\tau)}.$$

For $\tau \to 0$

$$w_{DE}(\tau) = e^{i\pi}\left(\frac{3\pi}{2}\right)^{2/3}\left\{1 - \frac{\tau^2}{2} + O(\tau^3)\right\}$$

and for $\tau \to \infty$

$$w_{DE} = e^{i\frac{2\pi}{3}}\left(\frac{3}{4}\tau\right)^{2/3} e^{2\tau/3}\left\{\left(1 - \frac{2}{3\tau} + \frac{2\pi i}{3\tau} + O\left(\frac{1}{\tau^2}\right)\right\}.$$

The equation of the curve CF has the form

$$w_{CF}(\tau) = \left(\frac{3}{2}\right)^{2/3}\left[e^{i\frac{\pi}{2}}\tau\sinh\tau + \frac{\pi}{2}e^{i\pi}\sinh\tau + e^{i\frac{3\pi}{2}}\cosh\tau\right]^{2/3}, \qquad \tau \geqslant 0,$$

$$w_{C'F'}(\tau) = \bar{w}_{CF}(\tau).$$

For $\tau \to 0$

$$w_{CF}(\tau) = \left(\frac{3}{2}\right)^{2/3} e^{i\pi}\left\{1 - \frac{\pi i}{3}\tau + O(\tau^2)\right\}$$

and for $\tau \to \infty$

$$w_{CF}(\tau) = e^{i\frac{\pi}{3}}\left(\frac{3}{4}\tau\right)^{2/3} e^{2\tau/3}\left\{1 - \frac{2}{3\tau} + \frac{\pi i}{3\tau} + O\left(\frac{1}{\tau^2}\right)\right\}.$$

Thus the function $w(\zeta)$ is regular in $\mathcal{D}_\zeta$ while the inverse function $\zeta(w)$ is regular in $\mathcal{D}_w$. Let us find the behavior of $\zeta(w)$ in a neighborhood of the point $w=0$. To this end we consider the function $\eta = \eta(w)$ in (1.13) for $|\eta| \leqslant \epsilon'$:

$$\eta\cosh\eta - \sinh\eta = \eta^3 \sum_{p=0}^{\infty} \frac{2(p+1)}{(2p+3)!}\eta^{2p} = \frac{2}{3}w^{3/2}. \tag{A.14}$$

We choose ϵ' so small that for $|\eta| \leqslant \epsilon'$

$$\left|\sum_{p=1}^{\infty} \frac{2(p+1)}{(2p+3)!}\eta^{2(p-1)}\right| \leqslant \frac{1}{3},$$

$$\left|\sum_{p=0}^{\infty} \frac{2(p+1)}{(2p+3)!}\eta^{2p}\right| \geqslant \frac{1}{3} - \frac{\varepsilon'^2}{3} \geqslant \frac{1}{6}.$$

Taking into account the latter inequality, we deduce from (A.14) the estimates

$$\left|\frac{\eta^3}{6}\right| \leqslant \left|\frac{2}{3} w^{3/2}\right|, \qquad |\eta| \leqslant 2^{2/3} |w|^{1/2}.$$

Consequently for $w \to 0$ equation (A.14) can be rewritten in the form

$$\frac{\eta^3}{3} = \frac{2}{3} w^{3/2} \{1 + O(w)\}. \tag{A.15}$$

Equation (A.14) reduces to an identity if one considers η as the function of ζ in (A.13) and w as the function of ζ in (A.6). The functions $\eta(\zeta) = \operatorname{arc\,cosh} \zeta$ and $w(\zeta)$ are positive for $\zeta > 1$ and tend to zero as $\zeta \to 1 + 0$; consequently (A.15) can be written in the form

$$\eta = 2^{1/3} w^{1/2} \{1 + O(w)\}. \tag{A.16}$$

Since $\zeta = \cosh \eta$, it follows from (A.16) that

$$\zeta(w) = 1 + 2^{-1/3} w + O(w^2). \tag{A.17}$$

Let us obtain the asymptotic behavior of $w(\zeta)$ for $\zeta \to \infty$. By using the well-known expansions 1.112 and 1.642 in [16], we can write equation (A.6) in the form

$$\frac{2}{3} w^{3/2} = \zeta \ln \frac{2\zeta}{e} + \zeta^{-1} \sum_{p=0}^{\infty} c_p \zeta^{-p},$$

where

$$0 < c_p = \frac{(2p+1)\,!!}{(2p+2)\,!!} - \frac{(2p+2)\,!}{2^{2p+3}(p+2)\,!\,(p+1)\,!} \leqslant 1.$$

Thus $\Sigma_0^\infty c_p \zeta^{-p}$ converges uniformly as $\zeta \to \infty$, and hence

$$\frac{2}{3} w^{3/2} = \zeta \ln \frac{2\zeta}{e} + O(\zeta^{-1}), \tag{A.18}$$

which implies

$$w(\zeta) = \left(\frac{3}{2} \zeta \ln \frac{2\zeta}{e}\right)^{2/3} + O(\zeta^{-1/3} \ln^{-1/3} \zeta). \tag{A.19}$$

Let us obtain, finally, the behavior of $\zeta(w)$ for $w \to \infty$. We note that $\zeta + \sqrt{\zeta^2 - 1}$ is bounded and does not vanish for $|\zeta| \leqslant C < \infty$, from which we conclude on the basis of (A.6) that $\zeta \to \infty$ as $w \to \infty$. Consequently we have (A.18) and the estimates

$$C_1 |w|^{3/2} \leqslant \left|\zeta \ln \frac{2\zeta}{e}\right| \leqslant C_2 |w|^{3/2}, \tag{A.20}$$

where C_1 and C_2 are positive constants. Since $|\ln 2\zeta/e| \leqslant |\zeta|^{1/2}$ for $\zeta \in \mathcal{D}_\zeta$ and $\zeta \to \infty$, we deduce from (A.20) the estimate $|w| \leqslant C_3 |\zeta|$, on the basis of which equation (A.18) can be rewritten for $w \to \infty$ in the form

$$\zeta \ln \frac{2\zeta}{e} = \frac{2}{3} w^{3/2} + O(w^{-1}). \tag{A.21}$$

Let

$$\zeta_1 = 2\zeta/e, \qquad \eta = 4w^{3/2}/3e + O(w^{-1}).$$

Then $\zeta_1 \ln \zeta_1 = \eta$, which implies the inequalities

$$|\zeta_1| \leqslant |\zeta_1 \ln \zeta_1| \leqslant |\eta|$$

for $\zeta_1 \to \infty$ and $\eta \to \infty$. Let $\zeta_1 = \zeta_2 \eta$. Then $\zeta_2(\ln \eta + \ln \zeta_2) = 1$ and $|\zeta_2| \leqslant 1$ for $\eta \to \infty$. From the latter inequality we obtain the estimate

$$\left|\zeta_2 - \frac{1}{\ln \eta}\right| \leqslant \frac{|\zeta_2 \ln \zeta_2|}{|\ln \eta|} \leqslant \frac{C_4}{|\ln \eta|}$$

for $\eta \to \infty$. The function $\zeta_3 = \zeta_2 - 1/\ln \eta$ satisfies the equation

$$\zeta_3 (\ln \eta - \ln \ln \eta) = \frac{\ln \eta}{\ln \ln \eta} - \frac{(1 + \zeta_3 \ln \eta) \ln (1 + \zeta_3 \ln \eta)}{\ln \eta} \tag{A.22}$$

and the inequality $|\ln \eta|\, |\zeta_3| \leqslant C_4$ for $\eta \to \infty$. From (A.22) we obtain the estimate

$$\left|\zeta_3 - \frac{\ln \ln \eta}{\ln \eta (\ln \eta - \ln \ln \eta)}\right| \leqslant \frac{C_5}{|\ln \eta|\,|\ln \eta - \ln \ln \eta|}$$

for $\eta \to \infty$; consequently, for $\eta \to \infty$

$$\zeta_3 \ln \eta = O\left(\frac{\ln \ln \eta}{\ln \eta}\right)$$

and

$$\ln (1 + \zeta_3 \ln \eta) = \zeta_3 \ln \eta + \left(\frac{\ln \ln \eta}{\ln \eta}\right)^2 O(1).$$

Substituting the latter estimate into (A.22), we get

$$\zeta_3 (\eta) = \frac{\ln \ln \eta}{\ln^2 \eta} + \frac{(\ln \ln \eta)^2}{\ln^3 \eta} O(1).$$

Returning to the variables w and $\zeta(w)$, we conclude that $\zeta(w)$ has the following asymptotic behavior for $w \in \mathcal{D}_w$ and $w \to \infty$:

$$\zeta(w) = \frac{\frac{2}{3} w^{3/2}}{\ln \frac{4w^{3/2}}{3e}} \left\{ 1 + \frac{\ln \ln \frac{4w^{3/2}}{3e}}{\ln \frac{4w^{3/2}}{3e}} + \left(\frac{\ln \ln w}{\ln w}\right)^2 O(1) \right\}. \tag{A.23}$$

Consider the functions $dw/d\zeta$ and $d^2w/d\zeta^2$ for $\zeta \in \mathcal{D}_\zeta$. On the basis of (A.6) they have the forms

$$\frac{dw}{d\zeta} = \frac{\ln (\zeta + \sqrt{\zeta^2 - 1})}{w^{1/2}(\zeta)},$$

$$\frac{d^2w}{d\zeta^2} = \frac{1}{w^{1/2}(\zeta)\sqrt{\zeta^2 - 1}} - \frac{1}{2} \frac{[\ln (\zeta + \sqrt{\zeta^2 - 1})]^2}{w^2(\zeta)},$$

which together with (A.7) and (A.23) directly imply the inequalities

$$C_1\frac{\ln(1+|w(\zeta)|^{1/2})}{|w(\zeta)|^{1/2}}\leqslant\left|\frac{dw(\zeta)}{d\zeta}\right|\leqslant C_2\frac{\ln(1+|w(\zeta)|^{1/2})}{|w(\zeta)|^{1/2}}, \tag{A.24}$$

$$\left|\frac{d^2w(\zeta)}{d\zeta^2}\right|\leqslant C_3\frac{\ln(1+|w(\zeta)|^2)}{|w(\zeta)|^2}, \tag{A.25}$$

where C_1, C_2 and C_3 are positive numbers.

B. Suppose $k\in N(\delta,\nu)$. Then the zeros $\xi(k)$ of the function $H'_{k\xi}(k)$ that lie in $M(\theta,\mu)$ satisfy uniformly in k the asymptotic equality

$$\mathrm{Ai}'\left(e^{i\frac{2\pi}{3}}k^{2/3}w(\xi)\right)+(1+|k^{2/3}w(\xi)|)^{1/4}\exp\left\{\frac{2}{3}|\operatorname{Re}kw^{3/2}(\xi)|\right\}O(|k\xi|^{-1})=0. \tag{B.1}$$

Let $\tilde{\xi}_n(k)$, $n=1,2,\cdots$, denote the roots of the equation

$$\mathrm{Ai}'\left(e^{i\frac{2\pi}{3}}k^{2/3}w(\xi)\right)=0 \tag{B.2}$$

that lie in the domain $M(\theta,\mu)$. On the basis of (A.10) we have

$$w(\xi_n)=e^{i\left(\frac{\pi}{3}-\frac{2\theta}{3}\right)}a_n|k|^{-2/3},\qquad n=1,\ 2,\ \ldots,\qquad \theta=\arg k, \tag{B.3}$$

i.e.

$$\xi_n=\zeta(x_n,\ \theta),\qquad x_n=a_n|k|^{-2/3},\qquad n=1,\ 2,\ \ldots, \tag{B.4}$$

where the function $\tilde{\zeta}(x,\theta)$ is determined for $0\leqslant x<\infty$ from the equation

$$w(\tilde{\zeta})=e^{i\left(\frac{\pi}{3}-\frac{2\theta}{3}\right)}x. \tag{B.5}$$

Let us establish some of the properties of the curve $\tilde{\zeta}(x,\theta)$. We assume everywhere below that ϵ' is a sufficiently small positive number. On the basis of (A.17) we have

$$\zeta(x,\ \theta)=1+2^{-1/3}e^{i\left(\frac{\pi}{3}-\frac{2\theta}{3}\right)}x+O(x^2) \tag{B.6}$$

for $0\leqslant x\leqslant\epsilon'$, while on the basis of (A.23) we have

$$\zeta(x,\ \theta)=\frac{2/3\,e^{i\left(\frac{\pi}{2}-\theta\right)}x^{3/2}}{\ln\frac{4x^{3/2}}{3e}}\left\{1+\frac{\ln\ln\frac{4x^{3/2}}{3e}-i\left(\frac{\pi}{2}-\theta\right)}{\ln\frac{4x^{3/2}}{3e}}+\left(\frac{\ln\ln x}{\ln x}\right)^2O(1)\right\} \tag{B.7}$$

for $x\geqslant 1/\epsilon'$. It follows that $\tilde{\zeta}(x,\theta)$ lies in $M(\theta,1)$ for $0\leqslant x\leqslant\epsilon'$ and $1/\epsilon'\leqslant x<\infty$. We will show that it also has this property for $\epsilon'<x<1/\epsilon'$ when $|\theta|\leqslant 2\delta'=2\delta'(\epsilon')$, where δ' is sufficiently small. From (B.6) we have the inequalities

$$0<\alpha(\varepsilon',\ \delta)\leqslant\frac{d\operatorname{Re}\zeta(x,\ \theta)}{dx},\qquad 0<\beta(\varepsilon',\ \delta)=\frac{d\operatorname{Im}\zeta(x,\ \theta)}{dx} \tag{B.8}$$

for $0\leqslant x\leqslant\epsilon'$ and $|\theta|\leqslant\delta<\pi/2$. We will show that analogous inequalities hold for

$\epsilon' < x < 1/\epsilon'$ and $|\theta| \leqslant 2\delta'$. Suppose first $\theta = 0$. Then (B.5) and (A.6) imply that the function $\zeta_0(x) = \zeta(x, 0)$ satisfies the equation

$$\zeta_0(x) \ln\left[\zeta_0(x) + \sqrt{\zeta_0^2(x) - 1}\right] - \sqrt{\zeta_0^2 - 1} = \frac{2}{3} i x^{3/2}. \tag{B.9}$$

Differentiating with respect to x, we obtain

$$\frac{d\zeta_0(x)}{dx} \ln\left[\zeta_0(x) + \sqrt{\zeta_0^2(x) - 1}\right] = i\sqrt{x}. \tag{B.10}$$

It follows from (B.8) that the functions $d \operatorname{Re} \zeta_0(x)/dx$ and $d \operatorname{Im} \zeta_0(x)/dx$ are positive for $0 \leqslant x \leqslant \epsilon'$. Moreover, they remain positive for $x > \epsilon'$. For suppose that at some point $x' > \epsilon'$, for example,

$$\left.\frac{d \operatorname{Re} \zeta_0(x)}{dx}\right|_{x=x'} = 0.$$

Then (B.10) implies

$$\left.\frac{d \operatorname{Im} \zeta_0(x)}{dx}\right|_{x=x'} \ln\left[\zeta_0(x') + \sqrt{\zeta_0^2(x') - 1}\right] = \sqrt{x'},$$

i.e.

$$\left.\frac{d \operatorname{Im} \zeta_0(x)}{dx}\right|_{x=x'} = \beta' \neq 0,$$

and hence

$$\ln\left[\zeta_0(x') + \sqrt{\zeta_0^2(x') - 1}\right] = \sqrt{x'}/\beta'.$$

Solving for ζ_0, we get $\zeta_0(x') = \cosh(\sqrt{x'}/\beta') \geqslant 1$. Substituting this in (B.9), we find that the left side of (B.9) is a real number, whereas the right side is an imaginary number; hence $d \operatorname{Re} \zeta_0(x)/dx$ does not vanish for $0 \leqslant x < \infty$ and we have

$$0 < d \operatorname{Re} \zeta_0(x)/dx, \qquad 0 \leqslant x < \infty. \tag{B.11}$$

Suppose, further, that at some point $x' > \epsilon'$

$$\left.\frac{d \operatorname{Im} \zeta_0(x)}{dx}\right|_{x=x'} = 0.$$

Then (B.10) implies

$$\left.\frac{d \operatorname{Re} \zeta_0(x)}{dx}\right|_{x=x'} \ln\left[\zeta_0(x') + \sqrt{\zeta_0^2(x') - 1}\right] = i\sqrt{x'},$$

i.e.

$$\left.\frac{d \operatorname{Re} \zeta_0(x)}{dx}\right|_{x=x'} = \alpha' \neq 0$$

and hence

$$\ln\left[\zeta_0(x') + \sqrt{\zeta_0^2(x') - 1}\right] = i\sqrt{x'}/\alpha'.$$

Solving the latter equation for ζ_0, we get

$$\zeta_0(x') = \cos(\sqrt{x'}/\alpha') = \operatorname{Re}\zeta_0(x') \leqslant 1.$$

Since $\operatorname{Re}\zeta_0(0) = 1$ (this follows from (B.6)), we conclude on the basis of (B.11) that $\operatorname{Re}\zeta_0(x') > 1$ for $0 < x' < \infty$; hence the supposition is impossible and we have the inequality

$$0 < d \operatorname{Im}\zeta_0(x)/dx, \qquad 0 \leqslant x < \infty. \tag{B.12}$$

By virtue of the continuous differentiability of $\tilde{\zeta}(x, \theta)$ with respect to x and θ and on the basis of (B.11)–(B.12) there exist positive numbers α', β', δ' depending on ϵ' such that $\tilde{\zeta}(x, \theta)$ satisfies the inequalities

$$0 < \alpha' \leqslant d \operatorname{Re}\tilde{\zeta}(x, \theta)/dx, \qquad 0 < \beta' \leqslant d \operatorname{Im}\tilde{\zeta}(x, \theta)/dx \tag{B.13}$$

for $0 \leqslant x \leqslant 1/\epsilon'$ and $|\theta| \leqslant 2\delta'$. It follows for $\epsilon' \leqslant x \leqslant 1/\epsilon'$ and $|\theta| \leqslant 2\delta'$ that $\operatorname{Re}\tilde{\zeta}(x, \theta)$ and $\operatorname{Im}\tilde{\zeta}(x, \theta)$ are monotonic in x and that the curve $\tilde{\zeta}(x, \theta)$ lies in the rectangle

$$\begin{gathered} 1 < a_1(\varepsilon') \leqslant \operatorname{Re}\tilde{\zeta}(x, \theta) \leqslant a_2(\varepsilon') < \infty, \\ 0 < b_1(\varepsilon') \leqslant \operatorname{Im}\tilde{\zeta}(x, \theta) \leqslant b_2(\varepsilon') < \infty \end{gathered} \tag{B.14}$$

in the domain $M(\theta, 1)$.

Thus we have shown that the roots $\tilde{\xi}_n(k)$, $n = 1, 2, \cdots$, of equation (B.2) lie on a curve $\tilde{\zeta}(x, \theta)$ having properties (B.6), (B.7), (B.13) and (B.14). Before proceeding with a study of the roots $\xi_n(k)$, $n = 1, 2, \cdots$, of equation (B.1), we will show that the neighborhoods

$$\tilde{T}_n = \{\xi\colon |\xi - \tilde{\xi}_n| < \tilde{C}|k|^{-5/6} a_n^{-1/4}\}, \qquad n = 1, 2, \ldots, \tag{B.15}$$

of the points $\tilde{\xi}_n$ do not intersect when $\tilde{C}$ is a sufficiently small positive number.

Consider the function

$$z(\xi) = e^{i\frac{2\pi}{3}} k^{2/3} w(\xi). \tag{B.16}$$

For $\xi \in \tilde{T}_n$ we have

$$\begin{aligned} z(\xi) = e^{i\frac{2\pi}{3}} k^{2/3} w(\tilde{\xi}_n) + e^{i\frac{2\pi}{3}} k^{2/3} (\xi - \tilde{\xi}_n) \left.\frac{dw}{d\xi}\right|_{\xi=\tilde{\xi}_n} \\ + k^{2/3} e^{i\frac{2\pi}{3}} \int_{\tilde{\xi}_n}^{\xi} (\xi - \eta) \frac{d^2 w(\eta)}{d\eta^2}\, d\eta. \end{aligned} \tag{B.17}$$

On the basis of (A.24), (A.25), (B.3) and (B.15) the images $\tilde{S}_n$, $n = 1, 2, \cdots$, of the neighborhoods $\tilde{T}_n$ in the z plane are contained in the disks

$$\begin{aligned} |z - a_n e^{i\pi}| &\leqslant \tilde{C} C_2 a_n^{-3/4} |k|^{-1/6} \ln(1 + a_n^{1/2}|k|^{-1/3}) + C_3 \tilde{C}^2 |k|^{-1} a_n^{-1/2} \\ &\leqslant \tilde{C} C_2 a_n^{-1/2} \max_{0 \leqslant x} [x^{-1/2} \ln(1 + x)] + C_3 \tilde{C}^2 |k|^{-1} a_n^{-1/2} \leqslant \tilde{C} C_4 a_n^{-1/2}, \end{aligned}$$

where the constant C_4 does not depend on k for sufficiently large values of $|k|$.

The estimates (A.10) of the numbers a_n imply the inequality

$$|a_{n+1} - a_n| > C_0 a_n^{-1/2},$$

where C_0 is a positive number. It follows that the neighborhoods $\widetilde{S}_n$, and hence the neighborhoods $\widetilde{T}_n$, do not intersect for $\widetilde{C} < C_0/4C_4$. Thus we have established the inequalities

$$|\xi_{n+1}(k) - \xi_n(k)| \geqslant 2\widetilde{C}\,|k|^{-5/6} a_n^{-1/4}, \qquad n = 1, 2, \ldots . \tag{B.18}$$

Let us show that the roots $\xi_n(k)$ of equation (B.1) lie in the neighborhoods

$$T_n = \{\xi : |\xi - \xi_n| < |k|^{-4/3} a_n^{-5/6}\}, \qquad n = 1, 2, \ldots, \tag{B.19}$$

for $k \in N(\delta', \nu)$ when the number $\nu = \nu(\epsilon')$ has been chosen sufficiently large. From (B.17), (A.24), (A.25) and (B.19) we have the estimates

$$|z - a_n e^{i\pi}| \geqslant |k|^{-2/3} a_n^{-5/6} \left| \frac{dw}{d\xi} \right|_{\xi=\xi_n} - |k|^{-2} a_n^{-5/3} \sup_{\xi \in T_n} \left| \frac{d^2 w}{d\xi^2} \right|$$
$$\geqslant C_1 |k|^{-1/3} a_n^{-4/3} \ln(1 + a_n^{1/2} |k|^{-1/3}) - C_3 |k|^{-2} a_n^{-5/3} \geqslant C_1 |k|^{-1/3} a_n^{-5/6} (a_n^{1/2} + |k|^{1/3})^{-1}$$
$$- C_3 |k|^{-2} a_n^{-5/3} \geqslant \frac{C_1}{2} |k|^{-1/3} a_n^{-5/6} (a_n^{1/2} + |k|^{1/3})^{-1}$$

for $\xi \in T_n$ and sufficiently large values of $|k|$ (we have made use of the inequality $\ln(1 + x) \geqslant x(1 + x)^{-1}, x \geqslant 0$).

Thus the images in the ξ plane of the disks

$$S_n = \left\{ z : |z - a_n e^{i\pi}| < \frac{C_1}{2} |k|^{-1/3} a_n^{-5/6} (a_n^{1/2} + |k|^{1/3})^{-1} \right\} \tag{B.20}$$

are contained in the neighborhoods T_n. In a disk S_n equation (B.1) can be rewritten in the form

$$\mathrm{Ai}'(z) + B_n(z) = 0, \tag{B. 21}$$

where $B_n(z)$ is an analytic function in S_n satisfying the inequality

$$|B_n(z)| \leqslant C_5 (1 + |z|)^{1/4} e^{\frac{2}{3} |\mathrm{Re}\, z^{3/2}|} |k\xi_n|^{-1}, \qquad n = 1, 2, \ldots . \tag{B.22}$$

We will make use of Rouché's theorem in estimating the roots of equation (B.21). To this end we establish a number of auxiliary inequalities.

Let us estimate the expression $|k\xi_n|$ from below. From (B.6) we deduce the inequality

$$|k\xi_n| \geqslant C_6(\varepsilon') |k| (1 + a_n |k|^{-1/3}) \geqslant C' (|k| + a_n^{3/2})$$

for those positive integral values of n such that $a_n \leqslant \epsilon' |k|^{2/3}$; from (B.14) we deduce the inequality

$$|k\xi_n| \geqslant C_7(\varepsilon') |k| \geqslant C' (|k| + a_n^{3/2}),$$

for those positive integral values of n such that $\epsilon'|k|^{2/3} < a_n < |k|^{2/3}/\epsilon'$ and a sufficiently small number $C' = C'(\epsilon')$; and from (B.7) we deduce the inequality

$$|k\xi_n| \geqslant C_8(\varepsilon')\, a_n^{3/2} [\ln (a_n^{3/2}/|k|)]^{-1} \geqslant C'(a_n^{3/2} + |k|) \ln^{-1}(1 + a_n) \tag{B.23}$$

for those positive integral values of n such that $a_n > |k|^{2/3}/\epsilon'$. It follows from the estimates (A.10) of the numbers a_n that the constants $C' = C'(\epsilon')$ can be chosen in such a way that (B.23) holds for all $n = 1, 2, \cdots$. On the basis of (B.20) we have the inequality

$$|{}^2/_3 z^{3/2} - {}^2/_3 a_n^{3/2} i| \leqslant C_9 |k|^{-1/3} a_n^{-1/3} (a_n^{1/2} + |k|^{1/3})^{-1} \tag{B.24}$$

for $z \in S_n$. Taking (B.22)–(B.24) into account, we obtain the estimates

$$|B_n(z)| \leqslant C_{10}(\varepsilon') (|k| + a_n^{3/2})^{-1} (1 + a_n)^{1/4} \ln (1 + a_n) \tag{B.25}$$

for the functions $B_n(z)$, $n = 1, 2, \cdots$, when $z \in S_n$. For $a_n e^{i\pi} + \eta = z \in S_n$ we introduce the function

$$\mathscr{F}_n(\eta) = \mathrm{Ai}'(a_n e^{i\pi} + \eta) + B_n(a_n e^{i\pi} + \eta) = \mathrm{Ai}''(a_n e^{i\pi})\, \eta$$
$$+ \left[\int_0^\eta \mathrm{Ai}'''(a_n e^{i\pi} + \eta_1)(\eta - \eta_1)\, d\eta_1 + B_n(a_n e^{i\pi} + \eta) \right] = F_n(\eta) + f_n(\eta),$$

in which

$$F_n(\eta) = a_n e^{i\pi} \mathrm{Ai}(a_n e^{i\pi})\, \eta,$$

$$f_n(\eta) = B_n(a_n e^{i\pi} + \eta) + \int_0^\eta (\eta - \eta_1) [(a_n e^{i\pi} + \eta_1) \mathrm{Ai}'(a_n e^{i\pi} + \eta_1) + \mathrm{Ai}(a_n e^{i\pi} + \eta_1)]\, d\eta_1$$

on the basis of (A.7). On the boundary of S_n we have

$$z = a_n e^{i\pi} + \eta = a_n e^{i(\pi + \alpha_n)}$$

and the estimate

$$|\alpha_n| \leqslant C_{11} |k|^{-1/3} a_n^{-11/6} (a_n^{1/2} + |k|^{1/3})^{-1},$$

which implies $|\alpha_n| < \pi/3$ for sufficiently large values of $|k|$. Taking into account (A.12), (B.20) and (B.25), we obtain the estimate

$$\begin{aligned} |f_n(\eta)| &\leqslant |B_n(a_n e^{i\pi} + \eta)| + |\eta|^2 (1 + a_n)^{5/4} e^{a_n^{3/2} |\alpha_n|} \\ &\leqslant C_{10} (|k| + a_n^{3/2})^{-1} (1 + a_n)^{1/4} \ln (1 + a_n) + C_{12} |k|^{-2/3} a_n^{-5/3} \\ &\quad \times (1 + a_n)^{5/4} (|k|^{1/3} + a_n^{1/2})^{-2} \leqslant 2C_{10} (|k| + a_n^{3/2})^{-1} (1 + a_n)^{1/4} \ln (1 + a_n) \\ &\leqslant C_{13}(\varepsilon') (|k| + a_n^{3/2})^{-1} a_n^{1/3} = \beta_n. \end{aligned}$$

On the basis of (A.10), (A.11) and (B.20) the following inequality is valid on the boundary of S_n:

$$|F_n(\eta)| \geqslant C_{14}|k|^{-1/3} a_n^{-1/2}(a_n^{1/2}+|k|^{1/3})^{-1}=\gamma_n.$$

Let us show that $\gamma_n > \beta_n$. We put $a_n^{1/2} = |k|^{1/3}\tau_n$. Then

$$\beta_n = C_{13}|k|^{-7/9}\tau_n^{2/9}(1+\tau_n^3)^{-1},$$

$$\gamma_n = C_{14}|k|^{-5/6}\tau_n^{-1/12}(1+\tau_n)^{-1}.$$

There exists a number $C > 0$ such that

$$\tau^{-1/12}(1+\tau)^{-1} \geqslant C\tau^{2/9}(1+\tau^3)^{-1}$$

for all $\tau \geqslant 0$. Consequently $\gamma_n > \beta_n$ if $|k| \geqslant \nu = \nu(\epsilon')$, where $\nu(\epsilon')$ is sufficiently great. Thus we have established the inequalities

$$|F_n(\eta)| > |f_n(\eta)|$$

on the boundaries of the disks S_n, $n = 1, 2, \cdots$, and since $F_n(\eta)$ vanishes at $\eta = 0$, it follows on the basis of Rouché's theorem that $\mathfrak{F}_n(\eta)$ vanishes for some $z_n = a_n e^{i\pi} + \eta_n \in S_n$; hence equation (B.1) has a root ξ_n in the disk T_n. We have thus obtained the following result.

Let ϵ' be an arbitrary positive number. It is possible to indicate a domain $N(\delta, \nu)$ with parameters δ and ν depending on ϵ' such that for any $k \subset N(\delta, \nu)$ the curve $\widetilde{\xi}(x, \arg k)$ determined from (B.5) can be "divided" into three parts by the straight lines $\operatorname{Im} \xi = b_1$ and $\operatorname{Im} \xi = b_2$ (see Figure 4), where b_1 and b_2 are positive numbers depending only on ϵ', the curve $\widetilde{\xi}(x, \arg k)$ having properties (B.6), (B.7), (B.13), (B.14) on the appropriate parts. The points $\widetilde{\xi}_n = \widetilde{\xi}(a_n|k|^{-2/3}, \arg k)$ $(\arg k = \theta)$ have the following properties:

(i) When $a_n|k|^{-2/3} \leqslant \epsilon'$

$$\xi_n = 1 + 2^{-1/3} e^{i\left(\frac{\pi}{2}-\frac{2\theta}{3}\right)} a_n|k|^{-2/3} + O(a_n^2|k|^{-4/3}), \tag{B.26}$$

(ii) When $\epsilon' < a_n|k|^{-2/3} < 1/\epsilon'$

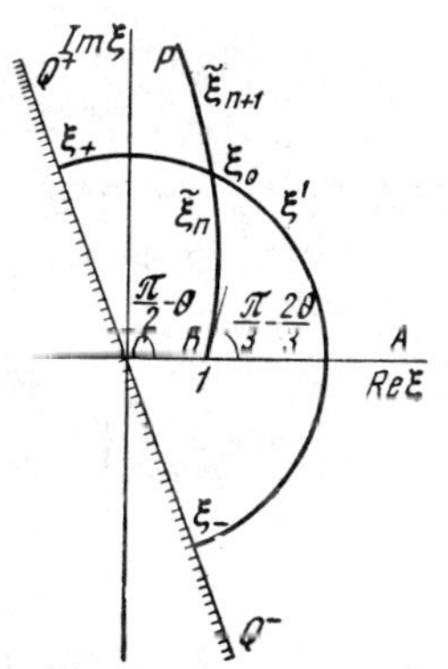

FIGURE 4

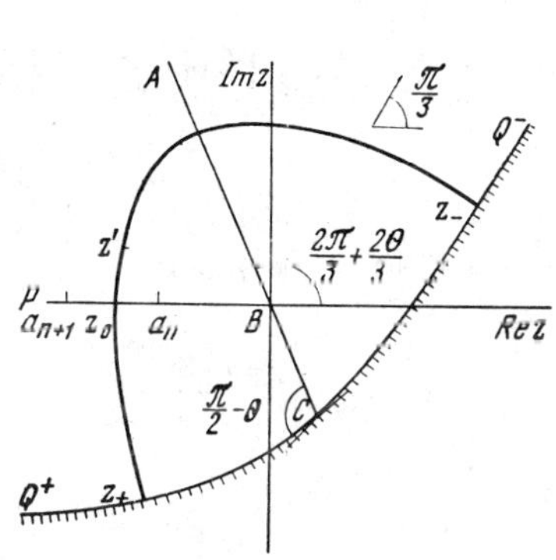

FIGURE 5

$$1 \leqslant a_1(\varepsilon') \leqslant \operatorname{Re} \xi_n < \operatorname{Re} \xi_{n+1} \leqslant a_2(\varepsilon') < \infty,$$
$$0 < b_1(\varepsilon') \leqslant \operatorname{Im} \xi_n < \operatorname{Im} \xi_{n+1} \leqslant b_2(\varepsilon') < \infty. \tag{B.27}$$

(iii) When $1/\epsilon' \leqslant a_n|k|^{-2/3}$

$$\xi_n = \frac{2e^{i\left(\frac{\pi}{2}-\theta\right)} a_n^{3/2}}{3|k|\ln\left(\frac{4a_n^{3/2}}{3|k|e}\right)}\left\{1 + \frac{\ln\ln\left(\frac{4a_n^{3/2}}{3|k|e}\right) - i\left(\frac{\pi}{2}-\theta\right)}{\ln\left(\frac{4a_n^{3/2}}{3|k|e}\right)} + \left(\frac{\ln\ln(a_n^{3/2}/|k|)}{\ln(a_n^{3/2}/|k|)}\right)^2 O(1)\right\}. \tag{B.28}$$

(iv) The number of points $\tilde{\xi}_n$ falling onto the first two parts of the curve $\tilde{\xi}(x, \theta)$ increases with increasing $|k|$.

(v) The distance between $\tilde{\xi}_n$ and $\tilde{\xi}_{n+1}$ $(n = 1, 2, \cdots)$ satisfies the estimate

$$|\tilde{\xi}_{n+1} - \tilde{\xi}_n| \geqslant 2\tilde{C}|k|^{-5/6} a_n^{-1/4}, \tag{B.29}$$

where C is a positive number depending only on ϵ'.

(vi) The neighborhood

$$|\xi - \tilde{\xi}_n| < |k|^{-4/3} a_n^{-5/6} \tag{B.30}$$

of each point $\tilde{\xi}_n$ contains a zero ξ_n of $H'_{k\xi}(k)$, all of which (for $n = 1, 2, \cdots$) belong to $M(\theta, 1)$, have properties (B.26)–(B.28) and satisfy the inequalities

$$|\xi_{n+1} - \xi_n| \geqslant \tilde{C}|k|^{-5/6} a_n^{-1/4}. \tag{B.31}$$

We conclude this subsection by showing that the function $H'_{k\xi}(k)$ does not have any other zeros in the halfplane $|\arg \xi + \theta| \leqslant \pi/2$ and all of the zeros ξ_n are simple.

Consider the mapping $z(\xi)$ (see (B.16)) of the halfplane $|\arg \xi + \theta| \leqslant \pi/2$ (Figures 4 and 5). The equation of the curve CQ^+ in the z plane has the form

$$z_{CQ^+}(\tau) = e^{i\left(\pi+\frac{2\theta}{3}\right)}\left(\frac{3}{2}|k|\right)^{2/3}\left\{\tau e^{-i\theta}\left[\frac{\pi}{2}e^{i\frac{\pi}{2}} + \ln\left(\tau e^{-i\theta} + \sqrt{1+\tau^2 e^{-2i\theta}}\right)\right] + e^{i\pi}\sqrt{1+\tau^2 e^{-2i\theta}}\right\}^{2/3},$$

where $\tau \in [0, \infty)$. For $\tau \to 0$

$$z_{CQ^+}(\tau) = e^{i\left(\frac{5\pi}{3}+\frac{2\theta}{3}\right)}\left(\frac{3}{2}|k|\right)^{2/3}\left\{1 + \frac{\pi}{3}e^{-i\left(\frac{\pi}{2}+\theta\right)}\tau + O(\tau^2)\right\}$$

and for $\tau \to \infty$

$$z_{CQ^+}(\tau) = e^{i\pi}\left(\frac{3}{2}|k|\,\tau \ln\frac{2\tau}{e}\right)^{2/3}\left\{1 + \frac{e^{i\frac{\pi}{2}}\left(\frac{\pi}{3}-\frac{2\theta}{3}\right)}{\ln\frac{2\tau}{e}} + O\left(1/\ln^2\frac{2\tau}{e}\right)\right\}.$$

The equation of the curve CQ^- in the z plane has the form

$$z_{CQ^-}(\tau)=e^{i\left(\frac{\pi}{3}+\frac{2\theta}{3}\right)}\left(\frac{3}{2}|k|\right)^{2/3}\left\{\tau e^{-i\theta}\left[\frac{\pi}{2}e^{-i\frac{\pi}{2}}+\ln\left(\tau e^{-i\theta}+\sqrt{1+\tau^2e^{-i2\theta}}\right)\right]+e^{-i\pi}\sqrt{1+\tau^2e^{-i2\theta}}\right\}^{2/3}.$$

For $\tau\to 0$

$$z_{CQ^-}(\tau)=e^{i\left(\frac{2\theta}{3}-\frac{\pi}{3}\right)}\left(\frac{3}{2}|k|\right)^{2/3}\left\{1+\frac{\pi}{3}e^{i\left(\frac{\pi}{2}-\theta\right)}\tau+O(\tau^2)\right\}$$

and for $\tau\to\infty$

$$z_{CQ^-}(\tau)=e^{i\frac{\pi}{3}}\left(\frac{3}{2}|k|\tau\ln\frac{2\tau}{e}\right)^{2/3}\left\{1+\frac{e^{-i\frac{\pi}{2}}\left(\frac{\pi}{3}+\frac{2\theta}{3}\right)}{\ln\frac{2\tau}{e}}+O\left(1/\ln^2\frac{2\tau}{e}\right)\right\}.$$

The curve $\tilde{\zeta}(x,\theta)$ (denoted by BP in Figure 4) is mapped into the negative real axis of the z plane.

Let k be an arbitrary fixed number in $N(\delta,\nu)$. On the curve $\tilde{\zeta}(x,\theta)$ we choose the point $\zeta_0=\zeta_0(n,k)$ such that

$$z_0=z(\xi_0)=\frac{1}{2}e^{i\pi}(a_{n+1}+a_n),\tag{B.32}$$

and draw the arc $\xi_-\xi_0\xi_+$ of radius $|\xi_0|$ (see Figure 4). Consider the closed contour $\xi_+\xi_-\xi_0\xi_+$ formed by the segments $[\xi_+,0]$, $[0,\xi_-]$ and the arc $\xi_-\xi_0\xi_+$. The contour $\xi_+\xi_-\xi_0\xi_+$ depends on k, which is fixed, and on n, which we will assume to be sufficiently large. Since $H'_{k\xi}(k)$ is an entire function of ξ, it follows from the argument principle that the number $p(n)$ (taking into account multiplicities) of zeros enclosed by the contour $\xi_+\xi_-\xi_0\xi_+$ is determined by the formula

$$p(n)=\frac{1}{2\pi}\Big[\operatorname*{var\,arg}_{[\xi_+,0]}+\operatorname*{var\,arg}_{[0,\xi_-]}+\operatorname*{var\,arg}_{\xi_-\xi_0\xi_+}\Big]H'_{k\xi}(k).$$

Since $\xi=e^{i(\pi/2-\theta)}y$ for $\xi\in[\xi_+,0]$ and $\xi=e^{-i(\pi/2+\theta)}y$ for $\xi\in[0,\xi_-]$ ($y\in[0,|\xi_0|]$), we have the equalities

$$\operatorname*{var\,arg}_{[\xi_+,0]}H'_{k\xi}(k)=-\operatorname*{var\,arg}_{y\in[0,|\xi_0|]}H'_{i|k|y}(k),\qquad \operatorname*{var\,arg}_{[0,\xi_-]}H'_{k\xi}(k)=\operatorname*{var\,arg}_{y\in[0,|\xi_0|]}H'_{-i|k|y}(k).$$

Taking into account the well-known identity 8.484 in [16]:

$$H'_{-i|k|y}(k)=e^{-\pi|k|y}=H'_{i|k|y}(k),$$

we conclude that

$$\arg H'_{i|k|y}(k)=\arg H'_{-i|k|y}(k).$$

Consequently

$$[\underset{[\xi_+,\,0]}{\text{var arg}} + \underset{[0,\,\xi_-]}{\text{var arg}}]\, H'_{k\xi}(k) = 0$$

and we get

$$p(n) = \frac{1}{2\pi} \underset{\xi_-\xi_0\xi_+}{\text{var arg}}\, H'_{k\xi}(k) = \frac{1}{2\pi}[\underset{\xi_-\xi'}{\text{var arg}} + \underset{\xi'\xi_0}{\text{var arg}} + \underset{\xi_0\xi_+}{\text{var arg}}]\, H'_{k\xi}(k),$$

where $\xi' = |\xi_0| e^{i(\pi/2-\theta-2\delta)}$.

By construction, $\xi_0 = \widetilde{\zeta}(x_0, \theta)$, where $x_0 > a_n|k|^{-2/3}$; hence $|\xi_0| \to \infty$ as $n \to \infty$, and from (A.4) and (B.16) we deduce the following asymptotic equality for $n \to \infty$:

$$p(n) = \frac{1}{4} + O(|\xi_0|^{-2}) + \frac{1}{2\pi}[\underset{\xi_-\xi'}{\text{var arg}} + \underset{\xi'\xi_0}{\text{var arg}} + \underset{\xi_0\xi_+}{\text{var arg}}]$$

$$\times \left\{ z^{-1/4}(\xi)\left[\mathrm{Ai}'(z(\xi)) + (1+|z(\xi)|)^{1/4} \exp\left[-\frac{2}{3} z^{3/2}(\xi)\right] O(|k\xi_0|^{-1})\right]\right\}. \quad \text{(B.33)}$$

Here when $\xi \in \xi'\xi_+$ the factor $\exp[-2z^{3/2}(\xi)/3]$ is replaced by $\exp|\mathrm{Re}\, 2z^{3/2}(\xi)/3|$. From (B.16) and (A.19) we get

$$z(\xi) = \left(\frac{3}{2} e^{i\pi} k\xi \ln \frac{2\xi}{e}\right)^{2/3} + O(k^{2/3}\xi^{-1/3} \ln^{-1/3} \xi) \quad \text{(B.34)}$$

for $\xi \to \infty$. When $\xi \in \xi_-\xi'$ $(-\pi/2 - \theta \leqslant \arg \xi \leqslant \pi/2 - \theta - \delta)$ relation (B.34) implies the inequalities

$$\frac{\pi}{3} - \delta < \arg z(\xi) < \pi - \delta/3,$$

which permit one to use (A.8) and obtain the asymptotic equality

$$\frac{1}{2\pi} \underset{\xi_-\xi'}{\text{var arg}} \left\{ z^{-1/4}(\xi)\left[\mathrm{Ai}'(z(\xi)) + (1+|z(\xi)|)^{1/4} \exp\left[-\frac{2}{3} z^{3/2}(\xi)\right] O(|k\xi_0|^{-1})\right]\right\}$$

$$= -\frac{1}{2\pi}\left(\mathrm{Im}\, \frac{2}{3} z'^{3/2} - \mathrm{Im}\, \frac{2}{3} z_-^{3/2}\right) + O_\delta(|z_0|^{-3/2} + |k\xi_0|^{-1}), \quad \text{(B.35)}$$

where $z' = z(\xi')$, $z_- = z(\xi_-)$ and $z_0 = z(\xi_0)$ (see Figure 5). When $\xi \in \xi'\xi_+$ $(\pi/2 - \theta - \delta \leqslant \arg \xi \leqslant \pi/2 - \theta)$ relation (B.34) implies the inequalities

$$\pi - \delta < \arg z(\xi) < \pi + \delta,$$

in which when $\xi \in \xi'\xi_0$

$$\pi - \delta < \arg z(\xi) \leqslant \pi \quad \text{(B.36)}$$

and when $\xi \in \xi_0\xi_+$

$$\pi \leqslant \arg z(\xi) < \pi + \delta. \quad \text{(B.37)}$$

With the use of (A.9) we obtain the asymptotic equalities

$$\frac{1}{2\pi}\operatorname*{var\,arg}_{\xi-\xi_0}\left\{z^{-1/4}(\xi)\left[\mathrm{Ai}'(z(\xi))+(1+|z(\xi)|)^{1/4}\exp\left|\operatorname{Re}\frac{2}{3}z^{3/2}(\xi)\right|O(|k\xi_0|^{-1})\right]\right\}$$
$$=\frac{1}{2\pi}\operatorname*{var\,arg}_{\xi\_\xi_0}\left\{\sin\left[\frac{2}{3}iz^{3/2}(\xi)-\frac{\pi}{4}\right]+\exp\left|\operatorname{Re}\frac{2}{3}z^{3/2}(\xi)\right|O(|z_0|^{-3/2}+|k\xi_0|^{-1})\right\}=p_1,$$
$$\frac{1}{2\pi}\operatorname*{var\,arg}_{\xi_0\xi+}\left\{z^{-1/4}(\xi)\left[\mathrm{Ai}'(z(\xi))+(1+|z(\xi)|)^{1/4}\exp\left|\operatorname{Re}\frac{2}{3}z^{3/2}(\xi)\right|O(|k\xi_0|^{-1})\right]\right\}$$
$$=\frac{1}{2\pi}\operatorname*{var\,arg}_{\xi_0\xi+}\left\{\sin\left[\frac{2}{3}iz^{3/2}(\xi)-\frac{\pi}{4}\right]+\exp\left|\operatorname{Re}\frac{2}{3}z^{3/2}(\xi)\right|O(|z_0|^{-3/2}+|k\xi_0|^{-1})\right\}=p_2.$$

On the basis of (B.36) and (B.37) we have the inequalities

$$\operatorname{Re}\frac{2}{3}z^{3/2}(\xi)\leqslant 0,\qquad \xi\in\xi'\xi_0,$$
$$0\leqslant\operatorname{Re}\frac{2}{3}z^{3/2}(\xi),\qquad \xi\in\xi_0\xi_+,$$

from which we obtain the representations

$$p_1=\frac{1}{2\pi}\operatorname*{var\,arg}_{\xi'\xi_0}\exp\left[-\frac{2}{3}z^{3/2}(\xi)\right]+\frac{1}{2\pi}\operatorname*{var\,arg}_{\xi'\xi_0}\left[1-i\exp\left[\frac{4}{3}z^{3/2}(\xi)\right]\right.$$
$$\left.+O(|z_0|^{-3/2}+|k\xi_0|^{-1}\right]=\frac{1}{2\pi}\left(\operatorname{Im}\frac{2}{3}z'^{3/2}-\operatorname{Im}\frac{2}{3}z_0^{3/2}\right)+p_1',$$
$$p_2=\frac{1}{2\pi}\operatorname*{var\,arg}_{\xi_0\xi+}\exp\left[\frac{2}{3}z^{3/2}(\xi)\right]+\frac{1}{2\pi}\operatorname*{var\,arg}_{\xi_0\xi+}\left[1+i\exp\left[-\frac{4}{3}z^{3/2}(\xi)\right]\right.$$
$$\left.+O(|z_0|^{-3/2}+|k\xi_0|^{-1})\right]=\frac{1}{2\pi}\left(\operatorname{Im}\frac{2}{3}z_+^{3/2}-\operatorname{Im}\frac{2}{3}z_0^{3/2}\right)+p_2'.$$

For $\xi\to\infty$ we have on the basis of (B.16) and (A.18) the relation

$$\frac{2}{3}z^{3/2}(\xi)=-k\xi\ln\frac{2\xi}{e}+O\left(\frac{k}{\xi}\right), \tag{B.38}$$

with the use of which we obtain the asymptotic equalities

$$\frac{2}{3}z'^{3/2}=\frac{2}{3}z^{3/2}\left(|\xi_0|e^{-i\left(\frac{\pi}{2}+\theta\right)}\right)=i|k\xi_0|\left[\ln\frac{2|\xi_0|}{e}-i\left(\frac{\pi}{2}+\theta\right)\right]+O\left(\frac{k}{\xi_0}\right),$$
$$\frac{2}{3}z_+^{3/2}=\frac{2}{3}z^{3/2}\left(|\xi_0|e^{i\left(\frac{\pi}{2}-\theta\right)}\right)=-i|k\xi_0|\left[\ln\frac{2|\xi_0|}{e}+i\left(\frac{\pi}{2}+\theta\right)\right]+O\left(\frac{k}{\xi_0}\right).$$

From (B.32) and (A.10) it follows that

$$\frac{2}{3}z_0^{3/2}=-ib_n,$$

where

$$b_n=\pi n\left[1-\frac{1}{4n}+O\left(\frac{1}{n^2}\right)\right].$$

Thus the expression (B.33) for $n\to\infty$ can be rewritten in the form

$$p(n)=n+p_1'+p_2'+O(n^{-1}+|k||\xi_0(n)|^{-1}).$$

The contour $\xi_+\xi_-\xi_0\xi_+$ has been chosen so that it encloses at least n zeros of the function $H'_{k\xi}(k)$, i.e. $n \leqslant p(n)$ and $|\xi_0(n)|^{-1} \leqslant |\widetilde{\xi}_n|^{-1}$. Using (B.28), (A.10) and the latter inequality, we get that

$$O(n^{-1}+|k||\xi_0(n)|^{-1})=O(|k|^2 n^{-1}\ln n)\to 0 \qquad \text{for} \qquad n\to\infty.$$

We will show below that also $p_1', p_2' \to 0$ as $n\to\infty$ and hence $p(n)=n$, so that the sequence $\{\xi_n\}_1^\infty$ contains all of the zeros of $H'_{k\xi}(k)$ lying in the halfplane $|\arg k\xi| \leqslant \pi/2$, these zeros being simple. Thus let us show that $p_1' \to 0$ as $n\to\infty$.

Suppose $\xi \in \xi'\xi_0$. Then

$$\xi=|\xi_0|e^{i\left(\frac{\pi}{2}-\theta-\psi\right)},\quad 0<\psi_0\leqslant\psi\leqslant\delta,$$

where ξ_0 is represented in the form $\xi_0 = |\xi_0|e^{i(\pi/2-\theta-\psi_0)}$ and, on the basis of (B.7), $\psi_0 \to 0$ as $n\to\infty$. From (B.16) and (B.3) we get

$$\frac{2}{3}z^{3/2}(\xi(\psi))=-|k\xi_0|\left[\sin\psi\ln\frac{2|\xi_0|}{e}-\cos\psi\left(\frac{\pi}{2}-\theta-\psi\right)\right]+O(|k||\xi_0|^{-1})$$
$$-i|k\xi_0|\left[\cos\psi\ln\frac{2|\xi_0|}{e}+\sin\psi\left(\frac{\pi}{2}-\theta-\psi\right)\right]+iO(|k||\xi_0|^{-1}). \qquad \text{(B.39)}$$

Let ϵ be an arbitrarily small positive number. Then for $n\geqslant N(\epsilon,k)$ we have

$$|O(|k|\xi_0|^{-1})|\leqslant\varepsilon.$$

Suppose $\xi(\psi)$ is such that $\operatorname{Re} 2z^{3/2}(\xi(\psi))/3 \geqslant -\epsilon$ (we recall that $\operatorname{Re} 2z^{3/2}(\xi)/3 \leqslant 0$ for $\xi\in\xi'\xi_0$). Then on the basis of (B.39) we have

$$|k\xi_0|\left[\sin\psi\ln\frac{2|\xi_0|}{e}-\cos\psi\left(\frac{\pi}{2}-\theta-\psi\right)\right]\leqslant 2\varepsilon.$$

Since $\operatorname{Re} 2z_0^{3/2}/3 = \operatorname{Re} 2z^{3/2}(\xi(\psi_0))/3 = 0$, it follows from (B.39) that

$$|k\xi_0|\left[\cos\psi_0\left(\frac{\pi}{2}-\theta-\psi_0\right)-\sin\psi_0\ln\frac{2|\xi_0|}{e}\right]\leqslant\varepsilon.$$

Adding the latter two inequalities, we obtain the estimate

$$|k\xi_0|\ln\frac{2|\xi_0|}{e}(\sin\psi-\sin\psi_0)$$
$$+|k\xi_0|\left[\left(\frac{\pi}{2}-\theta-\psi_0\right)\cos\psi_0-\left(\frac{\pi}{2}-\theta-\psi\right)\cos\psi\right]\leqslant 3\varepsilon. \qquad \text{(B.40)}$$

We note that the function $(\pi/2-\theta-\psi_0)\cos\psi$ is positive for $0\leqslant\psi\leqslant\delta\leqslant 1$ and monotonically decreases, since

$$[(\pi/2-\theta-\psi)\cos\psi]'=-(\pi/2-\theta-\psi)\sin\psi-\cos\psi\leqslant 0.$$

This permits us to deduce from (B.40) the inequality

$$|k\xi_0|\ln\frac{2|\xi_0|}{e}(\sin\psi-\sin\psi_0)\leqslant 3\varepsilon.$$

When $0 < \psi_0 \leqslant \psi \leqslant \delta$ we have the inequality

$$\frac{1}{2}(\psi - \psi_0) \leqslant (\psi - \psi_0)\cos\delta \leqslant \sin\psi - \sin\psi_0$$

(since $\delta = \delta(\epsilon')$ is a sufficiently small positive number), and hence when ψ is such that $\operatorname{Re} 2z^{3/2}(\xi(\psi))/3 \geqslant -\epsilon$ we have the estimate

$$\psi - \psi_0 \leqslant 6\varepsilon\left(|k\xi_0| \ln \frac{2|\xi_0|}{e}\right)^{-1}. \tag{B.41}$$

When $\psi \geqslant \psi_0$ and satisfies (B.41) we obtain from (B.39) the inequalities

$$\begin{aligned}\left|\operatorname{Im}\frac{2}{3}z^{3/2}(\xi(\psi)) - \operatorname{Im}\frac{2}{3}z_0^{3/2}\right| &\leqslant |k\xi_0| \ln \frac{2|\xi_0|}{e}|\cos\psi - \cos\psi_0| \\ &+ |k\xi_0|\left|\left(\frac{\pi}{2} - \theta - \psi\right)\sin\psi - \left(\frac{\pi}{2} - \theta - \psi_0\right)\sin\psi_0\right| + \varepsilon \leqslant 6\varepsilon \\ &+ 6\varepsilon\left(\frac{\pi}{2} + 1\right)\left(\ln\frac{2|\xi_0|}{e}\right)^{-1} + \varepsilon \leqslant 8\varepsilon\end{aligned}$$

where

$$\left(\max_{0\leqslant\psi\leqslant\delta}\left|\left[\left(\frac{\pi}{2} - \theta - \psi\right)\sin\psi\right]'\right|(\ln 2|\xi_0|/e)^{-1} \leqslant \left(\frac{\pi}{2} + 1\right)(\ln 2|\xi_0|/e)^{-1} < 1\right.$$

for sufficiently large n. Recalling that $\operatorname{Im} 2z_0^{3/2}/3 = -b_n = -\pi n + \pi/4 + O(1/n)$, we conclude that the relations

$$\begin{aligned}1 - i\exp&\left[\frac{4}{3}z^{3/2}(\xi(\psi))\right] + O(|z_0|^{-3/2} + |k\xi_0|^{-1}) \\ &= 1 - i\exp\left[-i2\pi n + i\frac{\pi}{2} + O(\varepsilon)\right] + O(|k|^2 n^{-1}\ln n) \\ &= 2 + O(\varepsilon + |k|^2 n^{-1}\ln n) = 2 + O(\varepsilon)\end{aligned} \tag{B.42}$$

hold for ψ satisfying (B.41) and $n \geqslant N = N(\epsilon, k)$. Thus when $-\epsilon \leqslant \operatorname{Re} 2z^{3/2}(\xi)/3 \leqslant 0$ we get

$$\arg\left(1 - i\exp\left[\frac{4}{3}z^{3/2}(\xi)\right] + O(|z_0|^{-3/2} + |k\xi_0|^{-1})\right) = O(\varepsilon).$$

When $\xi \in \xi'\xi_0$ and is such that $-\epsilon \leqslant \operatorname{Re} 2z^{3/2}(\xi)/3 \leqslant 0$ we get

$$\left|-i\exp\left[\frac{4}{3}z^{3/2}(\xi)\right] + O|z|^{-3/2} + |k\xi_0|^{-1})\right| \leqslant e^{-2\varepsilon} + \varepsilon \leqslant 1 - \frac{\varepsilon}{2},$$

so that

$$\left|1 - i\exp\left[\frac{4}{3}z^{3/2}(\xi)\right] + O(|z_0|^{-3/2} + |k\xi_0|^{-1})\right| \geqslant \frac{\varepsilon}{2} \tag{B.43}$$

and

$$\begin{aligned}p_1' &= \frac{1}{2\pi}\operatorname*{var\,arg}_{\xi'\xi_0}\left[1 - i\exp\left[\frac{4}{3}z^{3/2}(\xi)\right] + O(|z_0|^{-3/2} + |k\xi_0|^{-1})\right] \\ &= \frac{1}{2\pi}\arg[2 + O(\varepsilon)] - \frac{1}{2\pi}\arg\left[1 - \exp\left(i\frac{4}{3}z'^{3/2}\right) + O(\varepsilon)\right] = O(\varepsilon)\end{aligned}$$

for $n \geqslant N = N(\epsilon, k)$, i.e. $p_1' \to 0$ as $n \to \infty$. It can be shown analogously that $p_2' \to 0$ as $n \to \infty$.

REMARK. On the basis of (B.42) and (B.43) there exist positive numbers C' and $N = N(k)$ such that

$$\left|1 - i \exp\left[\frac{4}{3} z^{3/2}(\xi)\right] + O(|z_0|^{-3/2} + |k\xi_0|^{-1})\right| \geqslant C' \tag{B.44}$$

for $n \geqslant N(k)$ and $\xi \in \xi'\xi_0$. Analogously, when $n \geqslant N(k)$ and $\xi \in \xi_0\xi_+$

$$\left|1 + i \exp\left[-\frac{4}{3} z^{3/2}(\xi)\right] + O(|z_0|^{-3/2} + |k\xi_0|^{-1})\right| \geqslant C'. \tag{B.45}$$

C. We note that when $\operatorname{Im} k \geqslant 0$ and $\rho > 1$ the circuit relations for the functions $H_m(z)$ and $H_m'(z)$, $m = 0, 1, \cdots$ ([16], 8.746), and the convergence of the series (4.1) imply the equality

$$\overline{\Phi(\rho, \varphi, k)} = \Phi(\rho, \varphi, ke^{i\pi}),$$

which permits us to prove the assertion of the lemma for $k \in K_+ = K \cap \{\operatorname{Re} k > 0\}$. As we noted at the beginning of this section, the proof of the lemma makes use of equality (4.5). Before establishing this equality, let us represent the function $\Phi(\rho, \phi, k)$ given by (4.1) in the form of a contour integral. To this end we introduce the function

$$\Psi(\rho, \varphi, k, \xi) = \frac{1}{2\pi i} \frac{H_{k\xi}(k\rho)}{H_{k\xi}'(k)} \frac{\cos k\xi(\pi - |\varphi|)}{\sin k\xi\pi}, \tag{C.1}$$

where $\rho \geqslant 1$, $|\varphi| \leqslant \pi$, $k \in K_+ \cap \{\operatorname{Im} k \geqslant 0\}$ and $\xi \in \{|\arg k\xi| \leqslant \pi/2\}$. This function has poles at the points $\hat{\xi}_m = m/k$, $m = 0, 1, 2, \cdots$, at which, clearly,

$$2\pi i [\operatorname{Res} \Psi(\rho, \varphi, k, \xi)]_{\xi=\hat{\xi}_m} = \frac{1}{\pi k} \frac{H_m(k\rho)}{H_m'(k)} \cos m\varphi, \quad m = 0, 1, \ldots. \tag{C.2}$$

It also has poles at the zeros ξ_n, $n = 1, 2, \cdots$, of $H_{k\xi}'(k)$, which are separated from the poles $\{\hat{\xi}_m\}_0^\infty$ by the contour $\Pi_1\Pi_2$ in Figure 6. From (B.27) and (B.30) we deduce the inequalities

$$\operatorname{Im} \xi_1 = 2^{-1/3} \sin\left(\frac{\pi}{3} - \frac{2\theta}{3}\right) a_1 |k|^{-2/3} + O(|k|^{-4/3}) \geqslant 2^{-4/3} a_1 |k|^{-2/3},$$

$$\operatorname{Re} \xi_1 = 1 + 2^{-1/3} \cos\left(\frac{\pi}{3} - \frac{2\theta}{3}\right) a_1 |k|^{-2/3} + O(|k|^{-4/3}) \leqslant 1 + 2^{-1/3} a_1 |k|^{-2/3}$$

(for $|\theta| \leqslant \delta \ll 1$ and $|k| \geqslant \nu \gg 1$), from which it follows on the basis of (B.27), (B.28) and (B.30) that the roots ξ_n, $n = 1, 2, \cdots$, lie in the domain $a'|k|^{-2/3} < \arg \xi < \pi/2 - \theta$, where a' is a positive number. On the other hand, when $\theta \geqslant 0$ the points $\hat{\xi}_m$, $m = 0, 1, \cdots$, are situated on a straight line $\hat{\mathfrak{L}}$ (Figure 6) lying in the region $\arg \xi \leqslant 0$. This permits us, when $\theta \geqslant 0$, to construct in the domain $0 < \arg \xi < \pi/2 - \theta$ a contour $\Pi_1\Pi_2$ separating the poles ξ_n and $\hat{\xi}_m$ ($n = 1, 2, \cdots$; $m = 0, 1, \cdots$) of $\Psi(\rho, \phi, k, \xi)$, this contour being chosen so that "when $\xi \to \Pi_2$"

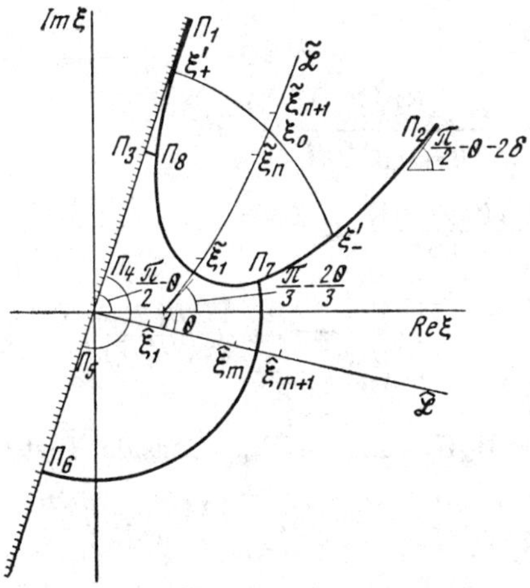

FIGURE 6

$$\Pi_1\Pi_2 = \{\xi\colon \xi - |\xi|\, e^{i(\pi/2-\theta-2\delta)} = o(|k\xi|^{-1})\} \tag{C.3}$$

and "when $\xi \to \Pi_1$"

$$\Pi_1\Pi_2 = \{\xi\colon \xi - |\xi|\, e^{i(\pi/2-\theta)} = o(|k\xi|^{-1}\}. \tag{C.4}$$

Let us show that the equality

$$\Phi(\rho, \varphi, k) = \int_{\Pi_1\Pi_2} \Psi(\rho, \varphi, k, \xi)\, d\xi \tag{C.5}$$

holds for $k \in N(\delta, \nu) \cap \{\operatorname{Im} k \geqslant 0\}$, $\rho > 1$ and $|\phi| \leqslant \pi$. Consider for $m \to \infty$ the integral

$$I_m = \oint_{\Pi_m} \Psi\, d\xi, \tag{C.6}$$

where the contour $\Pi_m = \Pi_3\Pi_4 \cdots \Pi_8\Pi_3$ (Figure 6) has been chosen so that the radius of the arc $\Pi_3\Pi_6$ is equal to $(m+\epsilon_2)|k|^{-1}$ and the radius of the arc $\Pi_4\Pi_5$ is equal to $\epsilon_1|k|^{-1}$, $0 < \epsilon_1, \epsilon_2 < 1$. Using the well-known identity 8.484 in [16]:

$$\frac{H_{-z}(z_1)}{H'_{-z}(z_2)} = \frac{H_z(z_1)}{H'_z(z_2)},$$

we obtain the equalities

$$\left\{\int_{\Pi_3\Pi_4} + \int_{\Pi_5\Pi_6}\right\}\Psi\, d\xi = \frac{1}{2\pi}\int_{\varepsilon_1|k|^{-1}}^{(m+\varepsilon_2)|k|^{-1}} e^{-i\theta}\left[\frac{H_{-i|k|y}(k\rho)}{H'_{-i|k|y}(k)} - \frac{H_{i|k|y}(k\rho)}{H'_{i|k|y}(k)}\right] \times \frac{\cos[i|k|y(\pi-|\varphi|)]}{\sin(i|k|y\pi)}\, dy = 0. \tag{C.7}$$

For $\xi \in \Pi_4\Pi_5$ we introduce the variable $\xi_1 = k\xi = \epsilon_1 e^{i\psi_1}$, $|\psi_1| \leqslant \pi/2$. Then

$$\int_{\Pi_4\Pi_5} \Psi\, d\xi = -\frac{\varepsilon_1}{2\pi k}\int_{-\pi/2}^{\pi/2} e^{i\psi_1}\frac{H_{\varepsilon_1 e^{i\psi}}(k\rho)}{H'_{\varepsilon_1 e^{i\psi}}(k)}\,\frac{\cos[\varepsilon_1 e^{i\psi_1}(\pi-|\varphi|)]}{\sin(\varepsilon_1 e^{i\psi_1}\pi)}\, d\psi$$

$$= -\frac{\varepsilon_1}{2\pi k}\int_{-\pi/2}^{\pi/2} e^{i(\psi_1+\pi)}\frac{H_{\varepsilon_1 e^{i(\psi_1+\pi)}}(k\rho)}{H'_{\varepsilon_1 e^{i(\psi_1+\pi)}}(k)}\,\frac{\cos[\varepsilon_1 e^{i(\psi_1+\pi)}(\pi-|\varphi|)]}{\sin(\varepsilon_1 e^{i(\psi_1+\pi)}\pi)}\, d(\psi_1+\pi)$$

$$= -\frac{\varepsilon_1}{2\pi k}\int_{\pi/2}^{3\pi/2} e^{i\psi_1}\frac{H_{\varepsilon_1 e^{i\psi_1}}(k\rho)}{H'_{\varepsilon_1 e^{i\psi_1}}(k)}\,\frac{\cos[\varepsilon_1 e^{i\psi_1}(\pi-|\varphi|)]}{\sin(\varepsilon_1 e^{i\psi_1}\pi)}\, d\psi_1.$$

Thus

$$\int_{\Pi_4\Pi_5} \Psi d\xi = -\frac{\varepsilon_1}{4\pi k}\int_{-\pi/2}^{3\pi/2} e^{i\psi_1}\frac{H_{\varepsilon_1 e^{i\psi_1}}(k\rho)}{H'_{\varepsilon_1 e^{i\psi_1}}(k)}\frac{\cos[\varepsilon_1 e^{i\psi_1}(\pi-|\varphi|)]}{\sin(\varepsilon_1 e^{i\psi_1}\pi)}d\psi_1$$
$$= -\frac{1}{4\pi k i}\oint_{|\xi_1|=\varepsilon_1}\frac{H_{\xi_1}(k\rho)}{H'_{\xi_1}(k)}\frac{\cos\xi_1(\pi-|\varphi|)}{\sin\xi_1\pi}d\xi_1$$
$$= -\frac{1}{2k}\operatorname{Res}\left[\frac{H_{\xi_1}(k\rho)}{H'_{\xi_1}(k)}\frac{\cos\xi_1(\pi-|\varphi|)}{\sin\xi_1\pi}\right]_{\xi_1=0} = -\frac{1}{2\pi k}\frac{H_0(k\rho)}{H'_0(k)}. \qquad (C.8)$$

We estimate the function Ψ on the arcs $\Pi_6\Pi_7$ and $\Pi_8\Pi_3$. Consider first the ratio $\cos k\xi(\pi-|\phi|)/\sin k\xi\pi$ on the semicircle $\Pi_3\Pi_6$: $k\xi=(m+\epsilon_2)e^{i\psi}$, where $|\psi|\leqslant\pi/2$. We have

$$k\xi\pi = (\pi m+\pi\varepsilon_2)\cos\psi + i(\pi m+\pi\varepsilon_2)\sin\psi = x(\psi)+iy(\psi).$$

For $m\to\infty$ and ψ such that

$$|y(\psi)| = |\pi m+\pi\varepsilon_2|\,|\sin\psi|\leqslant 1$$

we have the estimate

$$|\psi|\leqslant 2/\pi m;$$

hence

$$x(\psi) = \pi m+\pi\varepsilon_2+O(m^{-1}).$$

Since $0<\epsilon_2<1$, for $m\to\infty$

$$|\cos[2\pi\varepsilon_2+O(m^{-1})]|<1-C,$$

where $C=C(\epsilon_2)>0$. Therefore

$$|\sin(x+iy)|^2 = {}^1\!/_2\cosh 2y - {}^1\!/_2\cos 2x \geqslant {}^1\!/_2\, C\cosh 2y.$$

For $|y(\psi)|\geqslant 1$, clearly,

$$|\sin(x+iy)|^2\geqslant {}^1\!/_4\cosh 2y. \qquad (C.9)$$

Thus on the semicircle $\Pi_3\Pi_6$

$${}^1\!/_2 C(\varepsilon_2)\,e^{|\operatorname{Im}(k\xi\pi)|}\leqslant|\sin(k\xi\pi)|.$$

Taking into account the obvious inequality

$$|\cos(x+iy)|\leqslant(e^y+e^{-y})/2, \qquad (C.10)$$

we obtain on the arcs $\Pi_6\Pi_7$ and $\Pi_8\Pi_3$ the estimate

$$\left|\frac{\cos k\xi(\pi-|\varphi|)}{\sin k\xi\pi}\right|\leqslant\frac{2}{C(\varepsilon_2)}e^{-|\operatorname{Im}(k\xi\varphi)|}. \qquad (C.11)$$

We estimate the ratio $H_{k\xi}(k\rho)/H'_{k\xi}(k)$ on the arcs $\Pi_6\Pi_7$ and $\Pi_8\Pi_3$. For $\xi\in\Pi_6\Pi_7$ and $m\to\infty$ we have $|\xi|\to\infty$ and $-\pi/2\leqslant\arg k\xi\leqslant\pi/2-\delta$. Consequently on the basis of (A.18) and (A.19) the function

$$z(\xi, \rho) = e^{i\frac{2\pi}{3}}(k\rho)^{2/3} w(\xi/\rho) \tag{C.12}$$

satisfies the relations

$$z(\xi, \rho) = \left(\frac{3}{2} e^{i\pi}\xi k \ln \frac{2\xi}{\rho e}\right)^{2/3} + o_{\rho,k}(1), \tag{C.13}$$

$$\frac{2}{3} z^{3/2}(\xi, \rho) = -k\xi \ln \frac{2\xi}{\rho e} + o_{\rho,k}(1), \tag{C.14}$$

where $o_{\rho,k}(1) \to 0$ as $m \to \infty$ for each fixed $\rho \in [1, \infty)$ and each fixed $k \in N(\delta(\epsilon'), \nu(\epsilon')) \cap \{\operatorname{Im} k \geqslant 0\}$. From (C.13) it follows that for $m \to \infty$

$$\pi/3 - \delta \leqslant \arg(\xi/\rho) \leqslant \pi - \delta,$$

so that, using (A.3) and (A.8), we get

$$\frac{H_{k\xi}(k\rho)}{H'_{k\xi}(k)} = \frac{\exp[-2/3z^{3/2}(\xi/\rho) + 2/3z^{3/2}(\xi)]}{(\xi^2/\rho^2 - 1)^{1/4}(\xi^2 - 1)^{1/4}} \{1 + o_{\rho,k,\epsilon'}(1)\}.$$

Taking into account the fact that $k\xi = (m + \epsilon_2)e^{i\psi}$ for $\xi \in \Pi_6\Pi_7$, where $-\pi/2 \leqslant \psi \leqslant \pi/2 - \delta$, and using (C.14), we deduce from the latter asymptotic relation the estimate

$$\left|\frac{H_{k\xi}(k\rho)}{H'_{k\xi}(k)}\right| \leqslant \frac{C(\rho, k, \epsilon')}{m} e^{-m\cos\psi \ln \rho}. \tag{C.15}$$

For $\xi \in \Pi_8\Pi_3$ we have $k\xi = (m + \epsilon_2)e^{i\psi}$, where ψ satisfies by virtue of (C.3) the condition

$$\pi/2 - o(m^{-1}) \leqslant \psi \leqslant \pi/2. \tag{C.16}$$

It therefore follows from (C.13) that

$$\pi - o_{\rho,k}(1) \leqslant \arg z(\xi, \rho) \leqslant \pi + o_{\rho,k}(1),$$

and on the basis of (A.3), (A.4) and (A.8) that

$$\frac{H_{k\xi}(k\rho)}{H'_{k\xi}(k)} = \frac{1}{\left(\frac{\xi^2}{\rho^2} - 1\right)^{1/4}(\xi^2 - 1)^{1/4}} \times \frac{\cos\left[\frac{2}{3} z^{3/2}(\xi, \rho) - \frac{\pi}{4}\right] + \left[\exp\left|\operatorname{Re}\frac{2}{3} z^{3/2}(\xi, \rho)\right|\right] o_{\rho,k}(1)}{\sin\left[\frac{2}{3} z^{3/2}(\xi, 1) - \frac{\pi}{4}\right] + \left[\exp\left|\operatorname{Re}\frac{2}{3} z^{3/2}(\xi, 1)\right|\right] o_{\rho,k}(1)}. \tag{C.17}$$

Taking (C.16) into account, we can rewrite (C.14) when $\xi \in \Pi_8\Pi_3$ in the form

$$\frac{2}{3} z^{3/2}(\xi(\psi), \rho) = (m + \epsilon_2)\frac{\pi}{2} + o(1)\ln\frac{2(m + \epsilon_2)}{3\rho|k|e} - i(m + \epsilon_2)\ln\frac{2(m + \epsilon_2)}{3\rho|k|e} + io(1),$$

which implies that $\operatorname{Im}[2iz^{3/2}(\xi, \rho)/3 - \pi/4] \to \infty$ as $m \to \infty$ and hence that (C.17) has the form

$$\frac{H_{k\xi}(k\rho)}{H'_{k\xi}(k)}=\left(\frac{\xi^2}{\rho^2}-1\right)^{-1/4}(\xi^2-1)^{-1/4}\{1+o_{\rho,k}(1)\}.$$

Thus the estimate

$$\left|\frac{H_{k\xi}(k\rho)}{H'_{k\xi}(k)}\right|\leqslant\frac{C(\rho,k)}{m} \tag{C.18}$$

is valid on the arc $\Pi_8\Pi_3$ for $m\to\infty$. The following inequalities hold on the basis of (C.11), (C.15) and (C.18) for $\rho>1$, $|\phi|\leqslant\pi$ and $m\to\infty$:

$$\left|\int_{\Pi_8\Pi_3}\Psi d\xi\right|\leqslant C(\rho,k,\varepsilon_2)\int_{\pi/2-o(m^{-1})}^{\pi/2}d\psi\leqslant C(\rho,k,\varepsilon_2)\,o(m^{-1}),$$

$$\left|\int_{\Pi_6\Pi_7}\Psi d\xi\right|\leqslant C(\rho,k,\varepsilon_2,\varepsilon')\int_{-\pi/2}^{\pi/2-\delta}e^{-m\cos\psi\ln\rho}d\psi$$

$$\leqslant C(\rho,k,\varepsilon_2,\varepsilon')\left[\pi e^{-m\sin\delta\ln\rho}+\int_{-\pi/2}^{-\pi/2+\delta}e^{-m\cos\psi\ln\rho}d\psi\right]$$

$$\leqslant C(\rho,k,\varepsilon_2,\varepsilon')\left(\pi e^{-m\sin\delta\ln\rho}+\frac{2}{m\ln\rho}\right),$$

i.e. for $m\to\infty$, $\rho>1$ and $|\phi|\leqslant\pi$

$$\int_{\Pi_6\Pi_7}\Psi d\xi\to 0,\qquad\int_{\Pi_8\Pi_3}\Psi d\xi\to 0. \tag{C.19}$$

Using (C.2), (C.7) and (C.8), we obtain the equality

$$\oint_{\Pi_m}\Psi d\xi=\sum_{j=1}^{m}\frac{1}{\pi k}\frac{H_j(k\rho)}{H'_j(k)}\cos j\varphi=-\frac{1}{2\pi k}\frac{H_0(k\rho)}{H'_0(k)}+\left\{\int_{\Pi_6\Pi_7}+\int_{\Pi_7\Pi_8}+\int_{\Pi_8\Pi_3}\right\}\Psi d\xi,$$

which transforms into the form

$$\int_{\Pi_7\Pi_8}\Psi d\xi=\frac{1}{2\pi k}\left[\frac{H_0(k\rho)}{H'_0(k)}+2\sum_{j=1}^{m}\frac{H_j(k\rho)}{H'_j(k)}\cos j\varphi\right]+\left\{\int_{\Pi_6\Pi_7}+\int_{\Pi_8\Pi_3}\right\}\Psi d\xi.$$

For $\rho>1$, $|\phi|\leqslant\pi$ and $k\in N(\delta,\nu)\cap\{\operatorname{Im}k\geqslant 0\}$ the right side of the latter equality has a limit for $m\to\infty$, which on the basis of (4.1) and (C.19) is equal to $\Phi(\rho,\phi,k)$. Consequently when $\rho>1$, $|\phi|\leqslant\pi$ and $k\in N(\delta,\nu)\cap\{\operatorname{Im}k\geqslant 0\}$ we have

$$\int_{\Pi_1\Pi_2}\Psi(\rho,\varphi,k,\xi)\,d\xi=-\lim_{m\to\infty}\int_{\Pi_7\Pi_8(m)}\Psi(\rho,\varphi,k,\xi)\,d\xi=\Phi(\rho,\varphi,k). \tag{C.20}$$

Let us now prove equality (4.5). In §4.B we indicated a point $\xi_0=\xi_0(n,k)$

(see Figures 4 and 6) on the curve $\widetilde{\mathfrak{L}}$ such that the contour $\xi'_+\xi'_-\xi_0\xi'_+$ formed by the arc $\xi'_-\xi_0\xi'_+$ of radius $|\xi_0|$ and the appropriate part of the contour $\Pi_1\Pi_2$ encloses for sufficiently large n precisely n poles $\xi_1, \cdots, \xi_n$ of the function $\Psi(\rho, \phi, k, \xi)$. Since these poles are simple, the equality

$$\left\{\int_{\xi'_+\xi'_-} + \int_{\xi'_-\xi_0\xi'_+}\right\}\Psi d\xi = \sum_{j=1}^{n} \frac{H_{k\xi_j}(k\rho)}{\left[\frac{dH'_{k\xi}(k)}{d\xi}\right]_{\xi=\xi_j}} \frac{\cos k\xi_j(\pi - |\varphi|)}{\sin k\xi_j\pi} \tag{C.21}$$

holds for any fixed and sufficiently large n. We estimate the function Ψ on the arc $\xi'_-\xi_0\xi'_+$ for $n \to \infty$. Since when $\xi \in \xi'_-\xi_0\xi'_+$ we have $\pi/2 - 2\delta < \arg k\xi < \pi/2$ and $|\xi| \to \infty$, we also have $\operatorname{Im}(k\xi\pi) \to \infty$; hence inequalities (C.9) and (C.10) can be used to obtain the estimate

$$\left|\frac{\cos k\xi(\pi - |\varphi|)}{\sin k\xi\pi}\right| \leqslant 16e^{-|\operatorname{Im}(k\xi\varphi)|}. \tag{C.22}$$

Consider the ratio $H_{k\xi}(k\rho)/H'_{k\xi}(k)$ for $\xi \in \xi'_-\xi_0\xi'_+$ and $n \to \infty$. Consequently, similarly to (C.17), the following equality is valid on the arc $\xi'_-\xi_0\xi'_+$ for $n \to \infty$:

$$\frac{H_{k\xi}(k\rho)}{H'_{k\xi}(k)} = \left(\frac{\xi^2}{\rho^2} - 1\right)^{-1/4}(\xi^2 - 1)^{-1/4}$$
$$\times \frac{\cos\left[\frac{2}{3}iz^{3/2}(\xi, \rho) - \frac{\pi}{4}\right] + \left[\exp\left|\operatorname{Re}\frac{2}{3}z^{3/2}(\xi, \rho)\right|\right]o_{\rho,k,\delta}(1)}{\sin\left[\frac{2}{3}iz^{3/2}(\xi, 1) - \frac{\pi}{4}\right] + \left[\exp\left|\operatorname{Re}\frac{2}{3}z^{3/2}(\xi, 1)\right|\right]o_{k,\delta}(1)}, \tag{C.23}$$

where $o_{\rho,k,\delta}(1) \to 0$ and $o_{k,\delta}(1) \to 0$ for $n \to \infty$ and any fixed numbers $\rho \in [1, \infty)$, $\delta = \delta(\epsilon')$ and $k \in N(\delta(\epsilon'), \nu(\epsilon'))$. The arc $\xi'_-\xi_0\xi'_+$ is the union of the arcs $\xi'_-\xi_0$ and $\xi_0\xi'_+$. We consider on each of these arcs the denominator of the ratio in (C.23). In §4.B we showed that when $\xi \in \xi'_-\xi_0 \subset \xi_-\xi_0$ (Figures 4 and 5) we have the inequality $\operatorname{Re} 2z^{3/2}(\xi, 1)/3 \leqslant 0$, which implies

$$\sin\left[\frac{2}{3}iz^{3/2}(\xi, 1) - \frac{\pi}{4}\right] + \left[\exp\left|\operatorname{Re}\frac{2}{3}z^{3/2}(\xi, 1)\right|\right]o_{k,\delta}(1)$$
$$= \frac{1}{2i}\exp\left[-\frac{2}{3}z^{3/2}(\xi, 1) - i\frac{\pi}{4}\right]\left\{1 - i\exp\left[\frac{4}{3}z^{3/2}(\xi, 1)\right] + o_{k,\delta}(1)\right\}. \tag{C.24}$$

Analogously, when $\xi \in \xi_0\xi'_+ \subset \xi_0\xi_+$ we have $\operatorname{Re} 2z^{3/2}(\xi, 1)/3 \geqslant 0$, and hence

$$\sin\left[\frac{2}{3}iz^{3/2}(\xi, 1) - \frac{\pi}{4}\right] + \left[\exp\left|\operatorname{Re}\frac{2}{3}z^{3/2}(\xi, 1)\right|\right]o_{k,\delta}(1)$$
$$= \frac{i}{2}\exp\left[\frac{2}{3}z^{3/2}(\xi, 1) + i\frac{\pi}{4}\right]\left\{1 + i\exp\left[-\frac{4}{3}z^{3/2}(\xi, 1)\right] + o_{k,\delta}(1)\right\}. \tag{C.25}$$

From (B.34) and (B.35) it follows that the moduli of the expressions in square brackets in the right sides of (C.24) and (C.25) are bounded from below by a number $C' = C'(k, \delta)$ as $n \to \infty$, and hence for $\xi \in \xi'_-\xi_0\xi'_+$ and $n \to \infty$ we have

$$\sin\left[\frac{2}{3}iz^{3/2}(\xi,\ 1)-\frac{\pi}{4}\right]+\left[\exp\left|\operatorname{Re}\frac{2}{3}z^{3/2}(\xi,\ 1)\right|\right]o_{k,\,\delta}(1)\left|\geqslant C'\exp\left|\operatorname{Re}\frac{2}{3}z^{3/2}(\xi,1)\right|.\right.$$

Taking this into account, we deduce from (C.23) the estimate

$$\left|\frac{H_{k\xi}(k\rho)}{H'_{k\xi}(k)}\right|\leqslant\frac{C(\rho,k,\delta)}{|\xi_0|}\exp\left[\left|\operatorname{Re}\frac{2}{3}z^{3/2}(\xi,\rho)\right|-\left|\operatorname{Re}\frac{2}{3}z^{3/2}(\xi,\ 1)\right|\right],$$

which on the basis of (C.14) can be rewritten in the form

$$\left|\frac{H_{k\xi}(k\rho)}{H'_{k\xi}(k)}\right|\leqslant\frac{C(\rho,k,\delta)}{|\xi_0|}\exp\left[\left|\operatorname{Re}\frac{2}{3}z^{3/2}(\xi,\ 1)+\operatorname{Re}(k\xi)\ln\rho\right|-\left|\operatorname{Re}\frac{2}{3}z^{3/2}(\xi,\ 1)\right|\right].\quad\text{(C.26)}$$

From (C.14) it follows that for any $\rho\geqslant 1$ there exists a sufficiently large number $A(\rho)>0$ such that for $\xi\in\xi'_-\xi_0\xi'_+$ satisfying the condition

$$\arg k\xi<\pi/2-A(\rho)/\ln(2|\xi_0|/e)$$

and $|\xi_0|\to\infty$ (i.e. $n\to\infty$) we have

$$-\operatorname{Re}\frac{2}{3}z^{3/2}(\xi,\ 1)=|k\xi_0|\left\{\cos[\arg(k\xi)]\ln\frac{2|\xi_0|}{e}-\sin[\arg(k\xi)]\arg\xi\right\}+o_k(1)$$

$$\geqslant|k\xi_0|\left\{\sin\left(\frac{A(\rho)}{\ln\frac{2|\xi_0|}{e}}\right)\ln\left(\frac{2|\xi_0|}{e}\right)-\pi\right\}\geqslant|k\xi_0|\frac{A(\rho)}{4}>|k\xi_0|\ln\rho\geqslant\operatorname{Re}(k\xi)\ln\rho,$$

so that (3.26) can be written in the form

$$\left|\frac{H_{k\xi}(k\rho)}{H'_{k\xi}(k)}\right|\leqslant\frac{C}{|\xi_0|}\begin{cases}\exp\{\operatorname{Re}(k\xi)\ln\rho\}, & \frac{\pi}{2}-\frac{A}{\ln(2|\xi_0|/e)}<\arg k\xi\leqslant\frac{\pi}{2},\\ \exp\{-\operatorname{Re}(k\xi)\ln\rho\}, & \frac{\pi}{2}-\delta\leqslant\arg k\xi\leqslant\frac{\pi}{2}-\frac{A}{\ln(2|\xi_0|/e)}.\end{cases}$$

We have

$$\operatorname{Re}(k\xi)/\operatorname{Im}(k\xi)\to 0\quad\text{for}\quad\frac{\pi}{2}-\frac{A}{\ln(2|\xi_0|/e)}<\arg k\xi\leqslant\frac{\pi}{2}\quad\text{and}\quad|\xi_0|\to\infty,$$

which together with (C.22) implies when $\phi\neq 0$ and $|\xi_0|\to\infty$ that

$$|\Psi|\leqslant\begin{cases}\dfrac{C(\nu,\delta,k)}{|\xi_0|}, & \frac{\pi}{2}-\frac{A}{\ln(2|\xi_0|/e)}<\arg k\xi\leqslant\frac{\pi}{2},\\ \dfrac{C(\rho,k,\delta)}{|\xi_0|}\exp\{-\operatorname{Re}(k\xi)\ln\rho\}, & \frac{\pi}{2}-\delta\leqslant\arg k\xi\leqslant\frac{\pi}{2}-\frac{A}{\ln(2|\xi_0|/e)}.\end{cases}$$

It follows from these estimates when $\rho>1,\phi\neq 0$ and $n\to\infty$ ($|\xi_0|\to\infty$) that

$$\left|\int_{\xi'_-\xi_0\xi'_+}\Psi\,d\xi\right|\leqslant C_1(\rho,k,\delta)\int_{\frac{\pi}{2}-\frac{A}{\ln(2|\xi_0|/e)}}^{\frac{\pi}{2}}d\psi+C_2(\rho,k,\delta)\int_{\frac{\pi}{2}-\delta}^{\frac{\pi}{2}-\frac{A}{\ln(2|\xi_0/e)}}e^{-|k\xi_0|\ln\rho\cos\psi}d\psi$$

$$\leqslant\frac{C_1A}{\ln(2|\xi_0|/e)}+C_2\int_0^{2\delta}e^{-1/2|k\xi_0|\ln\rho\psi_1}d\psi_1\leqslant\frac{C_1A}{\ln(2|\xi_0|/e)}+\frac{2C_2}{|k\xi_0|\ln\rho}\to 0,$$

which for any fixed $\rho > 1, \phi \neq 0$ and $k \in N(\delta, \nu) \cap \{\operatorname{Im} k > 0\}$ permits us to pass to the limit as $n \to \infty$ in (C.21) (taking into account the convergence of the integral (C.20)) and thereby obtain the equality

$$\Phi(\rho, \varphi, k) = 2 \sum_{n=1}^{\infty} \frac{H_{k\xi_n}(k\rho)}{[dH'_{k\xi}(k)/d\xi]_{\xi=\xi_n}} \frac{\cos k\xi_n(\pi - |\varphi|)}{\sin k\xi_n \pi}. \tag{C.27}$$

Differentiating (C.27) termwise with respect to ρ and ϕ, we obtain the formal equalities

$$\frac{\partial \Phi}{\partial \rho} = 2 \sum_{n=1}^{\infty} \frac{kH'_{k\xi_n}(k\rho)}{[dH'_{k\xi}(k)/d\xi]_{\xi=\xi_n}} \frac{\cos k\xi_n(\pi - |\varphi|)}{\sin k\xi_n \pi}, \tag{C.28}$$

$$\frac{\partial \Phi}{\partial \varphi} = 2 \sum_{n=1}^{\infty} \frac{k\xi_n H_{k\xi_n}(k\rho)}{[dH'_{k\xi}(k)/d\xi]_{\xi=\xi_n}} \frac{\sin k\xi_n(\pi - |\varphi|)}{\sin k\xi_n \pi}. \tag{C.29}$$

D. Let us proceed with the proof of the lemma. To this end we will construct series majorizing (C.27)–(C.29). We first establish an estimate from below of the modulus of the expression $[dH'_{k\xi}(k)/d\xi]_{\xi=\xi_n}$ with the use of (A.5). The expressions entering into (A.5) will be successively estimated. Consider for $\xi \in T_n$ the derivative $dw/d\xi$, which can be represented in the form

$$\frac{dw(\xi)}{d\xi} = \frac{dw}{d\xi}\bigg|_{\xi=\xi_n} + \int_{\xi_n}^{\xi} \frac{d^2w(\eta)}{d\eta^2}\, d\eta. \tag{D.1}$$

Recalling that $w(\widetilde{\xi}_n) = e^{i(\pi/3 - 2\theta/3)} a_n |k|^{-2/3}$, taking into account (A.24), (A.25) and (B.30) and using (D.1), we get

$$C_1 a_n^{-1/2} |k|^{1/3} \ln(1 + a_n^{1/2} |k|^{-1/3}) - C_3 a_n^{-5/6} |k|^{-1/3} \leqslant |dw(\xi)/d\xi| \\ \leqslant C_2 a_n^{-1/2} |k|^{1/3} \ln(1 + a_n^{1/2}|k|^{-1/3}) + C_3 a_n^{-5/6} |k|^{-1/3}.$$

Since $a_{n+1} > a_n \geqslant a_1 > 0$ and $\ln(1+x) \geqslant x(1+x)^{-1}$ for $x \geqslant 0$, we get

$$C_1 a_n^{-1/2} |k|^{1/3} \ln(1 + a_n^{1/2} |k|^{-1/3}) \geqslant C_1 (1 + a_n^{1/2} |k|^{-1/3})^{-1} \\ \geqslant C_1 |k|^{1/3} a_n^{-1/2} \geqslant 2C_3 a_n^{-5/6} |k|^{-4/3},$$

if $|k| > \nu$ and ν is sufficiently great. Thus the following estimates hold for $\xi \in T_n$, $n = 1, 2, \cdots$, and $k \in N(\delta, \nu)$:

$$C_1 a_n^{-1/2} |k|^{1/3} \ln(1 + a_n^{1/2} |k|^{-1/3}) \leqslant |dw(\xi)/d\xi| \\ \leqslant C_2 a_n^{-1/2} |k|^{1/3} \ln(1 + a_n^{1/2} |k|^{-1/3}). \tag{D.2}$$

Consider now the expression $\xi_n^2 - 1$, which can be represented in the form $\xi_n^2 - 1 = (\widetilde{\xi}_n - 1 + \xi_n - \widetilde{\xi}_n)(\widetilde{\xi}_n + 1 + \xi_n - \widetilde{\xi}_n)$. To estimate the difference $\widetilde{\xi}_n - 1$ for those positive integral values of n such that $a_n|k|^{-2/3} < c'$ we note on the basis of (B.26) that

$$\xi_n - 1 = 2^{-1/3} e^{i(\pi/3 - 2\theta/3)} a_n |k|^{-2/3} \{1 + O(\varepsilon')\},$$

and hence

$$C_3 a_n |k|^{-2/3} \leqslant |\xi_n - 1| \leqslant C_4 a_n |k|^{-2/3}$$

(recall that ϵ' can be assumed to be arbitrarily small). For n such that $\epsilon' < a_n|k|^{-2/3} < 1/\epsilon'$ we deduce from (B.27) the estimates

$$C_3(\varepsilon') a_n |k|^{-2/3} \leqslant b_1(\varepsilon') \leqslant \operatorname{Im} \xi_n \leqslant |\xi_n - 1| \leqslant |\xi_n| \leqslant a_2(\varepsilon') + b_2(\varepsilon') \leqslant C_4(\varepsilon') a_n |k|^{-2/3}.$$

The constants C_3 and C_4 can be chosen so that (B.28) will imply the following estimates for n such that $a_n|k|^{-2/3} \geqslant 1/\epsilon'$:

$$C_3(\varepsilon') a_n^{3/2} |k|^{-1} \ln^{-1}(1 + a_n^{1/2} |k|^{-1/3}) \leqslant |\xi_n - 1| \leqslant C_4(\varepsilon') a_n^{3/2} |k|^{-1} \ln^{-1}(1 + a_n^{1/2} |k|^{-1/3}). \quad \text{(D.3)}$$

Since the inequality $1 \leqslant x \ln^{-1}(1 + x) \leqslant 1/\epsilon'$ holds for $0 \leqslant x \leqslant 1/\epsilon'$, we conclude upon setting $x = x_n = a_n^{1/2}|k|^{-1/3}$ that estimates (D.3) will be valid for all $n = 1, 2, \cdots$ under an appropriate choice of the constants $C_3(\epsilon')$ and $C_4(\epsilon')$.

In exactly the same way we establish the estimates

$$C_3(\varepsilon') a_n^{1/2} (a_n + |k|^{2/3}) |k|^{-1} \ln^{-1}(1 + a_n^{1/2} |k|^{-1/3}) \leqslant |1 + \xi_n| \leqslant C_4(\varepsilon') a_n^{1/2} (a_n + |k|^{2/3}) |k|^{-1} \ln^{-1}(1 + a_n^{1/2} |k|^{-1/3}). \quad \text{(D.4)}$$

Since $x^3 \ln^{-1}(1 + x)$ and $x(1 + x^2) \ln^{-1}(1 + x)$ are monotonically increasing on $(0, \infty)$, we obtain upon setting $x = x_n = a_n^{1/2}|k|^{-1/3} \geqslant a_1^{1/2}|k|^{-1/3}$ the estimates

$$|\xi_n - 1| \geqslant C_5(\varepsilon') |k|^{-2/3}, \qquad |\xi_n + 1| \geqslant C_5(\varepsilon').$$

On the other hand,

$$2|\xi_n - \xi_n| \leqslant 2a_n^{-5/6} |k|^{-4/3} \leqslant 2a_1^{-5/6} |k|^{-4/3} \leqslant C_5 |k|^{-2/3} \leqslant C_5$$

for $k \in N(\delta, \nu)$, where $\nu = \nu(\epsilon')$ is sufficiently great. Taking into account (D.3), (D.4) and the latter inequalities, we obtain the following estimates for all $n = 1, 2, \cdots$ and $k \in N(\delta, \nu)$:

$$C_3(\varepsilon') a_n^2 |k|^{-2} (a_n + |k|^{2/3}) \ln^{-2}(1 + a_n^{1/2} |k|^{-1/3}) \leqslant |\xi_n^2 - 1| \leqslant C_4(\varepsilon') a_n^2 |k|^{-2} (a_n + |k|^{2/3}) \ln^{-2}(1 + a_n^{1/2} |k|^{-1/3}). \quad \text{(D.5)}$$

From the equality

$$w(\xi_n) = e^{i(\pi/3 - 2\theta/3)} a_n |k|^{-2/3} + \int_{\xi_n}^{\xi} \frac{dw(\eta)}{d\eta}\, d\eta,$$

and estimates (B.30) and (D.2) we obtain the estimates

$$C_5 a_n |k|^{-2/3} \leqslant |w(\xi_n)| \leqslant C_6 a_n |k|^{-2/3}. \quad \text{(D.6)}$$

By repeating the arguments used in §4.B to estimate the functions f_n from below, we obtain the estimates

$$C_7 a_n^{3/4} \leqslant |\operatorname{Ai}''(z(\xi_n)) + (1+|z(\xi_n)|)^{3/4} e^{-2/3 z^{3/2}(\xi_n)} O(|k\xi_n|^{-1})|. \tag{D.7}$$

From (D.2) and (D.5)–(D.7) we deduce the estimates

$$C a_n^{1/2} (a_n + |k|^{2/3})^{1/4} \ln^{1/2}(1 + a_n^{1/2}|k|^{-1/3}) \leqslant |dH'_{k\xi}(k)/d\xi|_{\xi=\xi_n}. \tag{D.8}$$

Let us now estimate from above the moduli of the ratios

$$\cos \nu\xi_n(\pi - |\varphi|)/\sin \nu\xi_n\pi, \quad \sin \nu\xi_n(\pi - |\varphi|)/\sin \nu\xi_n\pi.$$

To this end we first establish the inequality

$$C_1(\varepsilon') a_n^{3/2} \ln^{-1}(1 + a_n^{1/2}|k|^{-1/3}) \leqslant \operatorname{Im}(k\xi_n), \tag{D.9}$$

which is valid in a smaller domain than $N(\delta, \nu)$. In order to determine this domain we consider the expression $k\tilde{\xi}_n$. For those positive integral values of n such that $a_n|k|^{-2/3} \leqslant \epsilon'$ we obtain from (B.26) the relation

$$k\xi_n = |k| e^{i\theta} + 2^{-1/3} e^{i(\pi/3+\theta/3)} a_n |k|^{1/3}\{1 + O(\varepsilon')\},$$

from which, when ϵ' and $\delta = \delta(\epsilon')$ are sufficiently small positive numbers (recall that $|\theta| \leqslant \delta$), we deduce the estimate

$$\operatorname{Im}(k\xi_n) \geqslant |k| \sin\theta + 1/4 a_n |k|^{1/3}.$$

Thus for $k = |k|e^{i\theta}$ satisfying the conditions

$$1 \leqslant \nu(\varepsilon') \leqslant |k|, \quad -1/8\, a_1 |k|^{-2/3} < \theta < \delta(\varepsilon'), \tag{D.10}$$

and those positive integral n such that $a_n|k|^{-2/3} \leqslant \epsilon'$ we have the inequality

$$1/8\, a_n |k|^{1/3} \leqslant \operatorname{Im}(k\xi_n). \tag{D.11}$$

Further, for n such that $\epsilon' < a_n|k|^{-2/3} < 1/\epsilon'$ we deduce from (B.27) the inequalities

$$0 < C_1(\varepsilon') \leqslant |\xi_n|, \quad 0 < \psi_1(\varepsilon') \leqslant \arg\xi_n < \pi/2;$$

which for k satisfying (B.14) imply the estimates $\psi_1/2 \leqslant \arg(\tilde{\xi}_n + \theta) \leqslant \pi/2$ and hence the inequalities

$$\operatorname{Im}(k\xi_n) = |k\xi_n| \sin(\arg\xi_n + \theta) \geqslant C_2'(\varepsilon')|k| \geqslant C_1(\varepsilon') a_n |k|^{1/3}. \tag{D.12}$$

For n such that $a_n|k|^{-2/3} \geqslant 1/\epsilon'$ we readily obtain from (B.28) the inequality

$$\operatorname{Im}(k\xi_n) \geqslant C_1 a_n^{3/2} \ln^{-1}(1 + a_n^{1/2}|k|^{-1/3}). \tag{D.13}$$

On the basis of (D.11)–(D.13) we conclude that (D.9) will be valid for all $n = 1, 2, \cdots$ under an appropriate choice of the constant $C_1(\epsilon')$. Again using the monotonicity of $x^3 \ln^{-1}(1+x)$ and setting $x = x_n = a_n^{1/2}|k|^{-1/3}$, we get

$$2|k(\xi_n - \tilde{\xi}_n)| \leqslant 2a_1^{-5/6}|k|^{-1} \leqslant C_1(\varepsilon') a_1^{3/2} \ln^{-1}(1 + a_1^{1/2}|k|^{-1/3}) \leqslant \operatorname{Im}(k\tilde{\xi}_n)$$

for $|k| \geqslant \nu(\epsilon')$, where $\nu(\epsilon')$ is sufficiently great, which implies the estimate

$$\operatorname{Im}(k\xi_n) \geqslant \operatorname{Im}(k\tilde{\xi}_n) - |k(\xi_n - \tilde{\xi}_n)| \geqslant \frac{1}{2} C_1(\varepsilon') a_n^{3/2} \ln^{-1}(1 + a_n^{1/2}|k|^{-1/3}). \tag{D.14}$$

In exactly the same way we establish the estimates

$$C_1(\varepsilon')\, a_n^{1/2}(a_n + |k|^{2/3}) \ln^{-1}(1 + a_n^{1/2}|k|^{-1/3}) \\ \leqslant |k\xi_n| \leqslant C_2(\varepsilon')\, a_n^{1/2}(a_n + |k|^{2/3}) \ln^{-1}(1 + a_n^{1/2}|k|^{-1/3}). \quad \text{(D.15)}$$

Let $K_+(\beta, \kappa)$ denote a domain of the complex k plane of the form

$$K_+(\beta, \varkappa) = \{k\colon \varkappa < \operatorname{Re} k,\ |\operatorname{Im} k| < \beta(\operatorname{Re} k)^{1/3}\}, \quad \text{(D.16)}$$

where β and κ are positive numbers. Clearly it is possible to choose $\beta = \beta(a_1)$ and $\kappa = \kappa(\epsilon')$ so that $K_+(\beta, \kappa(\epsilon')) \subset N(\delta(\epsilon'), \nu(\epsilon'))$ and conditions (D.10) are satisfied for $k \in K_+(\beta, \kappa)$. Thus all of the inequalities obtained above are valid for $k \in K_+(\beta, \kappa(\epsilon'))$.

It follows from (D.14) that $\operatorname{Im}(k\xi_n) \geqslant \operatorname{Im}(k\xi_1) \geqslant C_1(\epsilon')|k|^{1/3} \to \infty$ as $k \to \infty$, and hence (C.9) and (C.10) can be used when $k \in K_+(\beta, \kappa(\epsilon'))$ to obtain the estimates

$$\left|\frac{\cos k\xi_n(\pi - |\varphi|)}{\sin k\xi_n\pi}\right| + \left|\frac{\sin k\xi_n(\pi - |\varphi|)}{\sin k\xi_n\pi}\right| \leqslant 16 e^{-\operatorname{Im}(k\xi_n)|\varphi|}. \quad \text{(D.17)}$$

Let us estimate, finally, the functions $H_{k\xi_n}(k\rho)$ and $H'_{k\xi_n}(k\rho)$, which have the asymptotic representations (A.3) and (A.4). We first establish estimates of the functions $\operatorname{Ai}(z(\xi, \rho))$ and $\operatorname{Ai}'(z(\xi, \rho))$ in these representations, where

$$z(\xi, \rho) = e^{i\frac{2\pi}{3}}(k\rho)^{2/3} w(\xi/\rho). \quad \text{(D.18)}$$

Using (A.6), we get

$$\begin{aligned} \frac{2}{3} z^{3/2}(\xi_n, \rho) &= e^{i\pi} k\left[\xi_n \ln\left(\frac{\xi_n}{\rho^2} + \sqrt{\frac{\xi_n^2}{\rho^2} - 1}\right) - \sqrt{\xi_n^2 - \rho^2}\right] \\ &= e^{i\pi} k\left[\xi_n \ln\left(\frac{\xi_n}{\rho} + \sqrt{\frac{\xi_n^2}{\rho} - 1}\right) - \sqrt{\xi_n^2 - \rho^2}\right] \\ &\quad + e^{i\pi} k \int_{\xi_n}^{\xi_n} \ln\left(\frac{\eta}{\rho} + \sqrt{\frac{\eta^2}{\rho^2} - 1}\right) d\eta = \frac{2}{3} z^{3/2}(\xi_n, \rho) + g_n(\rho, k). \end{aligned} \quad \text{(D.19)}$$

We recall that $\tilde{\xi}_n = \tilde{\zeta}(a_n|k|^{-2/3}, \theta)$, $n = 1, 2, \cdots$, where the curve $\tilde{\zeta}(x, \theta)$ satisfies the equation

$$\zeta \ln\left(\zeta + \sqrt{\zeta^2 - 1}\right) - \sqrt{\zeta^2 - 1} = \frac{2}{3} e^{i(\pi/2 - \theta)} x^{3/2} \quad \text{(D.20)}$$

for all $x \geqslant 0$ and has properties (B.6) and (B.7).

We show that the inequality

$$0 \leqslant d \operatorname{Re} 2/3\, z^{\frac{2}{3}}(\zeta(x, \theta), \rho)/dx \quad \text{(D.21)}$$

holds for $0 \leqslant x \leqslant 1/\epsilon'$ and $\rho \geqslant 1$. Differentiating the function $2z^{3/2}(\tilde{\zeta}(x, \theta), \rho)/3$ with respect to x and using (D.20), we get

$$d\,\frac{2}{3}\,z^{3/2}(\tilde{\zeta}(x,\theta),\rho)/dx = e^{i\pi}kd\tilde{\zeta}(x,\theta)/dx\,\ln\left(\tilde{\zeta}(x,\theta)/\rho+\sqrt{\tilde{\zeta}^2(x,\theta)/\rho^2-1}\right)$$
$$=-i\,|k|\,x^{1/2}\ln\left(\tilde{\zeta}(x,\theta)/\rho+\sqrt{\tilde{\zeta}^2(x,\theta)/\rho^2-1}\right)$$
$$\times\ln^{-1}\left(\tilde{\zeta}(x,\theta)+\sqrt{\tilde{\zeta}^2(x,\theta)-1}\right). \tag{D.22}$$

Suppose $0\leqslant\arg\eta\leqslant\pi/2$ and consider the function $\chi(\eta)=\operatorname{arc\,cosh}\eta=\ln(\eta+\sqrt{\eta^2-1})$. The quadrant $0\leqslant\arg\eta\leqslant\pi/2$ is mapped into the strip $0\leqslant\operatorname{Re}\chi=\chi_1\leqslant\infty$, $0\leqslant\operatorname{Im}\chi=\chi_2\leqslant\pi/2$ in the χ plane. Here $\eta=\cosh\chi$ and

$$\eta_1=\operatorname{Re}\eta=\cosh\chi_1\cos\chi_2,\quad \eta_2=\operatorname{Im}\eta=\sinh\chi_1\sin\chi_2. \tag{D.23}$$

We show that the inequality $\arg h(\eta,\rho)\geqslant 0$, where

$$h(\eta,\rho)=\chi(\eta/\rho)[\chi(\eta)]^{-1}, \tag{D.24}$$

holds for $\rho\geqslant 1$. The equalities

$$\rho_1\cosh\chi_1(\eta/\rho_1)\cos\chi_2(\eta/\rho_1)=\rho_2\cosh\chi_1(\eta/\rho_2)\cos\chi_2(\eta/\rho_2), \tag{D.25}$$

$$\rho_1\sinh\chi_1(\eta/\rho_1)\sin\chi_2(\eta/\rho_2)=\rho_2\sinh\chi_1(\eta/\rho_2)\sin\chi_2(\eta/\rho_2) \tag{D.26}$$

follow from (D.23) for $0\leqslant\arg\eta\leqslant\pi/2$ and $\rho_1,\rho_2\in(0,\infty)$. On the basis of (D.25) and (D.26) we conclude that $\chi_1(\eta/\rho_1)\leqslant\chi_1(\eta/\rho_2)$ for $\rho_1\geqslant\rho_2$. In fact, since $\chi_1(\eta/\rho_1)\geqslant 0$ and $\chi_2(\eta/\rho_2)\geqslant 0$, the inequalities $\cosh\chi_1(\eta/\rho_1)>\cosh\chi_1(\eta/\rho_2)$ and $\sinh\chi_1(\eta/\rho_1)>\sinh\chi_1(\eta/\rho_2)$ will hold if one assumes that $\chi_1(\eta/\rho_1)>\chi_1(\eta/\rho_2)$ for $\rho_1\geqslant\rho_2$ (considering $\chi_1(\eta/\rho)$ and $\chi_2(\eta/\rho)$ as functions of ρ for any fixed η). But then the inequalities

$$\cos\chi_2(\eta/\rho_1)<\cos\chi_2(\eta/\rho_2),\quad \sin\chi_2(\eta/\rho_1)<\sin\chi_2(\eta/\rho_2)$$

must hold simultaneously, which is impossible since $0\leqslant\chi_2(\eta/\rho_1)\leqslant\pi/2$ and $0\leqslant\chi_2(\eta/\rho_2)\leqslant\pi/2$.

Thus we have established that

$$d\chi_1(\eta/\rho)\,d\rho\leqslant 0 \tag{D.27}$$

for $0\leqslant\arg\eta\leqslant\pi/2$. Differentiating with respect to ρ the expression for $\tanh\chi_2(\eta/\rho)$ obtained from (D.25) and (D.26) by setting $\rho_1=1,\rho_2=\rho$ and taking the ratio of these equalities, we get

$$d\chi_2(\eta/\rho)/d\rho=-\tanh\chi_1(\eta)\tan\chi_2(\eta)\cos^2\chi_2(\eta/\rho)\sinh^{-2}\chi_1(\eta/\rho)\,d\chi_1(\eta/\rho)/d\rho\geqslant 0. \tag{D.28}$$

It follows from (D.27) and (D.28) that

$$0\leqslant\arg\chi(\eta)\leqslant\arg\chi(\eta/\rho)\leqslant\pi/2$$

for $\rho\geqslant 1$. Consequently

$$0\leqslant\arg h(\eta,\rho)\leqslant\pi/2,\qquad \operatorname{Im}h(\eta,\rho)\geqslant 0\quad(\rho\geqslant 1).$$

But

$$\operatorname{Im} h(\zeta(x,\theta),\rho)=\operatorname{Im}\left\{\ln\left(\zeta/\rho+\sqrt{\zeta^2/\rho^2-1}\right)\ln^{-1}\left(\zeta+\sqrt{\zeta^2-1}\right)\right\},$$

and hence

$$0\leqslant|k|\,x^{1/2}\operatorname{Im} h(\zeta(x,\theta),\rho)=d\operatorname{Re}\frac{2}{3}z^{3/2}(\zeta(x,\theta),\rho)/dx$$

for $0\leqslant x\leqslant 1/\epsilon'$, since then $0\leqslant\arg\tilde{\zeta}(x,\theta)<\pi/2$. On the basis of the above inequality we conclude that when $0\leqslant x\leqslant 1/\epsilon'$ and $\rho\geqslant 1$

$$\operatorname{Re}{}^2/_3\,z^{3/2}(1,\rho)\leqslant\operatorname{Re}{}^2/_3\,z^{3/2}(\zeta(x,\theta),\rho),$$

i.e.

$$-\operatorname{Im} k\left(\sqrt{\rho^2-1}-\arccos 1/\rho\right)\leqslant\operatorname{Re}\frac{2}{3}z^{3/2}(\zeta(x,\theta),\rho). \tag{D.29}$$

We show in addition that when $0\leqslant x\leqslant 1/\epsilon'$ and $\rho\geqslant 1$

$$\operatorname{Re}\frac{2}{3}z^{3/2}(\zeta(x,\theta),\rho)\leqslant\operatorname{Im}(k\zeta(x,\theta))\arccos 1/\rho-\operatorname{Im} k\sqrt{\rho^2-1}. \tag{D.30}$$

The function $2z^{3/2}(\tilde{\zeta}(x,\theta),\rho)/3$ is represented in the form

$$\begin{aligned}{}^2/_3\,z^{3/2}(\zeta(x,\theta),\rho)&=-k\left[\zeta\ln\left(\zeta/\rho+\sqrt{\zeta^2/\rho^2-1}\right)-\sqrt{\zeta^2-\rho^2}\right]\\&=-k\left[\zeta\ln\left(1/\rho+\sqrt{1/\rho^2-1}\right)+\zeta\int_1^{\zeta}d\eta/\sqrt{\eta^2-\rho^2}-\sqrt{1-\rho^2}\right.\\&\left.-\int_1^{\zeta}\eta\,d\eta/\sqrt{\eta^2-\rho^2}\right]=-ik\zeta\arccos 1/\rho+ik\sqrt{\rho^2-1}+f(x,k,\rho),\end{aligned} \tag{D.31}$$

where

$$f(x,k,\rho)=-k\int_1^{\zeta(x,\theta)}\frac{(\zeta(x,\theta)-\eta)}{\sqrt{\eta^2-\rho^2}}\,d\eta.$$

We calculate the derivative df/dx:

$$\begin{aligned}\frac{df}{dx}&=-k\frac{d\zeta}{dx}\int_1^{\zeta(x,\theta)}\frac{d\eta}{\sqrt{\eta^2-\rho^2}}=-ix^{1/2}|k|\ln^{-1}\left(\zeta+\sqrt{\zeta^2-1}\right)\\&\times\int_1^{\zeta(x,\theta)}\frac{d\eta}{\sqrt{\eta^2-\rho^2}}=-ix^{1/2}|k|\int_1^{\zeta(x,\theta)}\frac{d\eta}{\sqrt{\eta^2-\rho^2}}\left(\int_1^{\zeta(x,\theta)}\frac{d\eta}{\sqrt{\eta^2-1}}\right)^{-1}.\end{aligned} \tag{D.32}$$

We will show that the conditions $0\leqslant x\leqslant 1/\epsilon'$ and $\rho\geqslant 1$ imply $d\operatorname{Re} f/dx\leqslant 0$. We can then conclude that $\operatorname{Re} f(x,k,\rho)\leqslant\operatorname{Re} f(0,k,\rho)=0$, since $\tilde{\zeta}(0,\theta)\equiv 1$, and the desired inequality (D.30) will follow directly from (D.31).

Thus let us establish the inequality $d\operatorname{Re} f/dx\leqslant 0$. For each point $\tilde{\zeta}(x,\theta)$ we determine the curve of integration $\eta=\eta(\tilde{\zeta},y)$ from the equation

$$\eta(\zeta,y)=\cosh(y\operatorname{arc\,cosh}\zeta),\qquad 0\leqslant y\leqslant 1. \tag{D.33}$$

The expression (D.32) can then be rewritten in the form

$$\frac{df}{dx} = -ix^{1/2}|k|\int_0^1 (\eta^2(\tilde{\zeta}, y) - 1)^{1/2}(\eta^2(\tilde{\zeta}, y) - \rho^2)^{-1/2}\,dy. \tag{D.34}$$

The mapping $\chi = \operatorname{arc\,cosh} \tilde{\zeta}$ takes the quadrant $0 \leqslant \arg \tilde{\zeta} \leqslant \pi/2$ into the strip $0 \leqslant \operatorname{Re} \chi$, $0 \leqslant \operatorname{Im} \chi \leqslant \pi/2$; consequently when $0 \leqslant x \leqslant 1/\epsilon'$ and $0 \leqslant y \leqslant 1$ the points $y \operatorname{arc\,cosh} \tilde{\zeta}$ lie in this strip while the points $\cosh(y \operatorname{arc\,cos} \tilde{\zeta}) = \eta(\tilde{\zeta}, y)$ lie in the quadrant $0 \leqslant \arg \eta \leqslant \pi/2$. But then $0 \leqslant \arg \eta^2(\tilde{\zeta}, y) \leqslant \pi$, and hence when $0 \leqslant y \leqslant 1$ and $\rho \geqslant 1$ we have the inequality

$$-\pi/2 \leqslant \arg\{(\eta^2(\tilde{\zeta}, y) - 1)^{1/2}(\eta^2(\tilde{\zeta}, y) - \rho^2)^{-1/2}\} \leqslant 0,$$

which together with (D.34) implies $d \operatorname{Re} f/dx \leqslant 0$.

Since $\sqrt{\rho^2 - 1} \geqslant \operatorname{arc\,cos} 1/\rho$ for $\rho \geqslant 1$ and $\operatorname{Im} k\tilde{\zeta}(x, \theta) \geqslant 0$ for $0 \leqslant x \leqslant \infty$, it follows from (D.19) and (D.20) that

$$|\operatorname{Re} 2/3\, z^{3/2}(\tilde{\zeta}(x, \theta), \rho)| \leqslant \operatorname{Im}(k\tilde{\zeta}(x, \theta)) \operatorname{arc\,cos} 1/\rho + |\operatorname{Im} k|\sqrt{\rho^2 - 1} \tag{D.35}$$

for $0 \leqslant x \leqslant 1/\epsilon'$ and $\rho \geqslant 1$.

Now suppose $x \geqslant 1/\epsilon'$. Since $\operatorname{Re} 2z^{3/2}(\tilde{\zeta}(x, \theta), 1)/3 \equiv 0$ (see (D.19) and (D.20)), we have the relation

$$\operatorname{Re} \frac{2}{3} z^{3/2}(\zeta, \rho) = \operatorname{Re}\left[\frac{2}{3} z^{3/2}(\zeta, \rho) - \frac{2}{3} z^{3/2}(\zeta, 1)\right]$$

$$= \operatorname{Re} \int_1^\rho \frac{d}{d\rho_1}\left[\frac{2}{3} z^{3/2}(\zeta, \rho_1)\right] d\rho_1 = \operatorname{Re}\left[k \int_1^\rho \rho_1^{-1} \sqrt{\zeta^2 - \rho_1^2}\, d\rho_1\right]. \tag{D.36}$$

On the basis of (B.7) we can indicate for any $\rho \in [1, \infty)$ a number $\epsilon' = \epsilon'(\rho)$ (and hence numbers $\delta(\epsilon') = \delta(\rho)$ and $\nu(\epsilon') = \nu(\rho)$) such that the following inequalities will hold for $x \geqslant 1/\epsilon'$ and $k \in N(\delta, \nu)$:

$$\rho|\zeta(x, \theta)|^{-1} \leqslant 1/2, \qquad (\rho^2 - 1)|\zeta(x, \theta)|^{-1} \leqslant 1/4 \operatorname{arc\,cos} 1/\rho.$$

Then

$$\sqrt{\zeta^2(x, \theta) - \rho_1^2} = \zeta(x, \theta)[1 + \rho_1^2\zeta^{-2}(x, \theta) O(1)]$$

and (D.36) implies the inequality

$$|\operatorname{Re} {}^2/_3\, z^{3/2}(\zeta(x, \theta), \rho)| \leqslant \operatorname{Re}(k\zeta(x, \theta)) \int_1^\rho \rho_1^{-1} d\rho_1$$

$$+ |k\zeta(x, \theta)| \left|\int_1^\rho |\zeta(x, \theta)|^{-2} \rho_1 d\rho_1\right| \leqslant \operatorname{Re}(k\zeta(x, \theta)) \ln \rho + {}^1/_4 |k\zeta(x, \theta)| \operatorname{arc\,cos} 1/\rho. \tag{D.37}$$

It follows from (B.7) that we can choose for any $\sigma > 0$ a number $\epsilon' = \epsilon'(\sigma)$ so small that in addition to (D.36) the following inequality is satisfied for $x \geqslant 1/\epsilon'$ and $k \in N(\delta(\rho, \sigma), \nu(\rho, \sigma))$:

$$\operatorname{Re}(k\zeta(x, \theta)) \leqslant \sigma \operatorname{Im}(k\zeta(x, \theta)).$$

We then obtain the estimate

$$|\operatorname{Re} \tfrac{2}{3} z^{3/2}(\zeta(x, \theta), \rho)| \leqslant \operatorname{Im}(k\xi(x, \theta)) \arccos 1/\rho\, [\sigma\rho^2 + 1/4(1+\sigma)]$$

by taking into account the inequality $\ln \rho \leqslant \rho^2 \arccos 1/\rho$, $1 \leqslant \rho$. Setting $\sigma = \rho^2/2$ for $k \in N(\delta(\rho), \nu(\rho))$ and $n = 1, 2, \cdots$, we deduce from (D.35) and (D.37) the estimate

$$|\operatorname{Re} \tfrac{2}{3} z^{3/2}(\xi_n, \rho)| \leqslant \operatorname{Im}(k\xi_n) \arccos 1/\rho + |\operatorname{Im} k| \sqrt{\rho^2 - 1},$$

which implies, on the basis of (D.19) and (B.30),

$$|\operatorname{Re} \tfrac{2}{3} z^{3/2}(\xi_n, \rho)| \leqslant \operatorname{Im}(k\xi_n) \arccos 1/\rho + |\operatorname{Im} k| \sqrt{\rho^2 - 1} + |k|^{-1/3} a_n^{-5/6} [1 + 4\pi + 2 \ln(1 + 2|\xi_n| + 2|k|^{-4/3} a_n^{-5/6}].$$

Using (D.15) and the inequality $\ln^{-1}(1+x) \leqslant x^{-1}(1+x)$ $(x \geqslant 0)$, we obtain

$$|\xi_n| \leqslant C(\varepsilon')|k|^{-1/3}(a_n + |k|^{2/3})(a_n^{1/2} + |k|^{1/3})^{-1} \leqslant C(\varepsilon')|k|^{-1/3}(a_n^{1/2} + |k|^{1/3}) \leqslant C(\varepsilon')(1 + a_n^{1/2}),$$

so that

$$\ln(1 + |\xi_n| + 2|k|^{-4/3} a_n^{-5/6}) \leqslant C(\rho) \ln(1 + a_n^{1/2})$$

(recall that $\epsilon' = \epsilon'(\rho)$) and when $k \in N(\delta(\rho), \nu(\rho))$, where $\nu(\rho)$ is sufficiently large, the following estimate holds:

$$|\operatorname{Re} \tfrac{2}{3} z^{3/2}(\xi_n, \rho)| \leqslant \operatorname{Im}(k\xi_n) \arccos 1/\rho + |\operatorname{Im} k| \sqrt{\rho^2 - 1} + 1. \tag{D.38}$$

Using the well-known inequalities

$$|d^p \operatorname{Ai}(z)/dz^p| \leqslant C_p (1 + |z|)^{p/2 - 1/4} e^{|\operatorname{Re} 2/3 z^{3/2}|}, \quad p = 0, 1,$$

which are valid for any z, as well as the representations (A.3) and (A.4), we obtain the estimates

$$|H_{k\xi_n}(k\rho)| \leqslant C |k\rho|^{-1/3} |w(\xi_n/\rho)|^{1/4} |\xi_n^2/\rho^2 - 1|^{-1/4} \times (1 + |k|^{2/3}\rho^{2/3} |w(\xi_n/\rho)|)^{-1/4} \exp\{\operatorname{Im}(k\xi_n) \arccos 1/\rho + |\operatorname{Im} k| \sqrt{\rho^2 - 1}\}, \tag{D.39}$$

$$|H'_{k\xi_n}(k\rho)| \leqslant C |k\rho|^{-2/3} |w(\xi_n/\rho)|^{-1/4} |\xi_n^2/\rho^2 - 1|^{1/4} \times (1 + |k|^{2/3}\rho^{2/3} |w(\xi_n/\rho)|)^{1/4} \exp\{\operatorname{Im}(k\xi_n) \arccos 1/\rho + |\operatorname{Im} k| \sqrt{\rho^2 - 1}\}. \tag{D.40}$$

The inequalities

$$C_1 |k\rho|^{-1/6} (1 + |\zeta|)^{-1/2} \leqslant |w(\zeta)|^{1/4} |\zeta^2 - 1|^{1/4} (1 + |k\rho|^{2/3} |w(\zeta)|)^{-1/4} \leqslant C_2 (1 + |\zeta|)^{-1/2}$$

follow from (A.17) and (A.18) for $\zeta \in \mathcal{D}_\zeta$ and together with (D.39) and (D.40) imply the estimates

$$|H_{k\xi_n}(k\rho)| \leqslant C |k\rho|^{1/6} |k\xi_n|^{-1/2} \exp\{\operatorname{Im}(k\xi_n) \arccos 1/\rho + |\operatorname{Im} k| \sqrt{\rho^2 - 1}\}, \tag{D.41}$$

$$|H'_{k\xi_n}(k\rho)| \leqslant C |k|^{-1} \rho^{-1/2} |k\xi_n|^{1/2} \exp\{\operatorname{Im}(k\xi_n) \arccos 1/\rho + |\operatorname{Im} k| \sqrt{\rho^2 - 1}\}, \tag{D.42}$$

which are valid for any $n = 1, 2, \cdots, \rho \geqslant 1$ and $k \in K(\beta, \kappa(\rho))$.

On the basis of (D.8), (D.9), (D.15), (D.17) and (D.41) we conclude that the series (C.27) is majorized by the series

$$C(\rho)\exp\{|\operatorname{Im} k|\sqrt{\rho^2-1}\}|k|^{1/6}\sum_{n=1}^{\infty} a_n^{-3/4}(a_n+|k|^{2/3})^{-3/4}$$
$$\times \exp\{-C_1(|\varphi|-\arccos 1/\rho)a_n^{3/2}\ln^{-1}(1+a_n^{1/2}|k|^{-1/3})\}, \quad \text{(D.43)}$$

in which the dependence of $C(\rho)$ is such that $C(\rho_1) \geqslant C(\rho_2)$ for $\rho_1 \geqslant \rho_2$. The representations (A.10) of the numbers a_n imply the estimates $C'n^{2/3} \leqslant a_n \leqslant C''n^{2/3}$. Therefore, by introducing the variable $x(n) = n^{1/3}|k|^{-1/3}$ for $n \geqslant 1$, we can majorize the series (D.43) by the integral

$$C_2(\rho)|k|^{1/6}\exp\{|\operatorname{Im} k|\sqrt{\rho^2-1}\}$$
$$\times \int_{|k|^{-1/3}}^{\infty} x^{1/2}(1+x^2)^{-3/4}\exp\{-C_3|k|(|\varphi|-\arccos 1/\rho)x^3\ln^{-1}(1+x)\}\,dx$$
$$\leqslant C_2(\rho)|k|^{1/6}\exp\{|\operatorname{Im} k|\sqrt{\rho^2-1}\}$$
$$\times\left[\int_{|k|^{-1/3}}^{1} x^{1/2}\exp\{-C_3|k|(|\varphi|-\arccos 1/\rho)x^2\}\,dx\right.$$
$$\left.+\int_1^{\infty} x^{-1}\exp\{-C_3|k|(|\varphi|-\arccos 1/\rho)x^3\ln^{-1}(1+x)\}\,dx\right]. \quad \text{(D.44)}$$

Let us estimate the last integral in (D.44). To this end we note that $y(x) = x^3\ln^{-1}(1+x)$ is monotonically increasing for $x \geqslant 0$ and that

$$dy = [3x^2\ln^{-1}(1+x) - x^3(1+x)^{-1}\ln^{-2}(1+x)]\,dx$$
$$\geqslant 2x^2\ln^{-1}(1+x)\,dx \geqslant 2x^{-1}y\,dx$$

by virtue of the inequality $x(1+x)^{-1}\ln^{-1}(1+x) \leqslant 1$ $(x \geqslant 0)$. Consequently

$$\int_1^{\infty} x^{-1}\exp\{-C_3|k|(|\varphi|-\arccos 1/\rho)x^3\ln^{-1}(1+x)\}\,dx$$
$$\leqslant \int_1^{\infty} y^{-1}\exp\{-C_3|k|(|\varphi|-\arccos 1/\rho)y\}\,dy$$
$$\leqslant C_4\left|\ln|k|\left(|\varphi|-\arccos\frac{1}{\rho}\right)\right|\exp\left\{-\frac{C_3}{2}|k|\left(|\varphi|-\arccos\frac{1}{\rho}\right)\right\}. \quad \text{(D.45)}$$

On the basis of (D.45) the integrals in (D.44) converge for any $\phi \neq 0$ and $\rho \in [1, \infty)$ such that $|\phi| - \arccos 1/\rho > 0$, and for $k \in K(\beta, \kappa(\rho))$. Therefore the series (C.27) is absolutely convergent and its sum $\Phi(\rho, \phi, k)$ satisfies the estimate

$$|\Phi| \leqslant C(\rho)\exp\{|\operatorname{Im} k|\sqrt{\rho^2-1}\}|k|^{1/6}$$
$$\times\Bigg[\int\limits_{|k|^{-1/3}}^{1} \tau^{1/2}\exp\Big\{-B|k|\Big(|\varphi|-\arccos\frac{1}{\rho}\Big)\tau^2\Big\}\,d\tau$$
$$+\Big|\ln|k|\Big(|\varphi|-\arccos\frac{1}{\rho}\Big)\Big|\exp\Big\{-B|k|\Big(|\varphi|-\arccos\frac{1}{\rho}\Big)\Big\}. \qquad \text{(D.46)}$$

Here B is a certain positive number, while the dependence of the constants $C(\rho)$ and $\kappa(\rho)$ is such that $C(\rho_1) \geqslant C(\rho_2)$ and $\kappa(\rho_1) \geqslant \kappa(\rho_2)$ for $\rho_1 \geqslant \rho_2$. It follows from estimate (D.46) that we can indicate for any $(\rho_0, \phi_0) \in \omega$ and $k_0 \in K(\beta, \kappa(\rho+1))$ neighborhoods ω_0 $(\bar\omega_0 \in \omega)$ and K_0 $(\bar K_0 \in K)$ of the points (ρ_0, ϕ_0) and k_0 respectively such that for all $(\rho, \phi) \in \bar\omega_0$ and $k \in \bar K_0$ the series (C.27) converges absolutely and uniformly, i.e. the function Φ is continuous in ω_R (for any $R > 0$) and admits an analytic continuation into $K(\beta, \kappa_R)$ with estimate (4.2).

With the use of analogous arguments we conclude that the series (C.29) is majorized by the following series and integral:

$$C(\rho)\exp\{|\operatorname{Im} k|\sqrt{\rho^2-1}\}|k|^{1/6}\sum_{n=1}^{\infty} a_n^{-1/4}(a_n+|k|^{2/3})^{1/4}\ln^{-1}(1+a_n^{1/2}|k|^{-1/3})$$
$$\times\exp\Big\{-C_1\Big(|\varphi|-\arccos\frac{1}{\rho}\Big)a_n^{3/2}\ln^{-1}(1+a_n^{1/2}|k|^{-1/3})\Big\}$$
$$\leqslant C_2(\rho)\exp\{|\operatorname{Im} k|\sqrt{\rho^2-1}\}|k|^{7/6}$$
$$\times\Bigg[\int\limits_{|k|^{-1/3}}^{1} x^{1/2}\exp\Big\{-C_3|k|\Big(|\varphi|-\arccos\frac{1}{\rho}\Big)x^2\Big\}\,dx$$
$$+\int\limits_1^{\infty} x^2\ln^{-1}(1+x)\exp\Big\{-C_3|k|\Big(|\varphi|-\arccos\frac{1}{\rho}\Big)x^3\ln^{-1}(1+x)\Big\}\,dx\Bigg]$$
$$\leqslant C_2(\rho)\exp\{|\operatorname{Im} k|\sqrt{\rho^2-1}\}|k|^{7/6}\Bigg[\Big(|k|\Big(|\varphi|-\arccos\frac{1}{\rho}\Big)\Big)^{-1}$$
$$\times\exp\Big\{-\frac{C_3}{2}|k|^{1/3}\Big(|\varphi|-\arccos\frac{1}{\rho}\Big)\Big\}\max_{0\leqslant y}(ye^{-C_3y})$$
$$\times\int\limits_{|k|^{-1/3}}^{1} x^{-3/2}dx+\frac{1}{2}\int\limits_1^{\infty}\exp\Big\{-C_3|k|\Big(|\varphi|-\arccos\frac{1}{\rho}\Big)y\Big\}\,dy\Bigg]$$
$$\leqslant C_4(\rho)|k|^{1/3}\Big(|\varphi|-\arccos\frac{1}{\rho}\Big)^{-1}\exp\Big\{-\frac{C_3}{2}|k|^{1/3}\Big(|\varphi|-\arccos\frac{1}{\rho}\Big)\Big\}.$$

It follows that for any $(\rho_0, \phi_0) \in \omega$ and $k_0 \in K(\beta, \kappa(\rho+1))$ the series (C.29) converges absolutely and uniformly for all $(\rho, \phi) \in \bar\omega_0$ and $k \in \bar K_0$. Consequently the function Φ is differentiable with respect to ϕ in ω, and its derivative $\partial\Phi/\partial\phi$ is defined by the series (C.29), is continuous in ω_R, is analytic in $K(\beta, \kappa_R)$ and satisfies inequality (4.3).

Let us show that in ω the function Φ has a continuous partial derivative $\partial\Phi/\partial\rho$, and let us estimate it. With the use of the relations

$$H'_{k\xi_n}(k\rho)=0, \qquad H''_{k\xi_n}(k\rho)+\frac{1}{k\rho}H'_{k\xi_n}(k\rho)+\left(1-\frac{\xi_n^2}{\rho^2}\right)H_{k\xi_n}(k\rho)=0$$

we obtain the inequality

$$|H'_{k\xi_n}(k\rho)|=|k|\left|\int_1^\rho H''_{k\xi_n}(k\rho_1)\,d\rho_1\right|\leqslant\int_1^\rho\frac{1}{\rho_1}|H'_{k\xi_n}(k\rho_1)|\,d\rho_1$$

$$+\,|k|\int_1^\rho|1-\xi_n^2/\rho^2|\,|H_{k\xi_n}(k\rho_1)|\,d\rho_1\leqslant(\rho-1)\max_{1\leqslant\rho_1\leqslant\rho}|H'_{k\xi_n}(k\rho_1)|$$

$$+\,|k|\,[(\rho^2-1)^2+|\xi_n^2-1|\,(\rho-1)]\max_{1\leqslant\rho_1\leqslant\rho}|H_{k\xi_n}(k\rho_1)|.$$

From (D.5), (D.8), (D.9), (D.15), (D.17), (D.41) and (D.42) we conclude that the series (C.28) is majorized by the sum of the following three series:

$$1)\ C(\rho)(\rho-1)^2\exp\left\{|\operatorname{Im}k|\sqrt{\rho^2-1}\right\}|k|^{13/6}\sum_{n=1}^{\infty}a_n^{-3/4}(a_n+|k|^{2/3})^{-3/4}$$

$$\times\exp\left\{-C_1\left(|\varphi|-\arccos\frac{1}{\rho}\right)a_n^{3/2}\ln^{-1}(1+a_n^{1/2}|k|^{-1/3})\right\}$$

$$\leqslant C_2(\rho)(\rho-1)^2\exp\left\{|\operatorname{Im}k|\sqrt{\rho^2-1}\right\}|k|^{17/6}\left[\int_{|k|^{-1/3}}^{1}\tau^{1/2}\exp\left\{-2B|k|\right.\right.$$

$$\left.\times\left(|\varphi|-\arccos\frac{1}{\rho}\right)\tau^2\right\}d\tau+\left|\ln|k|\left(|\varphi|-\arccos\frac{1}{\rho}\right)\right|$$

$$\left.\times\exp\left\{-2B|k|\left(|\varphi|-\arccos\frac{1}{\rho}\right\}\right]\right.$$

(since this series divided by $(\rho-1)^2|k|^2$ coincides with (D.43))

$$\leqslant C_3(\rho)(\rho-1)^2\exp\left\{|\operatorname{Im}k|\sqrt{\rho^2-1}-B|k|^{1/3}\left(|\varphi|-\arccos\frac{1}{\rho}\right)\right\}|k|^{13/6}$$

$$\times\left[\left(|k|\left(|\varphi|-\arccos\frac{1}{\rho}\right)\right)^{-4}\max_{0\leqslant y}(y^4e^{-By})\int_{|k|^{-1/3}}^{1}\tau^{-15/2}\,d\tau\right.$$

$$\left.+\left(|k|\left(|\varphi|-\arccos\frac{1}{\rho}\right)\right)^{-4}\max_{0\leqslant y}(y^4|\ln y|\,e^{-By})\right]\leqslant C(\rho)|k|^{1/3}$$

$$\times(\rho-1)^2\left(|\varphi|-\arccos\frac{1}{\rho}\right)^{-4}\exp\left\{|\operatorname{Im}k|\sqrt{\rho^2-1}\right.$$

$$\left.-B|k|^{1/3}\left(|\varphi|-\arccos\frac{1}{\rho}\right)\right\}; \tag{D.47}$$

$$2)\ C(\rho)(\rho-1)\exp\left\{|\operatorname{Im}k|\sqrt{\rho^2-1}\right\}|k|^{1/6}\sum_{n=1}^{\infty}a_n^{5/4}(a_n+|k|^{2/3})^{1/4}$$

$$\times\ln^{-2}(1+a_n^{1/2}|k|^{-1/3})\exp\left\{-C_1\left(|\varphi|-\arccos\frac{1}{\rho}\right)a_n^{3/2}\ln^{-1}(1+a_n^{1/2}|k|^{-1/3})\right\}\leqslant$$

$$\leqslant C_2(\rho)(\rho-1)|k|^{3/2}\exp\{\operatorname{Im} k\,|\sqrt{\rho^2-1}\}\int\limits_{|k|^{-1/3}}^{\infty} x^{5/2}(1+x^2)^{1/4}\ln^{-2}(1+x)$$

$$\times\exp\left\{-C_3|k|\left(|\varphi|-\arccos\frac{1}{\rho}\right)x^3\ln^{-1}(1+x)\right\}dx$$

$$\leqslant C_4(\rho)(\rho-1)|k|^{3/2}\exp\{|\operatorname{Im} k|\sqrt{\rho^2-1}\}$$

$$\times\left[\int\limits_{|k|^{-1/3}}^{1} x^{1/2}\exp\left\{-C_3|k|\left(|\varphi|-\arccos\frac{1}{\rho}\right)x^2\right\}dx\right.$$

$$\left.+\int\limits_1^{\infty} y^{1/3}\exp\left\{-C_3|k|\left(|\varphi|-\arccos\frac{1}{\rho}\right)y\right\}dy\right]\leqslant C_5(\rho)(\rho-1)|k|^{3/2}$$

$$\times\exp\{|\operatorname{Im} k|\sqrt{\rho^2-1}\}\left[\left(|k|\left(|\varphi|-\arccos\frac{1}{\rho}\right)\right)^{-2}\exp\left\{-\frac{1}{2}C_3|k|^{1/3}\right.\right.$$

$$\left.\times\left(|\varphi|-\arccos\frac{1}{\rho}\right)\right\}\int\limits_{|k|^{-1/3}}^{1} x^{-7/2}dx+\left(|k|\left(|\varphi|-\arccos\frac{1}{\rho}\right)\right)^{-4/3}$$

$$\left.\times\exp\left\{-\frac{1}{2}C_3|k|\left(|\varphi|-\arccos\frac{1}{\rho}\right)\right\}\right]$$

$$\leqslant C_6(\rho)|k|^{1/3}(\rho-1)\left(|\varphi|-\arccos\frac{1}{\rho}\right)^{-2}$$

$$\times\exp\left\{|\operatorname{Im} k|\sqrt{\rho^2-1}-B|k|^{1/3}\left(|\varphi|-\arccos\frac{1}{\rho}\right)\right\}; \tag{D.48}$$

3) $$C(\rho)(\rho-1)\exp\{|\operatorname{Im} k|\sqrt{\rho^2-1}\}\sum_{n=1}^{\infty} a_n^{-1/4}(a_n+|k|)^{1/4}\ln^{-1}(1+a_n^{1/2}|k|^{-1/3})$$

$$\times\exp\left\{-C_1\left(|\varphi|-\arccos\frac{1}{\rho}\right)a_n^{3/2}\ln^{-1}(1+a_n^{1/2}|k|^{-1/3})\right\}$$

$$\leqslant C_2(\rho)(\rho-1)|k|\exp\{|\operatorname{Im} k|\sqrt{\rho^2-1}\}\times$$

$$\times\left[\int\limits_{|k|^{-1/3}}^{1} x^{1/2}\exp\left\{-C_3|k|\left(|\varphi|-\arccos\frac{1}{\rho}\right)x^2\right\}dx\right.$$

$$\left.+\left|\ln|k|\left(|\varphi|-\arccos\frac{1}{\rho}\right)\right|\exp\left\{-\frac{1}{2}C_3|k|\left(|\varphi|-\arccos\frac{1}{\rho}\right)\right\}\right]$$

$$\leqslant C_4(\rho)|k|^{-1/6}(\rho-1)\left(|\varphi|-\arccos\frac{1}{\rho}\right)^{-2}$$

$$\times\exp\left\{|\operatorname{Im} k|\sqrt{\rho^2-1}-B|k|^{1/3}\left(|\varphi|-\arccos\frac{1}{\rho}\right)\right\}. \tag{D.49}$$

On the basis of (D.47)–(D.49) we conclude that for any $(\rho_0, \phi_0)\in\omega$ and $k_0\in K(\beta, \kappa(\rho+1))$ the series (C.28) converges absolutely and uniformly with respect to $(\rho, \phi)\in\bar\omega_0$ and $k\in\bar K_0$, where ω_0 and K_0 are sufficiently small neighborhoods of the points (ρ_0, ϕ_0) and k_0 respectively. Consequently (on the basis of (D.47)–(D.49)) in ω the function Φ has a continuous derivative with respect to ρ that satisfies (4.4).

BIBLIOGRAPHY

1. P. D. Lax, C. S. Morawetz and R. S. Phillips, *Exponential decay of solutions of the wave equation in the exterior of a star-shaped obstacle*, Comm. Pure Appl. Math. **16** (1963), 477–486. MR **27** #5033.

2. V. P. Mihaĭlov, *On the principle of limiting amplitude*, Dokl. Akad. Nauk SSSR **159** (1964), 750–752 = Soviet Math. Dokl. **5** (1964), 1599–1602. MR **30** #352.

3. C. S. Morawetz, *The limiting amplitude principle*, Comm. Pure Appl. Math. **15** (1962), 349–361. MR **27** #1696.

4. L. A. Muraveĭ, *The decrease of solutions of the second exterior boundary-value problem for the wave equation with two space variables*, Dokl. Akad. Nauk SSSR **193** (1970), 996–999 = Soviet Math. Dokl. **11** (1970), 1067–1071. MR **42** #3400.

5. ———, *Asymptotic behavior of the solutions of the second exterior boundary-value problem for the two-dimensional wave equation*, Differencial'nye Uravnenija **6** (1970), 2248–2262 = Differential Equations **6** (1970), 1709–1720. MR **43** #2366.

6. ———, *Stabilization of solutions of the second exterior boundary-value problem for wave equations with two space variables*, Candidate's Dissertation, Steklov Institute of Math., Moscow, 1971. (Russian)

7. V. M. Babič, *On the asymptotics of the Green's functions of certain wave problems*, Dokl. Akad. Nauk SSSR **199** (1971), 743–746 = Soviet Math. Dokl. **12** (1971), 1138–1142. MR **45** #3957.

8. F. Ursell, *On the short-wave asymptotic theory of the wave equation* $(\nabla^2 + k^2)\phi = 0$, Proc. Cambridge Philos. Soc. **53** (1957), 115–133. MR **18**, 848.

9. V. M. Babič and I. V. Olimpiev, *Estimates of the wave field in the umbral region for diffraction of a cylindrical wave by a bounded convex cylinder*, Third All-Union Sympos. on Wave Diffraction (Tbilisi, 1964), Abstracts of Reports, "Nauka", Moscow, 1964, pp. 104–105. (Russian)

10. R. Grimshaw, *High-frequency scattering by finite convex regions*, Comm. Pure Appl. Math. **19** (1966), 167–198. MR **33** #2107.

11. V. M. Babič, *Analytic continuation to the second sheet of the resolvent of exterior problems for the Laplace operator*, Teor. Funkciĭ Funkcional. Anal. i Priložen. Vyp. 3 (1966), 151–157. (Russian) MR **34** #6341.

12. R. C. MacCamy, *Low frequency acoustic oscillations*, Quart. Appl. Math. **23** (1965), 247–255. MR **32** #6823.

13. S. L. Sobolev, *Applications of functional analysis in mathematical physics*, Izdat. Leningrad. Gos. Univ., Leningrad, 1950; English transl., Transl. Math. Monographs, vol. 7, Amer. Math. Soc., Providence, R. I., 1963. MR **14**, 565; **29** #2624.

14. O. A. Ladyženskaja, *The mixed problem for a hyperbolic equation*, GITTL, Moscow, 1953. (Russian) MR **17**, 160.

15. I. N. Vekua, *On metaharmonic functions*, Trudy Tbiliss. Mat. Inst. **12** (1943), 105–174. (Russian) MR **6**, 154.

16. I. S. Gradšteĭn and I. M. Ryžik, *Tables of integrals, series and products*, 4th ed., Fizmatgiz, Moscow, 1963; English transl., Academic Press, New York, 1965. MR **28** #5198; **33** #5952.

17. S. N. Bernšteĭn, *Some a priori estimates in Dirichlet's generalized problem*, Dokl. Akad. Nauk SSSR **124** (1959), 735–738. (Russian) MR **21** #196.

18. O. A. Ladyženskaja and N. N. Ural'ceva, *Linear and quasi-linear equations of elliptic type*, "Nauka", Moscow, 1964; English transl., Academic Press, New York, 1968. MR **35** #1955; **39** #5941.

19. I. G. Petrovskiĭ, *Partial differential equations*, 3rd ed., Fizmatgiz, Moscow, 1961; English transl., Saunders, Philadelphia, Pa., 1967. MR **25** #2308; **35** #1906.

20. J. D. Tamarkin, *On Fredholm's integral equations whose kernels are analytic in a parameter*, Ann. of Math. (2) **28** (1927), 127–152.

21. F. W. J. Olver, *The asymptotic solution of linear differential equations of the second order for large values of a parameter*, Philos. Trans. Roy. Soc. London Ser. A **247** (1954), 307–327. MR **16**, 695.

Trudy Mat. Inst. Steklov.
126 (1973)

Proc. Steklov Inst. Math.
126 (1973)

ON THE STABILIZATION OF THE SOLUTION OF A BOUNDARY VALUE PROBLEM FOR A PARABOLIC EQUATION

UDC 517.947.4

D. V. PROKA

ABSTRACT. In this paper a necessary and sufficient condition is given on the boundary operator under which the solution of a boundary value problem for the equation $\partial u/\partial t = (-1)^p \partial^{2p} u/\partial x^{2p}$ for $x > 0$ is stabilized at $t \to \infty$ for any stabilizing boundary functions.

Bibliography: 5 items.

§1. Statement of the problem. Formulation of the results

Let $u(x, t)$, $x \geqslant 0$, $t \geqslant 0$, denote the bounded for $x \geqslant 0$, $0 \leqslant t \leqslant T$, $T > 0$, solution of the parabolic in the sense of I. G. Petrovskiĭ [1] equation

$$Lu = \frac{\partial u(x, t)}{\partial t} + (-1)^p \frac{\partial^{2p} u(x, t)}{\partial x^{2p}} = 0, \qquad x > 0, \qquad t > 0 \tag{1.1}$$

satisfying the initial condition

$$u(x, 0) = 0, \qquad x > 0 \tag{1.2}$$

and the boundary conditions

$$b^{(j)}(\partial/\partial x)\, u(x, t)\big|_{x=0} = \varphi_j(t), \qquad j = 1, 2, \ldots, p, \qquad t > 0, \tag{1.3}$$

where the $b^{(j)}(\partial/\partial x)$, $j = 1, \cdots, p$, are polynomials in $\partial/\partial x$ of arbitrary orders n_j with real constant coefficients satisfying the Lopatinskiĭ condition relative to the operator L and the $\varphi_j(t)$ are sufficiently smooth functions satisfying consistency conditions at $t = 0$.

Existence and uniqueness questions for the solution of problem (1.1)–(1.3) have been studied under more general assumptions in [2] and [3].

DEFINITION 1. We will say that the function $u(x, t)$ is *stabilized* for $t \to \infty$ if it has a finite limit as $t \to \infty$ for any $x \to 0$.

DEFINITION 2. A function $\varphi(t)$ is said to be *stabilizing* for $t \to \infty$ if it has a finite limit as $t \to \infty$.

AMS (MOS) subject classification (1975). Primary 35K35; Secondary 35B40.

Let z be a complex number and let $\omega_s = \exp [i(\pi/2p + (s-1)\pi/p)]$, $s = 1, 2, \cdots, p$, be the $2p$th roots of -1 with positive imaginary parts. Consider the matrix

$$B(z) = (b^{(j)}(i\omega_s z)), \qquad j,\ s = 1,\ 2,\ \ldots,\ p,$$

and let $P(z) = \det B(z)$. It will be proved below (Lemma 1) that the multiplicity of the root $z = 0$ of the polynomial $P(z)$ for any boundary operators $b^{(j)}(\partial/\partial x)$, $j = 1, 2, \cdots, p$, is not less than $p(p-1)/2$. Consequently the function $Q(z) = z^{-p(p-1)/2}P(z)$ is a polynomial.

The basic goal of the present paper is to give a necessary and sufficient condition on the operators $b^{(j)}(\partial/\partial x)$ under which the solution $u(x, t)$ of problem (1.1)–(1.3) is stabilized for any stabilizing functions $\varphi_j(t)$.

THEOREM 1. *In order for the solution* $u(x, t)$ *of problem* (1.1)–(1.3) *to be stabilized for* $t \to \infty$ *for any bounded functions* $\varphi_j(t)$, $j = 1, \cdots, p$, *having finite limits for* $t \to \infty$ *it is necessary and sufficient that the roots of the polynomial* $Q(z)$ *lie in the sector* $|\arg z| > \pi/4p$, $z \neq 0$.

THEOREM 2. *If the roots of the polynomial* $Q(z)$ *are distributed in the sector* $|\arg z| > \pi/4p$, $z \neq 0$, *the solution* $u(x, t)$ *of problem* (1.1)–(1.3) *tends to a polynomial in* x *of degree* $p-1$ *as* $t \to \infty$.

This result was obtained in the case when equation (1.1) is the heat equation ($p = 1$) by another method in [4]. The stabilization of the derivatives $\partial^\beta u(t, x)/\partial x^\beta$, of the solution, where $\beta \geqslant \alpha(B)$ $(\alpha(B) \geqslant p-1)$, has been studied in [5].

REMARK. We emphasize that Theorem 1 provides a criterion for the stabilization of the solution $u(x, t)$ for *any* stabilizing boundary functions. It is therefore impossible to use Theorem 1 to draw any conclusions concerning the distribution of the roots of the equation $Q(z) = 0$ from the fact that the solution $u(x, t)$ is stabilized for some particular stabilizing boundary functions. For example, if $\varphi_j(t) \equiv 0$, $j = 1, \cdots, p$, then $u(x, t) \equiv 0$ by virtue of the uniqueness theorem, i.e. the solution is stabilized independently of the distribution of the roots of the equation $Q(z) = 0$.

§2. Auxiliary constructions

The solution $u(x, t)$ of problem (1.1)–(1.3) will be sought in the form of the inverse Laplace transform of the bounded for $0 \leqslant x < \infty$ solution $v(x, \lambda)$, where $\operatorname{Re} \lambda > 0$, of the equation

$$\lambda v(x, \lambda) = (-1)^{p-1} \frac{d^{2p} v(x, \lambda)}{dx^{2p}}, \qquad \operatorname{Re} \lambda > 0, \tag{2.1}$$

satisfying the boundary conditions

$$b^{(j)}(d/dx)\,v(x,\ \lambda)\big|_{x=0}=\tilde{\varphi}_j(\lambda),\quad j=1,\ 2,\ \ldots,\ p. \tag{2.2}$$

Since the functions $\varphi_j(t)$ are bounded, their Laplace transforms

$$\tilde{\varphi}_j(\lambda)=\int_0^\infty e^{-\lambda t}\varphi_j(t)\,dt \tag{2.3}$$

are analytic in λ for $\operatorname{Re}\lambda > 0$, $j = 1,\cdots, p$.

Problem (2.1), (2.2) is obtained by applying the Laplace transform to problem (1.1)–(1.3). Any bounded solution of equation (2.1) can be written in the form

$$v(x,\ \lambda)=\sum_{s=1}^{p} C_s(\lambda)\,e^{i\tau_s x}, \tag{2.4}$$

where the τ_s, $s = 1, 2,\cdots, p$, are the roots with positive imaginary parts of the characteristic equation $\lambda + \tau^{2p} = 0$.

We denote by $\omega_s = \exp\,[i(\pi/2p + (s-1)\pi/p)]$, $s = 1, 2, \cdots, p$, the $2p$th roots of -1 with positive imaginary parts and by $z = \sqrt[2p]{\lambda}$ the $2p$th roots of λ that are real and positive for real $\lambda > 0$. Then $\tau_s = \omega_s z$, $s = 1, \cdots, p$. For the determination of the $C_s(\lambda)$ we have by virtue of the boundary conditions (2.2) the system of equations

$$\sum_{s=1}^{p} C_s(\lambda)\,b^{(j)}\left(i\omega_s \sqrt[2p]{\lambda}\right)=\tilde{\varphi}_j(\lambda),\qquad j=1,\ 2,\ \ldots,\ p. \tag{2.5}$$

We consider the matrix

$$B(z)=\begin{pmatrix} b^{(1)}(i\omega_1 z) & \ldots & b^{(1)}(i\omega_p z)\\ \cdot & \cdots & \cdot \\ b^{(p)}(i\omega_1 z) & \ldots & b^{(p)}(i\omega_p z)\end{pmatrix} \tag{2.6}$$

and denote its determinant by

$$P(z)=\det B(z). \tag{2.7}$$

By the Lopatinskiĭ condition for problem (1.1)–(1.3) we mean the condition

$$P(z)\not\equiv 0. \tag{2.8}$$

Since $P(z)$ is a polynomial in z having a finite number of roots, there exists a number $\sigma > 0$ such that $P(\sqrt[2p]{\lambda}) \neq 0$ for $\operatorname{Re}\lambda \geqslant 0$. For such λ, solving the system of equations (2.5), we get

$$C_s(\lambda)=\frac{1}{P\left(\sqrt[2p]{\lambda}\right)}\sum_{j=1}^{p}\tilde{\varphi}_j(\lambda)\,P_{js}\left(\sqrt[2p]{\lambda}\right),\qquad s=1,\ 2,\ \ldots,\ p, \tag{2.9}$$

where $P_{js}(z)$ is the cofactor of the element $b^{(j)}(i\omega_s z)$, $j, s = 1, \cdots, p$, of the matrix $B(z)$. Substituting these obtained values of the coefficients $C_s(\lambda)$ into (2.4), we get

$$v(x, \lambda) = \sum_{j=1}^{p} \tilde{\varphi}_j(\lambda) \tilde{G}\left(x, \sqrt[2p]{\lambda}\right), \tag{2.10}$$

where

$$\tilde{G}_j(x, z) = \frac{1}{P(z)} \sum_{s=1}^{p} P_{js}(z) e^{i\omega_s z x} = \frac{P^{(j)}(x, z)}{P(z)} \tag{2.11}$$

and

$$P^{(j)}(x, z) = \sum_{s=1}^{p} P_{js}(z) e^{i\omega_s z x}, \quad j = 1, 2, \ldots, p. \tag{2.12}$$

The solution of problem (1.1)–(1.3) can be written in the form

$$u(x, t) = \sum_{j=1}^{p} u_j(x, t), \tag{2.13}$$

where

$$u_j(x, t) = \int_0^t G_j(x, t - \tau) \varphi_j(\tau) \, d\tau, \quad j = 1, 2, \ldots, p, \tag{2.14}$$

in which the jth component of the Green vector is

$$G_j(x, t) = \frac{1}{2\pi i} \int_\gamma e^{\lambda t} \tilde{G}_j\left(x, \sqrt[2p]{\lambda}\right) d\lambda, \quad \gamma = \{\operatorname{Re} \lambda = \sigma, \ \sigma > 0\}. \tag{2.15}$$

LEMMA 1. *The multiplicity of the root* $z = 0$ *of the polynomial* $P(z)$ *is not less than* $p(p-1)/2$.

PROOF. Let

$$b^{(j)}(i\omega_s z) = (i\omega_s z)^{l_j} b_1^{(j)}(i\omega_s z),$$

where $j, s = 1, \cdots, p$, $l_j \geqslant 0$, and $b_1^{(j)}(i\omega_s z)$ is a polynomial with a nonzero constant term, i.e. $b_1^{(j)}(0) \neq 0$, $j = 1, \cdots, p$. We rearrange the rows of the determinant $P(z)$ so that $l_1 \leqslant l_2 \leqslant \cdots \leqslant l_p$. The resultant determinant

$$P_1(z) = \det\left((i\omega_s z)^{l_j} b_1^{(j)}(i\omega_s z)\right), \quad j, s = 1, 2, \ldots, p, \tag{2.16}$$

can differ only in sign from $P(z)$. By taking differences of the rows in $P_1(z)$, we can obtain a determinant of the same value for which $l_1 < l_2 < \cdots < l_p$, and l_j must satisfy the inequality $l_j \geqslant j - 1$, $j = 1, \cdots, p$. Thus $P(z)$ can be represented in the form

$$P(z)=(\pm 1)\det((i\omega_s z)^{l_j} b_1^{(j)}(i\omega_s z))=(\pm 1)\, z^{\sum_{j=1}^{p} l_j}\det((i\omega_s)^{l_j} b_1^{(j)}(i\omega_s z)),$$

where $j, s = 1, \cdots, p$ and $\Sigma_{j=1}^{p} l_j \geqslant \Sigma_{j=1}^{p} (j-1) = p(p-1)/2$. Lemma 1 is proved.

COROLLARY. *The function*

$$Q(z)=z^{-p(p-1)/2}P(z) \tag{2.17}$$

is a polynomial.

LEMMA 2. *The multiplicity of the root* $z = 0$ *of each of the functions* $P^{(j)}(x, z)$ *is not less than* $p(p-1)/2$.

PROOF. We note that the function $P^{(j)}(x, z)$ can be expressed in the form of a determinant:

$$P^{(j)}(x, z)=\det\begin{pmatrix} b^{(1)}(i\omega_1 z) & \cdots & b^{(1)}(i\omega_p z) \\ \cdots & \cdots & \cdots \\ b^{(j-1)}(i\omega_1 z) & \cdots & b^{(j-1)}(i\omega_p z) \\ e^{i\omega_1 zx} & \cdots & e^{i\omega_p zx} \\ b^{(j+1)}(i\omega_1 z) & \cdots & b^{(j+1)}(i\omega_p z) \\ \cdots & \cdots & \cdots \\ b^{(p)}(i\omega_1 z) & \cdots & b^{(p)}(i\omega_p z) \end{pmatrix}, \tag{2.18}$$

where $j = 1, \cdots, p$. The determinant (2.18) is obtained from $P(z)$ by replacing the jth row by $e^{i\omega_1 zx}, \cdots, e^{i\omega_p zx}$. We transfer the jth row of (2.18) to the pth place and rearrange the remaining $p-1$ rows

$$b^{(j)}(i\omega_s z)=(i\omega_s z)^{l_j} b_1^{(j)}(i\omega_s z),\qquad j=1, 2, \ldots, p-1,\qquad s=1, 2, \ldots, p,$$

so that $l_1 \leqslant l_2 \leqslant \cdots \leqslant l_{p-1}$.

In the same way as in the proof of Lemma 1, we obtain a determinant for which $l_1 < l_2 < \cdots < l_{p-1}$, and which we denote by $P_p^{(j)}(x, z)$. Clearly $P_p^{(j)}(x, z)$ can differ only in sign from the determinant (2.18):

$$P_p^{(j)}(x, z)=(\pm 1)\,P^{(j)}(x, z). \tag{2.19}$$

If it turns out that $l_1 \geqslant 1$, then

$$\sum_{j=1}^{p} l_j \geqslant \sum_{j=1}^{p-1} j = p(p-1)/2,$$

and Lemma 2 is proved. Suppose $l_1 = 0, \cdots, l_{k-1} = k-2$, $l_k > k, \cdots, l_{p-1} \geqslant p-1$, $k = 2, \cdots, p-1$. In this case we represent each element of the pth row of the determinant $P_p^{(j)}(x, z)$ in the form

$$e^{i\omega_s zx} = \sum_{m=0}^{k-1} \frac{1}{m!}(i\omega_s zx)^m + O(z^k), \tag{2.20}$$

where $s = 1, \cdots, p$, Then, by successively subtracting with the use of (2.20) linear combinations, in which the coefficients depend on x, of the first $k-1$ rows of $P_p^{(j)}(x, z)$ from the last row, we reduce each element of this row to the form

$$(i\omega_s z)^{k-1} M_{k-1}(x) + O(z^k), \tag{2.21}$$

where $s = 1, \cdots, p$ and $M_{k-1}(x)$ is a polynomial in x of degree $k-1$. We now transfer the pth row of $P_p^{(j)}(x, z)$ to the kth place and, taking into account expressions (2.19) and (2.20), reduce the determinant (2.18) to the form

$$P^{(j)}(x, z) = (\pm 1)\det \begin{pmatrix} (i\omega_1 z)^{l_1} b_1^{(1)}(i\omega_1 z) & \cdots & (i\omega_p z)^{l_1} b_1^{(1)}(i\omega_p z) \\ \cdots & \cdots & \cdots \\ (i\omega_1 z)^{l_{k-1}} b_1^{(k-1)}(i\omega_1 z) & \cdots & (i\omega_p z)^{l_{k-1}} b_1^{(k-1)}(i\omega_p z) \\ M_{k-1}(x)(i\omega_1 z)^{k-1} & \cdots & M_{k-1}(x)(i\omega_p z)^{k-1} \\ (i\omega_1 z)^{l_k} b_1^{(k)}(i\omega_1 z) & \cdots & (i\omega_p z)^{l_k} b_1^{(k)}(i\omega_p z) \\ \cdots & \cdots & \cdots \\ (i\omega_1 z)^{l_{p-1}} b_1^{(p-1)}(i\omega_1 z) & \cdots & (i\omega_p z)^{l_{p-1}} b_1^{(p-1)}(i\omega_p z) \end{pmatrix} + O\left(z^{k+\sum_{j=1}^{p-1} l_j}\right), \tag{2.22}$$

where $l_j = j - 1$ for $j = 1, \cdots, k-1$ and $l_s \geqslant s$ for $s = k, \cdots, p-1$. But then $k - 1 + \Sigma_{j=1}^{p-1} l_j \geqslant p(p-1)/2$, and Lemma 2 follows from the representation (2.22).

COROLLARY 1. *The function*

$$Q^{(j)}(x, z) = z^{-p(p-1)/2} P^{(j)}(x, z), \qquad j = 1, 2, \ldots, p, \tag{2.23}$$

is an entire function of z *for any* $x > 0$.

COROLLARY 2. *Equalities* (2.17), (2.23) *and* (2.11) *imply that*

$$\tilde{G}_j(x, z) = Q^{(j)}(x, z)/Q(z), \tag{2.24}$$

where $j = 1, \cdots, p$.

LEMMA 3. *Suppose* z *varies in an arbitrary compactum in which there are no roots of the polynomial* $Q(z)$, *and suppose* $|\arg z| = |\varphi/2p| \leqslant \pi/2p$. *Then for any* $x \geqslant 0$ *there exists a constant* $C'(x)$ *such that*

$$|\tilde{G}_j(x, z)| \leqslant C'(x), \qquad j = 1, 2, \ldots, p. \tag{2.25}$$

In addition, the inequality

$$|\tilde{G}_j(x, z)| \leqslant C'(x) e^{-\theta |z| x}, \quad x > 0, \quad j = 1, 2, \ldots, p, \tag{2.26}$$

is valid for all z *for which* $|z|$ *is sufficiently large and* $|\arg z| \leqslant (\pi - \theta_0)/2p$, *where* θ_0 *is a certain positive number and* $\theta = \frac{1}{2} \sin(\theta_0/2p) > 0$.

PROOF. Inequality (2.25) follows from equality (2.24) and the fact that $Q^{(j)}(x, z)$, $j = 1, \cdots, p$, is an entire function of z for any $x \geqslant 0$. Further, for any $x > 0$

$$\begin{aligned} |\exp i\omega_s zx| &= \Big|\exp\Big\{i|z|x\Big[\cos\Big(\frac{\varphi}{2p} + \frac{\pi}{2p} + \frac{(s-1)\pi}{p}\Big) \\ &\quad + i \sin\Big(\frac{\varphi}{2p} + \frac{\pi}{2p} + \frac{(s-1)\pi}{p}\Big)\Big]\Big\}\Big| \leqslant \exp\Big[-x|z| \sin \frac{\varphi + \pi(2s-1)}{2p}\Big] \\ &\leqslant \exp\Big(-x|z| \sin \frac{\theta_0}{2p}\Big), \quad s = 1, 2, \ldots, p. \end{aligned}$$

Hence

$$|\exp i\omega_s zx| \leqslant \exp(-\theta_1 |z| x) \quad \text{for} \quad |\arg z| \leqslant (\pi - \theta_0)/2p, \quad x > 0, \tag{2.27}$$

where $s = 1, \cdots, p$ and $\theta_1 = \sin(\theta_0/2p)$. Since

$$0 < \theta_0/2p \leqslant [\varphi + \pi(2s-1)]/2p \leqslant \pi - \theta_0/2p < \pi, \quad s = 1, 2, \ldots, p,$$

we have $\theta_1 = \sin(\theta_0/2p) > 0$, and consequently $\theta = \theta_1/2 > 0$. Inequality (2.26) follows from equality (2.11) and estimate (2.27). Lemma 3 is proved.

LEMMA 4. *If the roots of the polynomial* $Q(z)$ *are distributed in the sector* $|\arg z| > \pi/4p$, $z \neq 0$, *there exists for any* $x > 0$ *a constant* $C(x)$ *such that the Green vector* $G(x, t) = \{G_1(x, t), \cdots, G_p(x, t)\}$ *of problem* (1.1)–(1.3) *satisfies for any* $x > 0$ *the inequality*

$$|G(x, t)| = \sqrt{G_1^2(x, t) + \ldots + G_p^2(x, t)} \leqslant \frac{C(x)}{1 + t^{1+1/2p}}. \tag{2.28}$$

It clearly suffices to carry out the proof for any component $G_j(x, t)$ of the vector $G(x, t)$. Suppose the roots of the polynomial $Q(z)$ are distributed in the sector $|\arg z| > \pi/4p$, $z \neq 0$. Then the roots of the function $Q(\sqrt[2p]{\lambda})$ are distributed in the halfplane $\operatorname{Re} \lambda < 0$.

Inasmuch as the number of roots is finite, there exists a number $\delta > 0$ such that all of the roots of $Q(\sqrt[2p]{\lambda})$ will lie in the halfplane $\operatorname{Re} \lambda < \delta$. Since the function $\tilde{G}_j(x, \sqrt[2p]{\lambda})e^{\lambda t}$ is analytic with respect to λ in the halfplane $\operatorname{Re} \lambda \geqslant -\delta$ with the exception of the segment $[-\delta, 0]$ of the real axis, it follows in view of estimate (2.26) that the contour of integration $\gamma = \{\operatorname{Re} \lambda = \tau, \tau > 0\}$ in equality (2.15) can be replaced by the contour

$$\gamma' = \gamma_1 \cup \gamma_\varepsilon^+ \cup \gamma_\varepsilon \cup \gamma_\varepsilon^- \cup \gamma_2, \tag{2.29}$$

where

$$\gamma_1=\{\operatorname{Re}\lambda=-\delta,\ \operatorname{Im}\lambda\geqslant 0\},\qquad \gamma_\varepsilon^+=\{-\delta\leqslant\operatorname{Re}\lambda\leqslant-\varepsilon,\ \arg\lambda=\pi\},$$
$$\gamma_\varepsilon=\{|\lambda|=\varepsilon,\ -\pi\leqslant\arg\lambda\leqslant\pi\},\qquad \gamma_\varepsilon^-=\{-\delta\leqslant\operatorname{Re}\lambda\leqslant-\varepsilon,\ \arg\lambda=-\pi\},$$
$$\gamma_2=\{\operatorname{Re}\lambda=-\delta,\ \operatorname{Im}\lambda\leqslant 0\}.$$

i.e. for some $\epsilon\in(0,\delta)$

$$G_j(x,t)=\frac{1}{2\pi i}\int_\gamma e^{\lambda t}\tilde G_j\left(x,\sqrt[2p]{\lambda}\right)d\lambda$$
$$=\frac{1}{2\pi i}\left(\int_{\gamma_1\cup\gamma_2}+\int_{\gamma_\varepsilon}+\int_{\gamma_\varepsilon^+\cup\gamma_\varepsilon^-}\right)e^{-\lambda t}\tilde G_j\left(x,\sqrt[2p]{\lambda}\right)d\lambda. \qquad (2.30)$$

This assertion can be proved by making use of Cauchy's theorem and the fact that for fixed $x>0$ and $t\geqslant 0$

$$\left|\int_{Ri-\delta}^{Ri+\sigma} e^{\lambda t}\tilde G_j\left(x,\sqrt[2p]{\lambda}\right)d\lambda\right|\leqslant C_1(x,t)\,e^{-\theta R^{1/2p}x}\to 0 \qquad \text{as}\qquad R\to\infty$$

and, analogously,

$$\left|\int_{-Ri-\delta}^{-Ri+\sigma} e^{\lambda t}\tilde G_j\left(x,\sqrt[2p]{\lambda}\right)d\lambda\right|\leqslant C_2(x,t)\,e^{-\theta R^{1/2p}x}\to 0 \qquad \text{as}\qquad R\to\infty,$$

where $C_1(x,t)$ and $C_2(x,t)$ are constants depending on x and t.

Estimating the first summand of the right side of (2.30) with the use of (2.26) we get

$$\left|\frac{1}{2\pi i}\int_{\gamma_1\cup\gamma_2} e^{\lambda t}\tilde G_j\left(x,\sqrt[2p]{\lambda}\right)d\lambda\right|\leqslant C_1(x)\,e^{-\delta t},\qquad j=1,2,\ldots,p. \qquad (2.31)$$

We estimate the second summand of the right side of (2.30) by making use of (2.25):

$$\left|\frac{1}{2\pi i}\int_{\gamma_\varepsilon} e^{\lambda t}\tilde G_j\left(x,\sqrt[2p]{\lambda}\right)d\lambda\right|\leqslant C'(x)\,e^{\varepsilon t}\varepsilon,\qquad j=1,2,\ldots,p.$$

It follows that

$$\lim_{\varepsilon\to 0}\left|\frac{1}{2\pi i}\int_{\gamma_\varepsilon} e^{\lambda t}\tilde G_j\left(x,\sqrt[2p]{\lambda}\right)d\lambda)\right|=0. \qquad (2.32)$$

Further, using (2.24), we get

$$\lim_{\varepsilon\to 0}\left|\frac{1}{2\pi i}\int_{\gamma_\varepsilon^+\cup\gamma_\varepsilon^-} e^{\lambda t}\tilde{G}_j\left(x, \sqrt[2p]{\lambda}\right)d\lambda\right|$$

$$=\left|\frac{1}{2\pi i}\int_0^\delta e^{-yt}\left[\tilde{G}_j\left(x, \sqrt[2p]{ye^{i\pi}}\right)-\tilde{G}_j\left(x, \sqrt[2p]{ye^{-i\pi}}\right)\right]dy\right|$$

$$=\frac{1}{2\pi}\left|\int_0^\delta e^{-yt}\left[\frac{Q^{(j)}\left(x, \sqrt[2p]{y}\,e^{i\pi/2}\right)}{Q\left(\sqrt[2p]{y}\,e^{i\pi/2p}\right)}-\frac{Q^{(j)}\left(x, \sqrt[2p]{y}\,e^{-i\pi/2p}\right)}{Q\left(\sqrt[2p]{y}\,e^{-i\pi/2p}\right)}\right]dy\right|$$

$$=\frac{1}{2\pi}\left|\int_0^\delta e^{-yt}\left\{\left[\frac{1}{Q(0)}+o\left(\sqrt[2p]{y}\right)\right]\left[Q^{(j)}(x, 0)+o\left(\sqrt[2p]{y}\right)\right]\right.\right.$$

$$\left.\left.-\left[\frac{1}{Q(0)}+o\left(\sqrt[2p]{y}\right)\right]\left[Q^{(j)}(x, 0)+o\left(\sqrt[2p]{y}\right)\right]\right\}dy\right|$$

$$=\left|\int_0^\delta e^{-yt}o\left(\sqrt[2p]{y}\right)dy\right|\leqslant\left|\int_0^\infty e^{-yt}o\left(\sqrt[2p]{y}\right)dy\right|\leqslant\frac{C_2(x)}{t^{1+1/2p}},\qquad j=1, 2, \ldots, p,$$

where $C_1(x)$ and $C_2(x)$ are constants not depending on t. Hence

$$\lim_{\varepsilon\to 0}\left|\frac{1}{2\pi i}\int_{\gamma_\varepsilon^+\cup\gamma_\varepsilon^-} e^{\lambda t}\tilde{G}_j\left(x, \sqrt[2p]{\lambda}\right)d\lambda\right|\leqslant\frac{C_2(x)}{t^{1+1/2p}},\qquad j=1, 2, \ldots, p. \tag{2.33}$$

From (2.31), (2.33) and (2.32) it follows that

$$|G_j(x, t)|\leqslant\frac{C_3(x)}{t^{1+1/2p}},\qquad j=1, 2, \ldots, p, \tag{2.34}$$

where $C_3(x) = \max\{C'(x),\ C_1(x),\ C_2(x)\}$.

We now prove that for any $T > 0$ there exists a constant $C_4(x)$ such that

$$|G_j(x, t)|\leqslant C_4(x) \tag{2.35}$$

for $0 \leqslant t \leqslant T$. In fact, by virtue of inequality (2.26)

$$|G_j(x, t)|=\left|\frac{1}{2\pi i}\int_{\substack{\mathrm{Re}\,\lambda=\sigma\\ \sigma>0}} e^{\lambda t}\tilde{G}_j\left(x, \sqrt[2p]{\lambda}\right)d\lambda\right|\leqslant\frac{e^{\sigma T}}{2\pi}\int_{\substack{\mathrm{Re}\,\lambda=\sigma\\ \sigma>0}}\left|\tilde{G}_j\left(x, \sqrt[2p]{\lambda}\right)\right|d\lambda=C_4(x).$$

Inequality (2.28) follows from (2.34) and (2.35). Lemma 4 is proved.

§3. Proof of Theorems 1 and 2

THEOREM 1. *In order for the solution* $u(x, t)$ *of problem* (1.1)–(1.3) *to be stabilized for* $t\to\infty$ *for any stabilizing boundary functions* $\varphi_j(t)$, $j = 1, \cdots, p$, *it is necessary and sufficient that the roots of the polynomial* $Q(z)$ *lie in the sector* $|\arg z| > \pi/4p,\ z \neq 0$.

PROOF. NECESSITY. Suppose the solution $u(x, t)$ of problem (1.1)–(1.3) is stabilized for any boundary functions $\varphi_j(t),\ j = 1, \cdots, p$, having a finite limit for $t \to \infty$. We will show that the roots of the polynomial $Q(z)$ are distributed in the sector $|\arg z| > \pi/4p,\ z \neq 0$, or, what is the same thing, that the roots of $Q(\sqrt[2p]{\lambda})$ are distributed in the halfplane $|\arg \lambda| > \pi/2,\ \lambda \neq 0$.

1. Let λ_0 be such that $\operatorname{Re} \lambda_0 > 0$ and $z_0 = \sqrt[2p]{\lambda_0}$ is a root of order $\kappa \geqslant 1$ of the polynomial $Q(z)$ or, what is the same thing inasmuch as $\lambda_0 \neq 0$, of the polynomial $P(z)$. Since the solution $u(x, t)$ of problem (1.1)–(1.3) is stabilized for $t \to \infty$ for any stabilizing boundary functions $\varphi_j(t)$, and is consequently bounded, its Laplace transform $v(x, \lambda)$ is for any fixed $x > 0$ an analytic function of λ for $\operatorname{Re} \lambda > 0$. Since equality (2.10) holds for any analytic for $\operatorname{Re} \lambda > 0$ functions $\widetilde{\varphi}_j(\lambda),\ j = 1, \cdots, p$, that are the Laplace transforms of functions with finite limits for $t \to \infty$, each of the functions $\widetilde{G}_j(x, \sqrt[2p]{\lambda})$ is for any $x > 0$ analytic with respect to λ in the halfplane $\operatorname{Re} \lambda > 0$ and hence in a neighborhood of λ_0. From equality (2.11) it then follows that $z_0 = \sqrt[2p]{\lambda_0}$ is for any $x > 0$ a root of multiplicity not less than $\kappa \geqslant 1$ for each of the functions $P^{(j)}(x, z)$. Equality (2.12) in turn shows that z_0 is a root of multiplicity not less than $\kappa \geqslant 1$ for each of the polynomials $P_{js}(z),\ j, s = 1, \cdots, p$, inasmuch as for any $x > 0$ the functions $e^{i\omega_s zx}$ are linearly independent as functions of x. Therefore each of the functions

$$P_{js}(z)/P(z), \qquad j,\ s = 1,\ 2,\ \ldots,\ p, \tag{3.1}$$

is an analytic function of z in a neighborhood of the point $z_0 = \sqrt[2p]{\lambda_0}$. But these functions are the elements of the inverse $B^{-1}(z)$ of the matrix $B(z)$, and hence the function $\det B^{-1}(z)$ is also analytic in this neighborhood, so that $\det B^{-1}(z) = B_0$, where B_0 is a complex number. On the other hand, by assumption, $P(z_0) = \det B(z_0) = 0$. We therefore have $\det B(z_0) \det B^{-1}(z_0) = 0$, which is impossible since $\det B(z_0) \det B^{-1}(z_0)$ must be equal to 1.

Consequently the function $Q(\sqrt[2p]{\lambda})$ cannot have any roots in the halfplane $\operatorname{Re} \lambda > 0$, or, equivalently, the polynomial $Q(z)$ does not have any roots in the sector $|\arg z| < \pi/4p$.

2. Let $\lambda_1, \cdots, \lambda_m$ denote all of the roots of $Q(\sqrt[2p]{\lambda})$ or, equivalently, let $z_1 = \sqrt[2p]{\lambda_1}, \cdots, z_m = \sqrt[2p]{\lambda_m}$ denote all of the roots of the polynomial $Q(z)$. Their multiplicities are denoted by $\kappa_1, \cdots, \kappa_m$ respectively. We will show that none of the roots $\lambda_1, \cdots, \lambda_m$ can be purely imaginary. Suppose, on the contrary, that $\lambda_m = i\rho_m$, where ρ_m is a real number not equal to zero. We will prove that $z_m = \sqrt[2p]{\lambda_m}$ is then a root of multiplicity not less than κ_m for each of the polynomials $P_{js}(z),\ j, s = 1, \cdots, p$. To this end we fix a natural number $j \in [1, p]$ and

set $\varphi_k(t) \equiv 0,\ k \neq j$, and

$$\varphi_j(t) = \int_\gamma e^{\lambda t} \tilde{\varphi}_j(\lambda)\, d\lambda, \tag{3.2}$$

where $\gamma = \{\operatorname{Re} \lambda = \sigma,\ \sigma > 0\}$ and

$$\tilde{\varphi}_j(\lambda) = \exp\left(-\sqrt[2p]{\lambda}\right) \frac{Q\left(\sqrt[2p]{\lambda}\right)}{\sqrt[2p]{\lambda} - \sqrt[2p]{\lambda_m}} = \exp\left(-\sqrt[2p]{\lambda}\right) N_j\left(\sqrt[2p]{\lambda}\right), \tag{3.3}$$

in which

$$N_j\left(\sqrt[2p]{\lambda}\right) = \frac{Q\left(\sqrt[2p]{\lambda}\right)}{\sqrt[2p]{\lambda} - \sqrt[2p]{\lambda_m}} \tag{3.4}$$

is, according to assumption, a polynomial in $\sqrt[2p]{\lambda}$. The function $\varphi_j(t)$ is infinitely differentiable, satisfies the consistency conditions of [3] and is such that $\lim_{t\to\infty} \varphi_j(t) = 0$.

For suppose λ varies in the halfplane $\operatorname{Re} \lambda \geqslant -\delta$ with the exception of the segment $[-\delta, 0]$ of the real axis, where δ is an arbitrary positive number, and let θ_0 denote a positive number from the interval $(0, \pi/2)$. For sufficiently large $|\lambda|$ we have $|\arg \lambda| \leqslant \pi - \theta_0$, so that

$$\left|\arg \sqrt[2p]{\lambda}\right| \leqslant (\pi - \theta_0)/2p \quad \text{and} \quad 0 < \cos\left[(\pi - \theta_0)/2p\right] = \theta_1 \leqslant \cos\left(\arg \sqrt[2p]{\lambda}\right).$$

Hence

$$\left|\exp\left(-\sqrt[2p]{\lambda}\right)\right| \leqslant \exp\left[-\sqrt[2p]{|\lambda|}\cos\left(\arg \sqrt[2p]{\lambda}\right)\right] \leqslant \exp\left(-\theta_1 \sqrt[2p]{|\lambda|}\right). \tag{3.5}$$

From (3.3), taking into account (3.5), we get that for sufficiently large $|\lambda|$

$$|\tilde{\varphi}_j(\lambda)| \leqslant \text{const} \cdot \exp\left(-\frac{\theta_1}{2} \sqrt[2p]{|\lambda|}\right), \quad \theta_1 > 0. \tag{3.6}$$

It follows from (3.6) that the integral

$$\int_\gamma e^{\lambda t} \lambda^k \tilde{\varphi}_j(\lambda)\, d\lambda, \qquad \gamma = \{\operatorname{Re} \lambda = \sigma,\ \sigma > 0\}$$

converges uniformly with respect to $t \in [0, T]$, where T is an arbitrary positive number, for any natural number k. But this means that the function $\varphi_j(t)$ is infinitely differentiable for $t \geqslant 0$. We now show that the functions $\varphi_j(t)$ satisfy the consistency conditions of [3]. For this purpose it suffices to establish that

$$\left.\frac{d^k \varphi_j(t)}{dt^k}\right|_{t=0} = \lim_{t\to+0} \frac{d^k \varphi_j(t)}{dt^k} = \int_\gamma \lambda^k \tilde{\varphi}_j(\lambda)\, d\lambda = 0$$

for an arbitrary natural number $k > 0$. Suppose $\lambda = |\lambda| e^{i\varphi}$ and $|\varphi| < \pi/2$. Since $\lambda^k \tilde{\varphi}_j(\lambda)$ is an analytic function of λ in the halfplane $\operatorname{Re} \lambda > 0$, it follows from

Cauchy's theorem in view of (3.6) that

$$\left|\int_{\substack{\operatorname{Re}\lambda=\sigma\\ \sigma>0}} \lambda^k \tilde{\varphi}_j(\lambda)\,d\lambda\right| = \lim_{|\lambda|\to\infty}\left|\int_{-\frac{\pi}{2}+\arcsin\frac{\sigma}{|\lambda|}}^{\frac{\pi}{2}-\arcsin\frac{\sigma}{|\lambda|}} (|\lambda|e^{i\varphi})^k\,\tilde{\varphi}_j(|\lambda|e^{i\varphi})\,|\lambda|\,ie^{i\varphi}d\varphi\right|$$

$$\leqslant \lim_{|\lambda|\to\infty} \text{const}\cdot|\lambda|^{k+1}e^{-\frac{\theta_1}{2}\sqrt[2p]{|\lambda|}} = 0.$$

We show, finally that $\lim_{t\to\infty}\varphi_j(t) = 0$. To this end we replace the contour γ of the integral in (3.2) by the contour γ' in the same way as was done in estimating the integral in (2.15). This is permissible inasmuch as the function $e^{\lambda t}\tilde{\varphi}_j(\lambda)$ is analytic and estimate (3.6) holds in the halfplane $\operatorname{Re}\lambda \geqslant -\delta$ with the exception of the segment $[-\delta, 0]$ of the real axis. Thus

$$\varphi_j(t) = \frac{1}{2\pi i}\left(\int_{\gamma_1\cup\gamma_2} + \int_{\gamma_\varepsilon} + \int_{\gamma_\varepsilon^+\cup\gamma_\varepsilon^-}\right) e^{\lambda t}\tilde{\varphi}_j(\lambda)\,d\lambda. \tag{3.7}$$

It immediately follows from (3.6) that

$$\left|\int_{\gamma_1\cup\gamma_2} e^{\lambda t}\tilde{\varphi}_j(\lambda)\,d\lambda\right| \leqslant \text{const}\cdot e^{-\delta t}, \qquad 1\leqslant j\leqslant p. \tag{3.8}$$

In the same way as in (2.32), we have

$$\lim_{\varepsilon\to 0}\left|\int_{\gamma_\varepsilon} e^{\lambda t}\tilde{\varphi}_j(\lambda)\,d\lambda\right| \leqslant \lim_{\varepsilon\to 0}\text{const}\cdot\varepsilon = 0. \tag{3.9}$$

We now estimate the third integral in (3.7):

$$\lim_{\varepsilon\to 0}\int_{\gamma_\varepsilon^+\cup\gamma_\varepsilon^-} (\exp\lambda t)\,\tilde{\varphi}_j(\lambda)\,d\lambda = \int_0^\delta [\exp(-yt)]\left\{\left[\exp\left(-\sqrt[2p]{y}\exp\frac{i\pi}{2p}\right)\right]\right.$$

$$\left.\times N_j\left(\sqrt[2p]{y}\exp\frac{i\pi}{2p}\right) - \left[\exp\left(-\sqrt[2p]{y}\exp\left(-\frac{i\pi}{2p}\right)\right)\right]N_j\left(\sqrt[2p]{y}\exp\left(-\frac{i\pi}{2p}\right)\right)\right\}dy$$

$$= \int_0^\delta [\exp(-yt)]\left\{\left[1+O\left(\sqrt[2p]{y}\right)\right]\left[N_j(0)+O\left(\sqrt[2p]{y}\right)\right] - \left[1+O\left(\sqrt[2p]{y}\right)\right]\right.$$

$$\left.\times\left[N_j(0)+O\left(\sqrt[2p]{y}\right)\right]\right\}dy = \int_0^\delta [\exp(-yt)]\,O\left(\sqrt[2p]{y}\right)dy.$$

But

$$\left|\int_0^\delta e^{-yt} O\left(\sqrt[2p]{y}\right) dy\right| \leqslant \left|\int_0^\infty e^{-yt} O\left(\sqrt[2p]{y}\right) dy\right| \leqslant \frac{\text{const}}{t^{1+1/2p}}.$$

Hence

$$\lim_{\varepsilon\to 0} \int_{\gamma_\varepsilon^+ \cup \gamma_\varepsilon^-} e^{\lambda t} \tilde{\varphi}_j(\lambda)\, d\lambda = O\left(\frac{1}{t^{1+1/2p}}\right). \tag{3.10}$$

and the equality

$$\lim_{t\to\infty} \varphi_j(t) = 0$$

follows from (3.7)–(3.10).

Substituting the functions $\widetilde{\varphi}_k(\lambda) = 0$, $k \neq j$, and the function $\widetilde{\varphi}_j(\lambda)$ given by (3.3) into (2.10) and taking (2.24) into account, we get

$$v(x, \lambda) = \frac{\left[\exp\left(-\sqrt[2p]{\lambda}\right)\right] Q^{(j)}\left(x, \sqrt[2p]{\lambda}\right)}{\sqrt[2p]{\lambda} - \sqrt[2p]{\lambda_m}} = \tilde{\varphi}_j(\lambda) \tilde{G}_j\left(x, \sqrt[2p]{\lambda}\right), \tag{3.11}$$

the inverse Laplace transform of which has the form

$$u(x, t) = \frac{1}{2\pi i} \int_\gamma e^{\lambda t} v(x, \lambda)\, d\lambda, \qquad \gamma = \{\operatorname{Re} \lambda = \sigma,\ \sigma > 0\}. \tag{3.12}$$

Estimates (2.25) and (3.6) imply that the following estimate holds for any fixed $x > 0$ and $|\lambda| \to \infty$ in the sector $|\arg \lambda| \leqslant \pi - \theta_0$:

$$|v(x, \lambda)| \leqslant C(x) \exp\left(-\theta \sqrt[2p]{|\lambda|}\, x\right), \tag{3.13}$$

where $C(x)$ is a constant depending on x, and θ_0 and θ are certain positive constants. Inasmuch as for any $x > 0$ the function $v(x, \lambda)$ has a pole at λ_m but is otherwise analytic with respect to λ in the halfplane $\operatorname{Re} \lambda \geqslant -\delta$ (δ is a certain positive number) with the exception of the segment $[-\delta, 0]$ of the real axis, we can replace the contour $\gamma = \{\operatorname{Re} \lambda = \sigma,\ \sigma > 0\}$ in equality (3.12) with the use of Cauchy's theorem and inequality (3.13) by the contour γ' given by (2.29):

$$u(x, t) - \frac{1}{2\pi i} \int_{\gamma'} e^{\lambda t} v(x, \lambda)\, d\lambda + \operatorname*{res}_{\lambda=\lambda_m} e^{\lambda t} v(x, \lambda). \tag{3.14}$$

From (3.13) we get

$$\left|\frac{1}{2\pi i} \int_{\gamma_1 \cup \gamma_2} e^{\lambda t} v(x, \lambda)\, d\lambda\right| \leqslant C_1(x) e^{-\delta t}, \tag{3.15}$$

where $C_1(x)$ is a constant not depending on t. Since $v(x, \lambda)$ has a pole only at the point $\lambda_m \neq 0$, it is bounded by a constant $C(x)$ in a sufficiently small neighborhood of the origin. Hence

$$\lim_{\varepsilon\to 0}\left|\frac{1}{2\pi i}\int_{\gamma_\varepsilon} e^{\lambda t} v(x, \lambda)\, d\lambda\right| \leqslant \lim_{\varepsilon\to 0} C(x) e^{\varepsilon t}\varepsilon = 0. \tag{3.16}$$

We estimate the summand

$$\lim_{\varepsilon\to 0}\frac{1}{2\pi i}\int_{\gamma_\varepsilon^+\cup\gamma_\varepsilon^-} e^{\lambda t} v(x, \lambda)\, d\lambda = \frac{1}{2\pi i}\int_0^\delta e^{-yt}\,[v(x, ye^{-i\pi})$$

$$-v(x, ye^{-i\pi})]\, dy = \frac{1}{2\pi}\int_0^\delta e^{-yt}\left\{\frac{\left[\exp\left(-\sqrt[2p]{y}\exp\frac{i\pi}{2p}\right)\right] Q^{(j)}\left(x, \sqrt[2p]{y}\exp\frac{i\pi}{2p}\right)}{\sqrt[2p]{y}\exp\frac{i\pi}{2p} - \sqrt[2p]{\lambda_m}}\right.$$

$$\left.-\frac{\left[\exp\left(-\sqrt[2p]{y}\exp\left(-\frac{i\pi}{2p}\right)\right)\right] Q^{(j)}\left[x, \sqrt[2p]{y}\exp\left(-\frac{i\pi}{2p}\right)\right]}{\sqrt[2p]{y}\exp\left(-\frac{i\pi}{2p}\right) - \sqrt[2p]{\lambda_m}}\right\} dy = \frac{1}{2\pi}\int_0^\delta e^{yt}$$

$$\times\left\{\left[1+o\left(\sqrt[2p]{y}\right)\right]\left[Q^{(j)}(x, 0)+o\left(\sqrt[2p]{y}\right)\right]\left[1+o\left(\sqrt[2p]{y}\right)\right]\left(-\frac{1}{\sqrt[2p]{\lambda_m}}\right)\right.$$

$$\left.-\left[1+o\left(\sqrt[2p]{y}\right)\right]\left[Q^{(j)}(x, 0)+o\left(\sqrt[2p]{y}\right)\right]\left[1+o\left(\sqrt[2p]{y}\right)\right]\left(-\frac{1}{\sqrt[2p]{\lambda_m}}\right)\right\} dy$$

$$=\int_0^\delta e^{-yt} o\left(\sqrt[2p]{y}\right) dy.$$

Consequently

$$\left|\lim_{\varepsilon\to 0}\frac{1}{2\pi i}\int_{\gamma_\varepsilon^+\cup\gamma_\varepsilon^-} e^{\lambda t} v(x, \lambda)\, d\lambda\right| \leqslant \left|\int_0^\delta e^{-yt} O\left(\sqrt[2p]{y}\right) dy\right| \leqslant \frac{C_2(x)}{t^{1+1\,2p}},$$

where $C_2(x)$ is a constant not depending on t. But this means that

$$\lim_{\varepsilon\to 0}\frac{1}{2\pi i}\int_{\gamma_\varepsilon^+\cup\gamma_\varepsilon^-} e^{\lambda t} v(x, \lambda)\, d\lambda = O\left(\frac{1}{t^{1+1/2p}}\right). \tag{3.17}$$

and the equality

$$u(x, t) = \operatorname*{res}_{\lambda=\lambda_m} e^{\lambda t} v(x, \lambda) + O\left(\frac{1}{t^{1+1/2p}}\right). \tag{3.18}$$

follows from (3.14)–(3.17). Since, by assumption, $u(x, t)$ is stabilized for $t\to\infty$ and $\lambda_m = ip_m \neq 0$, equality (3.18) implies

$$\operatorname*{res}_{\lambda=\lambda_m} e^{\lambda t} v(x, \lambda) = 2p\lambda_m^{1-1/2p}\exp[\lambda_m t]\left[\exp\left(-\sqrt[2p]{\lambda_m}\right)\right] Q^{(j)}\left(x, \sqrt[2p]{\lambda_m}\right) = 0. \tag{3.19}$$

From (3.19) it follows that $z_m = \sqrt[2p]{\lambda_m}$ is for any $x > 0$ a root of multiplicity at least unity of the function $Q^{(j)}(x, \sqrt[2p]{\lambda})$, or, equivalently, of the function $P^{(j)}(x, \sqrt[2p]{\lambda})$. Since j is any natural number from the interval $[1, p]$, the number z_m is a root of multiplicity at least unity of each of the functions $P^{(j)}(x, z)$. Taking into account the facts that $\lambda_m \neq 0$ and the functions $\exp i\omega_s z_m x$, $s = 1, \cdots, p$, are linearly independent as functions of x $(x > 0)$, we get from (2.12) that z_m is a root of multiplicity at least unity of each of the polynomials $P_{js}(z)$. If $\kappa_m > 1$, we divide the numerator and denominator of the right side of equality (2.11) by the common factor $\sqrt[2p]{\lambda} - \sqrt[2p]{\lambda_m}$, in place of the function (3.3) we take the function

$$\tilde{\varphi}_j(\lambda) = \exp\left(-\sqrt[2p]{\lambda}\right) \frac{Q\left(\sqrt[2p]{\lambda}\right)}{\left(\sqrt[2p]{\lambda} - \sqrt[2p]{\lambda_m}\right)^2}$$

and, with the use of the same arguments, we get that z_m is a root of multiplicity not less than two of each of the polynomials $p_{js}(z)$, etc. Consequently z_m is a root of multiplicity not less than κ_m of each of the polynomials $P_{js}(z)$. It then follows in the same way as in case 1 that $\det B(z_m) \det B^{-1}(z_n) = 0$, which is impossible.

3. We will show, finally, that $z = 0$ also cannot be a root of the polynomial $Q(z)$. Suppose, on the contrary, that $z = 0$ is a root of multiplicity $\kappa \geqslant 1$ of $Q(z)$, i.e.

$$Q(z) = z^{\kappa} R(z), \tag{3.20}$$

where $R(z)$ is a polynomial such that $R(0) \neq 0$. Then, by virtue of Lemma 1,

$$P(z) = z^{\kappa_1} R(z), \tag{3.21}$$

where $\kappa_1 = p(p-1)/2 + \kappa$.

We first show that among the functions $P^{(j)}(x, z)$ there exists at least one function $P^{(\beta_p)}(x, z)$, $1 \leqslant \beta_p \leqslant p$, for which the multiplicity of the root $z = 0$ is strictly less than κ_1. To this end we consider the determinant $P(z)$. We represent each of the polynomials $b^{(j)}(i\omega_s z)$, $j = 1, 2, \cdots, p$, in the form $(i\omega_s z)^{l_j} b_1^{(j)}(i\omega_s z)$, $j, s = 1, \cdots, p$, where $b_1^{(j)}(i\omega_s z)$ is a polynomial in $i\omega_s z$ such that $b_1^{(j)}(0) \neq 0$. Rearranging the rows of $P(z)$, we reduce it to the form

$$P(z) = (\pm 1) \det\left((i\omega_s z)^{l_j} b_1^{(j)}(i\omega_s z)\right), \tag{3.22}$$

where $l_1 \leqslant l_2 \leqslant \cdots \leqslant l_p$. If some $l_j = l_1$, $2 \leqslant j \leqslant p$, we subtract the first row multiplied by an appropriate constant from the jth row and obtain a determinant for which l_j, $j = 2, \cdots, p$, is greater than l_1. If it then turns out that some l_j, $2 \leqslant j \leqslant p$, is congruent to l_1 modulo $2p$, i.e. $l_j = l_1 + 2pm_j$ for some integer m_j, we increase l_j by multiplying the first row by $\lambda^{m_j} = (i\omega_s z)^{2pm_j}$, $s = 1, \cdots, p$, times an appropriate constant and subtracting it from the jth row. If l_j is again congruent to l_1 modulo $2p$, we repeat this operation. Each such operation produces an increase of the exponent l_j, while $\Sigma_{j=1}^{p} l_j \leqslant \kappa_1$, as can be seen from (3.22). Applying the indicated operation a sufficient number of times, we get that $l_j > l_1$ and $l_j \not\equiv l_1 \pmod{2p}$, $j = 2, \cdots, p$. We then transfer a row whose exponent $l_j \leqslant l_k$ for $k = 2, \cdots, p$ to the second place and change the notation of this exponent to l_2. Proceeding in the same way as for l_1, we obtain a determinant for which $l_j > l_2$ and $l_j \not\equiv l_1 \pmod{2p}$ $l_j \neq l_2 \pmod{2p}$, $j = 3, \cdots, p$, etc.

Thus it can be assumed that the determinant (3.22) is such that $l_1 < l_2 < \cdots < l_p$ and $l_j \not\equiv l_s \pmod{2p}$ for $j \neq s$. The transformations used in obtaining this new determinant can change only the sign of $P(z)$. Therefore the multiplicity of the root $z = 0$ of $P(z)$, which according to our assumption is equal to $\kappa_1 = p(p-1)/2 + \kappa$, is equal to the multiplicity of this root of the new determinant, while the latter is equal to $\Sigma_{j=1}^{p} l_j$ (the numbers l_j refer to the new determinant) since

$$P(z) = z^{\sum\limits_{j=1}^{p} l_j} \left[\text{const det} \begin{pmatrix} (i\omega_1)^{l_1} & (i\omega_2)^{l_1} & \dots & (i\omega_p)^{l_1} \\ (i\omega_1)^{l_2} & (i\omega_2)^{l_2} & \dots & (i\omega_p)^{l_2} \\ \cdot & \cdot & \cdot & \cdot \\ (i\omega_1)^{l_p} & (i\omega_2)^{l_p} & \dots & (i\omega_p)^{l_p} \end{pmatrix} + O(z) \right]$$

and

$$\det \begin{pmatrix} (i\omega_1)^{l_1} & (i\omega_2)^{l_1} & \dots & (i\omega_p)^{l_1} \\ (i\omega_1)^{l_2} & (i\omega_2)^{l_2} & \dots & (i\omega_p)^{l_2} \\ \cdot & \cdot & \cdot & \cdot \\ (i\omega_1)^{l_p} & (i\omega_2)^{l_p} & \dots & (i\omega_p)^{l_p} \end{pmatrix} = (i\omega_1)^{\sum\limits_{j=1}^{p} l_j} \det \begin{pmatrix} 1 & \left(e^{\frac{i\pi}{p}}\right)^{l_1} & \dots & \left(e^{\frac{i(p-1)\pi}{p}}\right)^{l_1} \\ 1 & \left(e^{\frac{i\pi}{p}}\right)^{l_2} & \dots & \left(e^{\frac{i(p-1)\pi}{p}}\right)^{l_2} \\ \cdot & \cdot & \cdot & \cdot \\ 1 & \left(e^{\frac{i\pi}{p}}\right)^{l_p} & \dots & \left(e^{\frac{i(p-1)\pi}{p}}\right)^{l_p} \end{pmatrix}$$

$$= (i\omega_1)^{\sum\limits_{j=1}^{p} l_j} \det \begin{pmatrix} 1 & 1 & \dots & 1 \\ e^{\frac{i\pi l_1}{p}} & e^{\frac{i\pi l_2}{p}} & \dots & e^{\frac{i\pi l_p}{p}} \\ \cdot & \cdot & \cdot & \cdot \\ \left(e^{\frac{i\pi l_1}{p}}\right)^{p-1} & \left(e^{\frac{i\pi l_2}{p}}\right)^{p-1} & \dots & \left(e^{\frac{i\pi l_p}{p}}\right)^{p-1} \end{pmatrix}$$

$$= (i\omega_1)^{\sum\limits_{j=1}^{p} l_j} \prod_{1 \leqslant j < s \leqslant p} \left(e^{\frac{i\pi l_s}{p}} - e^{\frac{i\pi l_j}{p}} \right) \neq 0.$$

Thus

$$\varkappa_1 = \sum_{j=1}^{p} l_j = \frac{p(p-1)}{2} + \varkappa. \tag{3.23}$$

We note in this connection that the row ending up as the last row in the new determinant was not used in obtaining the other rows.

Suppose this last row of the new determinant for $P(z)$ was the β_pth row of the original determinant $P(z)$. We transform the determinant $P^{(\beta_p)}(x, z)$ (see (2.18)) into a new determinant (which we will continue to denote by $P^{(\beta_p)}(x, z)$) with the use of precisely the same operations on its rows as were used in obtaining the new detreminant for $P(z)$. Thus the new determinant for $P^{(\beta_p)}(x, z)$ will differ at most in sign from the original determinant, and the β_pth row of the original determinant goes over into the last row of the new determinant:

$$P^{(\beta_p)}(x, z) = (\pm 1) \det \begin{pmatrix} (i\omega_1 z)^{l_1} b_1^{(1)}(i\omega_1 z) & \ldots & (i\omega_p z)^{l_1} b_1^{(1)}(i\omega_p z) \\ \cdot\ \cdot\ \cdot & \cdot\ \cdot\ \cdot & \cdot\ \cdot\ \cdot \\ (i\omega_1 z)^{l_{p-1}} b_1^{(p-1)}(i\omega_1 z) & \ldots & (i\omega_p z)^{l_{p-1}} b_1^{(p-1)}(i\omega_p z) \\ e^{i\omega_1 zx} & \ldots & e^{i\omega_p zx} \end{pmatrix}, \tag{3.24}$$

while the remaining rows of the new determinant coincide with the corresponding rows of the new determinant for $P(z)$, and consequently $l_1 < l_2 < \cdots < l_{p-1}$ and $l_j \not\equiv l_s \pmod{2p}$ for $j \neq s$, $j, s = 1, \cdots, p-1$.

We have the representation

$$l_j = \alpha_j + 2pm_j, \tag{3.25}$$

in which the α_j, $j = 1, \cdots, p$, are mutually distinct natural numbers from the interval $[0, 2p)$. Let n denote the least natural number from this interval that is not equal to an α_j. Clearly $n \in [0, p)$.

Further, from Taylor's theorem we have the representation

$$e^{i\omega_s zx} = \sum_{k=0}^{n} \frac{1}{k!}(i\omega_s zx)^k + O(z^{n+1}), \qquad s = 1, 2, \ldots, p, \tag{3.26}$$

which, when applied to (3.24), gives us the representation

$$P^{(\beta_p)}(x, z) = \sum_{k=0}^{n} M_k(x)\, z^{k + \sum_{j=1}^{p-1} l_j} + O\left(z^{n+1+\sum_{j=1}^{p} l_j}\right), \tag{3.27}$$

in which $M_k(x)$ is a polynomial in x of degree k:

$$M_k(x) = c_k^{(k)}x^k + c_{k-1}^{(k)}x^{k-1} + \dots + c_0^{(k)}$$

with coefficients $c_s^{(k)}$, $s = 0, 1, \cdots, k$, $k = 0, 1, \cdots, n$. The leading coefficients of the polynomials $M_k(x)$ can be written explicitly:

$$c_k^{(k)} = b_k \det \begin{pmatrix} \omega_1^{l_1} & \omega_2^{l_1} & \dots & \omega_p^{l_1} \\ \omega_1^{l_2} & \omega_2^{l_2} & \dots & \omega_p^{l_2} \\ \cdot & \cdot & \cdot & \cdot \\ \omega_1^{l_{p-1}} & \omega_2^{l_{p-1}} & \dots & \omega_p^{l_{p-1}} \\ \omega_1^k & \omega_2^k & \dots & \omega_p^k \end{pmatrix}, \tag{3.28}$$

where

$$b_k = i^{k+\sum_{j=1}^{p-1} l_j} \prod_{j=1}^{p-1} b_1^{(j)}(0) \neq 0, \qquad k = 0, 1, \dots, n.$$

Taking into account (3.25) and the fact that $\omega_s = \omega_1 e^{i(s-1)\pi/p}$, $s = 1, \cdots, p$, we get from (3.28) that

$$c_n^{(n)} = b_n \omega_1^{n+\sum_{j=1}^{p-1} l_j} \det \begin{pmatrix} 1 & \dots & 1 & 1 \\ e^{i\pi\alpha_1/p} & \dots & e^{i\pi\alpha_{p-1}/p} & e^{i\pi n/p} \\ (e^{i\pi\alpha_1/p})^2 & \dots & (e^{i\pi\alpha_{p-1}/p})^2 & (e^{i\pi n/p})^2 \\ \cdot & \cdot & \cdot & \cdot \\ (e^{i\pi\alpha_1/p})^{p-1} & \dots & (e^{i\pi\alpha_{p-1}/p})^{p-1} & (e^{i\pi n/p})^{p-1} \end{pmatrix}. \tag{3.29}$$

Since, by construction, $\alpha_j \neq n$, $j = 1, \cdots, p$, the determinant of (3.29) is a Vandermonde determinant and, consequently $c_n^{(n)} \neq 0$. But then (3.27) implies that the multiplicity of the root $z = 0$ for $P^{(\beta_p)}(x, z)$ is not greater than $n + \Sigma_{j=1}^{p-1} l_j$ for any $x > 0$. We will show that $n < l_p$. The desired conclusion that the multiplicity of the root $z = 0$ of $P^{(\beta_p)}(x, z)$ is strictly less than κ_1 will then follow from (3.23).

Suppose, on the contrary, that $l_p \leqslant n \leqslant p - 1$. Then the fact that $l_p > l_j$ for $j < p$ implies $\Sigma_{j=1}^{p} l_j = p(p-1)/2$, and since, on the other hand, $\Sigma_{j=1}^{p} l_j = p(p-1)/2 + \kappa$, we get $\kappa = 0$, which contradicts the assumption that $\kappa \geqslant 1$. This the multiplicity of the root $z = 0$ of $P^{(\beta_p)}(x, z)$ is strictly less than κ_1, i.e. if

$$P^{(\beta_p)}(x, z) = z^{\kappa_2} R^{(\beta_p)}(x, z), \tag{3.30}$$

where $R^{(\beta_p)}(x, 0) \not\equiv 0$, then $\kappa_2 < \kappa_1$. We will use this result to construct an unboundedly increasing for $t \to \infty$ solution of problem (1.1)–(1.3).

Suppose $\varphi_j(t) \equiv 0$ for $j \neq \beta_p$ while $\varphi_{\beta_p}(t) \in C^\infty([0, \infty))$, $\varphi_{\beta_p}(t) \equiv 0$ for

$0 \leqslant t \leqslant 1/2$ and $\varphi_{\beta_p}(t) \equiv 1$ for $t \geqslant 1$. Then

$$\tilde{\varphi}_{\beta_p}(\lambda) = \int_0^\infty e^{-\lambda t}\varphi_{\beta_p}(t)\,dt = \int_0^1 e^{-\lambda t}\varphi_{\beta_p}(t)\,dt$$

$$+ \int_1^\infty e^{-\lambda t}dt = \tilde{\varphi}^0_{\beta_p}(\lambda) + \frac{e^{-\lambda}}{\lambda} = \frac{\lambda\tilde{\varphi}^0_{\beta_p}(\lambda) + e^{-\lambda}}{\lambda} = \frac{\psi(\lambda)}{\lambda}, \tag{3.31}$$

where $\psi(\lambda)$ is an entire function of λ having the following property: for any $N > 0$ there exists a constant $C(N)$ such that

$$|\psi(\lambda)| \leqslant C(N) \tag{3.32}$$

for all $|\mathrm{Re}\,\lambda| \leqslant N$; in fact,

$$|\psi(\lambda)| \leqslant |e^{-\lambda}| + |\lambda| \left| \int_0^1 \varphi_{\beta_p}(t)\,e^{-\lambda t}dt \right| \leqslant e^N + |\lambda| \left| \frac{1}{\lambda} e^{-\lambda}\right.$$

$$\left. + \frac{1}{\lambda}\int_0^1 \varphi'_{\beta_p}(t)\,e^{-\lambda t}dt \right| \leqslant e^N + e^N + \max_{0 \leqslant t \leqslant 1} |\varphi_{\beta_p}(t)|\,e^N = C(N).$$

Substituting (3.31) into (2.10) and taking (2.11) into account, we get

$$v(x, \lambda) = \frac{P^{(\beta_p)}(x, \sqrt[2p]{\lambda})}{\lambda P(\sqrt[2p]{\lambda})}\,\psi(\lambda). \tag{3.33}$$

We rewrite this equality in the form

$$v(x, \lambda) = \frac{f(x, \sqrt[2p]{\lambda})\,\psi(\lambda)}{\lambda^{1+(\varkappa_1-\varkappa_2)/2p}}, \tag{3.34}$$

where

$$f(x, \sqrt[2p]{\lambda}) = \frac{R^{(\beta_p)}(x, \sqrt[2p]{\lambda})}{R(\sqrt[2p]{\lambda})}, \tag{3.35}$$

in which $R(\sqrt[2p]{\lambda})$ and $R^{(\beta_p)}(x, \sqrt[2p]{\lambda})$ are taken from (3.20) and (3.30). Since $Q(z) = z^\kappa R(z)$, $R(0) \neq 0$, and the roots of the polynomial $Q(z)$ can, as we have already proved, be distributed only in the sector $|\arg z| > \pi/4p$, $z \neq 0$, the roots of $R(\sqrt[2p]{\lambda})$ are distributed in the halfplane $\mathrm{Re}\,\lambda < 0$. Since the number of roots of $R(\sqrt[2p]{\lambda})$ is finite, there exists a number $\delta > 0$ such that the roots of $R(\sqrt[2p]{\lambda})$ will be distributed in the halfplane $\mathrm{Re}\,\lambda < -\delta$. Therefore the function $f(x, z)$ is analytic with respect to $z = \sqrt[2p]{\lambda}$ for $\mathrm{Re}\,\lambda \geqslant -\delta$ and consequently admits in a neighborhood of $z = \sqrt[2p]{\lambda} = 0$ the representation

$$f\left(x, \sqrt[2p]{\lambda}\right)=f_0(x)+f_1(x)\lambda^{1/2p}+\ldots+f_{\nu-2p-1}(x)\lambda^{(\nu-2p-1)/2p}$$
$$+f^*(x, \lambda^{1/2p})\lambda^{(\nu-2p)/2p}, \qquad (3.36)$$

in which $f_0(x)\not\equiv 0$, $\nu=2p+\kappa_1-\kappa_2\geqslant 2p+1$ (inasmuch as $\kappa_1\geqslant\kappa_2+1$) and $f^*(x, z)$ is analytic with respect to $z=\sqrt[2p]{\lambda}$ for $\operatorname{Re}\lambda\geqslant-\delta$.

Representing $\psi(\lambda)$ in the form

$$\psi(\lambda)=\lambda\tilde{\varphi}^0_{\beta_p}(\lambda)+e^{-\lambda}=1+C_1\lambda+C_2\lambda^2+\ldots,$$

we get from (3.36) that

$$f\left(x, \sqrt[2p]{\lambda}\right)\psi(\lambda)=f_0(x)+\psi_1(x)\lambda^{1/2p}+\ldots+\psi_{\nu-2p-1}(x)\lambda^{(\nu-2p-1)/2p}$$
$$+\psi^*\left(x, \sqrt[2p]{\lambda}\right)\lambda^{(\nu-2p)/2p}, \qquad (3.37)$$

where $\psi^*(x, z)$ is for any $x>0$ an analytic function of z for $\operatorname{Re}\lambda\geqslant-\delta$. Substituting (3.37) into (3.34), we get

$$v(x, \lambda)=\frac{f_0(x)}{\lambda^{\nu/2p}}+\frac{\psi_1(x)}{\lambda^{(\nu-1)/2p}}+\ldots+\frac{\psi_{\nu-2p-1}(x)}{\lambda^{(2p+1)/2p}}+\frac{\psi^*\left(x, \sqrt[2p]{\lambda}\right)}{\lambda}. \qquad (3.38)$$

Taking the inverse Laplace transform of both sides of this equality, we get

$$u(x, t)=\frac{1}{2\pi i}\int_\gamma e^{\lambda t}v(x, \lambda)\,d\lambda=\frac{f_0(x)}{2\pi i}\int_\gamma\frac{e^{\lambda t}}{\lambda^{\nu/2p}}\,d\lambda+\frac{\psi_1(x)}{2\pi i}\int_\gamma\frac{e^{\lambda t}}{\lambda^{(\nu-1)/2p}}\,d\lambda+$$
$$\ldots+\frac{\psi_{\nu-2p-1}(x)}{2\pi i}\int_\gamma\frac{e^{\lambda t}}{\lambda^{(2p+1)/2p}}\,d\lambda+\frac{1}{2\pi i}\int_\gamma\frac{e^{\lambda t}\psi^*\left(x, \sqrt[2p]{\lambda}\right)}{\lambda}\,d\lambda$$
$$=\frac{f_0(x)}{\Gamma\left(\frac{\nu}{2p}\right)}t^{(\nu-2p)/2p}+\frac{\psi_1(x)}{\Gamma\left(\frac{\nu-1}{2p}\right)}t^{(\nu-2p-1)/2p}+\ldots+\frac{\psi_{\nu-2p-1}(x)}{\Gamma\left(\frac{2p-1}{2p}\right)}t^{1/2p}$$
$$+\frac{1}{2\pi i}\int_\gamma\frac{e^{\lambda t}\psi^*\left(x, \sqrt[2p]{\lambda}\right)}{\lambda}\,d\lambda. \qquad (3.39)$$

From (3.37) and (3.35) it follows that $\psi^*(x, z)$ is for any $x>0$ an analytic function of $z=\sqrt[2p]{\lambda}$ for $\operatorname{Re}\lambda\geqslant-\delta$.

From (3.33), (2.26) and (3.32) it follows that for any $x>0$ and $|\lambda|\to\infty$ in the strip $|\operatorname{Re}\lambda|<N$

$$|v(x, \lambda)|\leqslant K(x)\exp\left(-\theta\sqrt[2p]{|\lambda|}\,x\right), \qquad (3.40)$$

where $K(x)$ is a certain positive constant depending on x and θ is a positive number.

From (3.38) and (3.40) we get for any $x>0$ and sufficiently large $|\lambda|$ that

$$\left|\frac{\psi^*\left(x, \sqrt[2p]{\lambda}\right)}{\lambda}\right| \leqslant \frac{K_1(x)}{|\lambda|^{(2p+1)/2p}}, \tag{3.41}$$

where $K_1(x)$ is a positive constant depending on x.

In addition, for any $x > 0$ the function $\psi^*(x, \sqrt[2p]{\lambda})$ is bounded with respect to λ in a neighborhood of the origin. We now estimate the integral

$$\int_\gamma \frac{e^{\lambda t}\psi^*\left(x, \sqrt[2p]{\lambda}\right)}{\lambda}\, d\lambda, \quad \gamma = \{\operatorname{Re}\lambda = \sigma,\ \sigma > 0\} \tag{3.42}$$

in the right side of (3.39). Cauchy's theorem together with (3.41) permits us to replace the contour of integration γ in (3.42) by the contour $\gamma' = \gamma_1 \cup \gamma_\epsilon^+ \cup \gamma_\epsilon \cup \gamma_\epsilon^- \cup \gamma_2$:

$$\int_\gamma \frac{e^{\lambda t}\psi^*\left(x, \sqrt[2p]{\lambda}\right)}{\lambda}\, d\lambda = \left(\int_{\gamma_1\cup\gamma_2} + \int_{\gamma_\varepsilon} + \int_{\gamma_\varepsilon^+\cup\gamma_\varepsilon^-}\right)\frac{e^{\lambda t}\psi^*\left(x, \sqrt[2p]{\lambda}\right)}{\lambda}\, d\lambda. \tag{3.43}$$

We estimate each of the summands in the right side of this equality separately. For the first summand, using (3.41), we get

$$\left|\int_{\gamma_1\cup\gamma_2} \frac{e^{\lambda t}\psi^*\left(x, \sqrt[2p]{\lambda}\right)}{\lambda}\, d\lambda\right| \leqslant K_2(x)\, e^{-\delta t}, \tag{3.44}$$

where $K_2(x)$ is a certain positive constant depending on x. For the second summand, making the substitution $\lambda = \epsilon e^{i\varphi}$, we get

$$\lim_{\varepsilon\to 0}\left|\int_{\gamma_\varepsilon} \frac{e^{\lambda t}\psi^*\left(x, \sqrt[2p]{\lambda}\right)}{\lambda}\, d\lambda\right| \leqslant 2\pi\psi^*(x, 0). \tag{3.45}$$

For the third summand, representing the function $\psi^*(x, \sqrt[2p]{\lambda})$ in the form $\psi^*(x, 0) + O(\sqrt[2p]{\lambda})$, we get

$$\begin{aligned}\lim_{\varepsilon\to 0}&\left|\int_{\gamma_\varepsilon^+\cup\gamma_\varepsilon^-} \frac{e^{\lambda t}\psi^*\left(x, \sqrt[2p]{\lambda}\right)}{\lambda}\, d\lambda\right| \\ &= \left|\int_0^\delta \frac{e^{-yt}}{y}\left[\psi^*\left(x, \sqrt[2p]{ye^{i\pi}}\right) - \psi^*\left(x, \sqrt[2p]{ye^{-i\pi}}\right)\right] dy\right| \\ &\leqslant \left|\int_0^\delta \frac{e^{-yt}}{y}\, O\left(\sqrt[2p]{y}\right) dy\right| \leqslant \frac{K_3(x)}{t^{1/2p}},\end{aligned} \tag{3.46}$$

where $K_3(x)$ is a constant not depending on t. It follows from (3.43)–(3.46) that

$$\left|\int\limits_{\gamma} \frac{e^{\lambda t}\psi^*(x, \sqrt[2p]{\lambda})}{\lambda}\, d\lambda\right| \leqslant 2\pi\psi^*(x, 0) + \frac{K_4(x)}{t^{1/2p}}, \tag{3.47}$$

where $K_4(x) = \max\{K_2(x), K_3(x)\}$.

Taking now a positive value of x such that $f_0(x) \neq 0$, from (3.39), (3.47) and the fact that $\nu \geqslant 2p + 1$ we get that $u(x, t)$ increases at least as fast as $t^{1/2p}$ as $t \to \infty$, which contradicts the assumption concerning the stabilization of $u(x, t)$ as $t \to \infty$ for any stabilizing boundary functions.

The necessity is proved.

SUFFICIENCY. Suppose the roots of the polynomial $Q(z)$ are distributed in the sector $|\arg z| > \pi/4p$, $z \neq 0$. We will show that the solution $u(x, t)$ of problem (1.1)–(1.3) is stabilized as $t \to \infty$ for any stabilizing boundary functions $\varphi_j(t)$, $j = 1, \cdots, p$.

We first assume that $\lim_{t\to\infty} \varphi_j(t) = 0$, and show that $\lim_{t\to\infty} u(x, t) = 0$ for any $x > 0$. Since $\lim_{t\to\infty} \varphi_j(t) = 0$, there exists for any $\epsilon > 0$ a number $T > 0$ such that $|\varphi_j(t)| < \epsilon$, $j = 1, \cdots, p$, as long as $t > T$. Further, the convergence of the integral

$$\int\limits_0^\infty \frac{d\tau}{1 + \tau^{1+1/2p}}$$

implies that for this ϵ there exists a number $M = M(\epsilon)$ such that

$$\int\limits_M^\infty \frac{d\tau}{1 + \tau^{1+1/2p}} < \varepsilon.$$

It therefore follows from (2.28) and (2.14) that for $t > T + M$

$$\begin{aligned}
|u_j(x, t)| &= \left|\int\limits_0^t G_j(x, t-\tau)\varphi_j(\tau)\, d\tau\right| \\
&= \left|\int\limits_0^T G_j(x, t-\tau)\varphi_j(\tau)\, d\tau + \int\limits_T^t G_j(x, t-\tau)\varphi_j(\tau)\, d\tau\right| \\
&\leqslant C(x) \max_{0\leqslant\tau\leqslant\infty} |\varphi_j(\tau)| \int\limits_0^T \frac{d\tau}{1 + (t-\tau)^{1+1/2p}} + \varepsilon C(x) \int\limits_T^t \frac{1}{1 + (t-\tau)^{1+1/2p}}\, d\tau \\
&= C(x) \max_{0\leqslant\tau<\infty} |\varphi_j(\tau)| \int\limits_{t-T}^t \frac{d\tau}{1 + \tau^{1+1/2p}} + \varepsilon C(x) \int\limits_0^{t-T} \frac{d\tau}{1 + \tau^{1+1/2p}} \leqslant
\end{aligned}$$

$$\leqslant C(x)\varepsilon\left(\max_{0\leqslant\tau<\infty}|\varphi_j(\tau)|+\int_0^\infty\frac{d\tau}{1+\tau^{1+1/2p}}\right)=C_1(x)\varepsilon.$$

This means that

$$\lim_{t\to\infty}u(x,\ t)=0. \tag{3.48}$$

REMARK. We note that the consistency conditions for the functions $\varphi_j(t)$ were not used in the above proof, and hence (3.48) holds for any bounded on $[0,\infty)$ functions $\varphi_j(t)$ tending to zero as $t\to\infty$, independently of whether or not they satisfy the consistency conditions at $t=0$.

Suppose now $\varphi_j(t)\to A_j$ as $t\to\infty$. We have the representation

$$\varphi_j(t)=A_j+\varphi_j^{(0)}(t), \tag{3.49}$$

in which the functions

$$\varphi_j^{(0)}(t)\to 0 \quad \text{as} \quad t\to\infty \tag{3.50}$$

and generally do not satisfy the consistency conditions but of course are bounded for all $t\in[0,\infty)$.

By virtue of (2.14), (3.48) and the above remark

$$\begin{aligned}u(x,\ t)&=\sum_{j=1}^{p}u_j(x,\ t)=\sum_{j=1}^{p}\int_0^t G_j(x,\ t-\tau)A_j\,d\tau\\&+\sum_{j=1}^{p}\int_0^t G_j(x,\ t-\tau)\varphi_j^{(0)}(\tau)\,d\tau=\sum_{j=1}^{p}\int_0^t G_j(x,\ t-\tau)A_j\,d\tau+\varepsilon(x,\ t)\\&=\sum_{j=1}^{p}G_j(x,\ t)*A_j+\varepsilon(x,\ t)=u_0(x,\ t)+\varepsilon(x,\ t),\end{aligned} \tag{3.51}$$

where $\epsilon(x,t)\to 0$ as $t\to\infty$ for any $x>0$.

The Laplace transform of the function $u_0(x,t)$ in (3.51) is equal to

$$v_0(x,\ \lambda)=\sum_{j=1}^{p}\frac{A_j}{\lambda}\tilde{G}_j\left(x,\ \sqrt[2p]{\lambda}\right)=\sum_{j=1}^{p}\frac{A_jQ^{(j)}\left(x,\ \sqrt[2p]{\lambda}\right)}{\lambda Q\left(\sqrt[2p]{\lambda}\right)} \tag{3.52}$$

in terms of which

$$u_0(x,\ t)=\frac{1}{2\pi i}\int_\gamma e^{\lambda t}v_0(x,\ \lambda)\,d\lambda,\quad \gamma=\{\operatorname{Re}\lambda=\sigma,\ \sigma>0\}. \tag{3.53}$$

From (3.52) and (2.25) we get that in a neighborhood of the origin of the λ plane

$$|v_0(x,\ \lambda)|\leqslant\frac{C(x)}{|\lambda|}. \tag{3.54}$$

Also, we get from (3.52) with the use of (2.26) that for $|\lambda| \to \infty$ in the sector $|\arg \lambda| \leqslant \pi - \theta_0$

$$|v_0(x, \lambda)| \leqslant C_1(x) e^{-\theta |\lambda|^{1/2p} x}, \tag{3.55}$$

where θ_0 and θ are certain positive constants.

Taking into account (3.52), (3.55) and the fact that $Q(\sqrt[2p]{\lambda})$ does not have any roots in the halfplane $\operatorname{Re} \lambda \geqslant -\delta$, where δ is a positive number, we replace the contour γ in (3.53) by the contour γ' given by (2.29) with the use of Cauchy's theorem and proceed, as before, to estimate the resultant summands. In this connection we get

$$\left| \frac{1}{2\pi i} \int\limits_{\gamma_1 \cup \gamma_2} e^{\lambda t} v_0(x, \lambda)\, d\lambda \right| \leqslant C_2(x) e^{-\delta t}, \tag{3.56}$$

where $C_2(x)$ is a constant not depending on t.

Further, taking into account (3.52) and the fact that for any $x > 0$

$$\frac{Q^{(j)}(x, \sqrt[2p]{\lambda})}{Q(\sqrt[2p]{\lambda})} = \frac{Q^{(j)}(x, 0)}{Q(0)} + o(\sqrt[2p]{\lambda}) \tag{3.57}$$

in the halfplane $\operatorname{Re} \lambda \geqslant -\delta$, we get

$$\lim_{\varepsilon \to 0} \frac{1}{2\pi i} \int\limits_{\gamma_\varepsilon} e^{\lambda t} v_0(x, \lambda)\, d\lambda = \lim_{\varepsilon \to 0} \frac{1}{2\pi i} \int\limits_{\gamma_\varepsilon} e^{\lambda t} \sum_{j=1}^{p} \frac{A_j}{\lambda} \tilde{G}_j(x, \sqrt[2p]{\lambda})\, d\lambda$$

$$= \lim_{\varepsilon \to 0} \frac{1}{2\pi i} \int\limits_{\gamma_\varepsilon} \frac{1}{\lambda Q(\sqrt[2p]{\lambda})} e^{\lambda t} \left[\sum_{j=1}^{p} A_j Q^{(j)}(x, \sqrt[2p]{\lambda}) \right] d\lambda$$

$$= \lim_{\varepsilon \to 0} \frac{1}{2\pi i Q(0)} \left[\sum_{j=1}^{p} A_j Q^{(j)}(x, 0) \right] \int\limits_{\gamma_\varepsilon} \frac{e^{\lambda t}}{\lambda}\, d\lambda$$

$$+ \lim_{\varepsilon \to 0} \int\limits_{\gamma_\varepsilon} \frac{e^{\lambda t} o(\sqrt[2p]{\lambda})}{\lambda}\, d\lambda = \frac{1}{Q(0)} \sum_{j=1}^{p} A_j Q^{(j)}(x, 0). \tag{3.58}$$

Finally, with the use of (3.57) we get

$$\lim_{\varepsilon \to 0} \left| \frac{1}{2\pi i} \int\limits_{\gamma_\varepsilon^+ \cup \gamma_\varepsilon^-} e^{\lambda t} v_0(x, \lambda)\, d\lambda \right| = \left| \frac{1}{2\pi i} \int_0^\delta e^{-yt} [v_0(x, ye^{i\pi}) - v_0(x, ye^{-i\pi})]\, dy \right|$$

$$= \frac{1}{2\pi} \left| \int_0^\delta \frac{e^{-yt}}{y} \left[\frac{e^{-i\pi}}{Q(\sqrt[2p]{ye^{i\pi}})} \sum_{j=1}^{p} A_j Q^{(j)}(x, \sqrt[2p]{y} e^{i\pi/2p}) \right.\right.$$

$$\left.\left. - \frac{e^{i\pi}}{Q(\sqrt[2p]{ye^{-i\pi}})} \sum_{j=1}^{p} A_j Q^{(j)}(x, \sqrt[2p]{y} e^{-i\pi/2p}) \right] dy \right|$$

$$= \left| \int_0^\delta \frac{e^{-yt}}{y} o(\sqrt[2p]{y})\, dy \right| \leqslant \left| \int_0^\infty \frac{e^{-yt}}{y} o(\sqrt[2p]{y})\, dy \right| \leqslant \frac{C_3(x)}{\sqrt[2p]{t}}, \tag{3.59}$$

where $C_3(x)$ is a constant not depending on t. From (3.56), (3.58) and (3.59) we conclude that

$$\lim_{t\to\infty} u_0(x, t) = \frac{1}{Q(0)} \sum_{j=1}^{p} A_j Q^{(j)}(x, 0).$$

Theorem 1 now follows from equality (3.51), with

$$\lim_{t\to\infty} u(x, t) = \frac{1}{Q(0)} \sum_{j=1}^{p} A_j Q^{(j)}(x, 0). \tag{3.60}$$

Theorem 1 is proved.

THEOREM 2. *If the roots of the polynomial* $Q(z)$ *are distributed in the sector* $|\arg z| > \pi/4p$, $z \neq 0$, *then the solution* $u(x, t)$ *of problem* (1.1)–(1.3) *for any stabilizing boundary functions* $\varphi_j(t)$, $j = 1, \cdots, p$, *tends to a polynomial in* x *of degree* $p-1$ *as* $t \to \infty$.

PROOF. From equality (3.60) it follows that it suffices to establish that $Q^{(j)}(x, 0)$, $j = 1, \cdots, p$, is a polynomial in x of degree $p-1$. Suppose the determinant $P^{(j)}(x, z) = z^{p(p-1)/2} Q^{(j)}(x, z)$ has already been reduced to the form

$$P^{(j)}(x, z) = \det \begin{pmatrix} (i\omega_1 z)^{l_1} b^{(1)}(i\omega_1 z) & \dots & (i\omega_p z)^{l_1} b^{(1)}(i\omega_p z) \\ (i\omega_1 z)^{l_2} b^{(2)}(i\omega_1 z) & \dots & (i\omega_p z)^{l_2} b^{(2)}(i\omega_p z) \\ \cdots & \cdots & \cdots \\ (i\omega_p z)^{l_{p-1}} b^{(p-1)}(i\omega_1 z) & \dots & (i\omega_p z)^{l_{p-1}} b^{(p-1)}(i\omega_p z) \\ e^{i\omega_1 z x} & \dots & e^{i\omega_p z x} \end{pmatrix} \tag{3.61}$$

where $l_1 < l_2 < \cdots < l_{p-1}$ and $b^{(j)}(0) \neq 0$, $j = 1, \cdots, p$.

We note that since $Q(0) \neq 0$ by assumption, only the cases $l_m \leqslant m$, $m = 1, \cdots, p$, are possible. If $l_m = m$, the determinant (3.61) can be written in the form

$$P^{(j)}(x, z) = z^{p(p-1)/2} \left[V \prod_{j=1}^{p-1} b^{(j)}(0) + O(z) \right], \tag{3.62}$$

where V is the Vandermonde determinant $\det((i\omega_s)^{j-1})$, $j, s = 1, \cdots, p$, i.e.

$$V = i^{p(p-1)/2} \prod_{1 \leqslant m < s \leqslant p} (\omega_s - \omega_m). \tag{3.63}$$

From (3.62), taking (2.23) into account, we get

$$Q^{(j)}(x,\ z) = V \prod_{j=1}^{p-1} b^{(j)}(0) + O(z),$$

i.e. $Q^{(j)}(x, 0) = V \prod_{j=1}^{p-1} b^{(j)}(0)$, in the case $l_m = m$, $m = 1, \cdots, p-1$.

Consider the case $l_1 = 0$, $l_2 = 1, \cdots, l_k = k-1$, $l_{k+1} = k+1, \cdots, l_{p-1} = p-1$ for some $k \in [1, p)$, and let

$$b^{(j)}(i\omega_s z) = b_0^{(j)}(i\omega_s z)^{l_j} c^{(j)}(i\omega_s z),$$

where $b_0^{(j)} = b^{(j)}(0) \neq 0$, $j = 1, \cdots, p$, and $c^{(j)}(i\omega_s z) = 1 + c_1^{(j)}(i\omega_s z) + \cdots + c_{n_j - l_j}^{(j)}(i\omega_s z)^{n_j - l_j}$, $j, s = 1, \cdots, p$. The determinant (3.61) then takes the form

$$P^{(j)}(x,\ z) = \prod_{m=1}^{p-1} b_0^{(m)}$$

$$\times \det \begin{pmatrix} c^{(1)}(i\omega_1 z) & \ldots & c^{(1)}(i\omega_p z) \\ (i\omega_1 z)\, c^{(2)}(i\omega_1 z) & \ldots & (i\omega_p z)\, c^{(2)}(i\omega_p z) \\ \cdots & \cdots & \cdots \\ (i\omega_1 z)^{k-1} c^{(k)}(i\omega_1 z) & \ldots & (i\omega_p z)^{k-1} c^{(k)}(i\omega_p z) \\ (i\omega_1 z)^{k+1} c^{(k+1)}(i\omega_1 z) & \ldots & (i\omega_p z)^{k+1} c^{(k+1)}(i\omega_p z) \\ \cdots & \cdots & \cdots \\ (i\omega_1 z)^{p-1} c^{(p-1)}(i\omega_1 z) & \ldots & (i\omega_p z)^{p-1} c^{(p-1)}(i\omega_p z) \\ \sum_{m=0}^{k} \frac{1}{m!}(i\omega_1 zx)^m + O(z^{k+1}) & \ldots & \sum_{m=0}^{k} \frac{1}{m!}(i\omega_p zx)^m + O(z^{k+1}) \end{pmatrix}. \tag{3.64}$$

We transform the last row of this determinant as follows. We begin by subtracting the first row from the last row, so that the sth element of the last row takes the form

$$(i\omega_s z)(x - c_1^{(1)}) + (i\omega_s z)^2\left(\frac{1}{2!}x^2 - c_2^{(1)}\right) + \\ \cdots + (i\omega_s z)^k\left(\frac{1}{k!}x^k - c_k^{(1)}\right) + O(z^{k+1}). \tag{3.65}$$

We next multiply the second row by the polynomial $M_1^{(j)}(x) = x - c_1^{(1)}$ and subtract the result from the new last row indicated by (3.65), so that the sth element of the last row next takes the form

$$(i\omega_s z)^2\left[\frac{1}{2!}x^2 - c_2^{(1)} - M_1^{(j)}c_1^{(2)}\right] +$$

$$\dots + (i\omega_s z)^k\left[\frac{1}{k!}x^k - c_k^{(1)} - M_1^{(j)}c_k^{(2)}\right] + O(z^{k+1}). \qquad (3.66)$$

Continuing in this manner, we multiply the third row by the polynomial $M_2^{(j)}(x) = x^2/2! - c_2^{(1)} - M_1^{(j)}(x)c_1^{(2)}$ and subtract the result from the new last row indicated by (3.66), etc. In this way we reduce each element of the last row to the form

$$M_k^{(j)}(x)(i\omega_s z)^k + O(z^{k+1}), \qquad s = 1, 2, \dots, p, \qquad (3.67)$$

in which

$$M_k^{(j)}(x) = \frac{1}{k!}x^k - c_k^{(1)} - c_k^{(2)}M_1^{(j)}(x) - \dots - c_k^{(k)}M_{k-1}^{(j)}(x), \qquad (3.68)$$

$k = 1, \cdots, p-1$ and $j = 1, \cdots, p$. We then transfer the new last row indicated by (3.68) to the $(k+1)$th place. This gives us

$$'P^{(j)}(x, z) = z^{p(p-1)/2}\left[VM_k^{(j)}(x)\prod_{m=1}^{p-1} b_0^{(m)} + O(z)\right], \qquad (3.69)$$

where V is the determinant (3.63). From (3.69) and (2.23) it follows that

$$Q^{(j)}(x, z) = M_k^{(j)}(x)V\prod_{m=1}^{p-1} b_0^{(m)} + O(z), \qquad (3.70)$$

where $j = 1, \cdots, p$ and $M_k^{(j)}(x)$ is the polynomial (3.68) of degree k.

Since k can take any of the values $1, \cdots, p-1$, the degree of the polynomial $M_k^{(j)}(x)$ is, in general, equal to $p-1$. We denote it by $M^{(j)}(x)$.

From (3.70) it follows that

$$Q^{(j)}(x, 0) = M^{(j)}(x)V\prod_{m=1}^{p-1} b_0^{(m)}. \qquad (3.71)$$

From (3.60) and (3.71) we have

$$\lim_{t\to\infty} u(x, t) = \frac{V}{Q(0)}\prod_{m=1}^{p-1} b_0^{(m)}\sum_{j=1}^{p} A_j M^{(j)}(x). \qquad (3.72)$$

Theorem 2 is proved.

§4. Examples

EXAMPLE 1. As an application of Theorems 1 and 2, we note the first boundary value problem, in which the boundary conditions (1.3) have the form

$$b^{(j)}\left(\frac{\partial}{\partial x}\right)u(x,\ t)\,|_{x=0}\equiv\frac{\partial^{j-1}u(x,\ t)}{\partial x^{j-1}}\bigg|_{x=0}=\varphi_j(t),\ j=1,\ 2,\ \ldots,\ p. \tag{4.1}$$

Here

$$P(z)=\det((i\omega_s z)^{j-1})=z^{p(p-1)/2}\det((i\omega_s)^{j-1})=z^{p(p-1)/2}V, \tag{4.2}$$

where V is the determinant (3.63). From (4.2) and (2.23) it follows that $Q(z) = V \neq 0$, so that $Q(0) = V \neq 0$.

According to Theorem 1, the solution $u(x, t)$ of problem (1.1), (1.2), (4.1) is stabilized for $t \to \infty$ for any stabilizing boundary functions $\varphi_j(t)$, $j = 1, \cdots, p$. In addition, if $\lim_{t\to\infty} \varphi_j(t) = A_j$, then

$$\lim_{t\to\infty} u(x,\ t)=\sum_{j=1}^{p}\frac{A_j}{(j-1)\,!}\,x^{j-1}. \tag{4.3}$$

In fact, (4.3) follows from (3.72) inasmuch as the boundary conditions (4.1) imply $Q(0) = V$, $\Pi_{m=1}^{p-1} b_0^{(m)} = 1$ and $M^{(j)}(x) = x^{j-1}/(j-1)!$, $j = 1, \cdots, p$.

EXAMPLE. 2. Consider the Nuemann problem for the heat equation

$$\frac{\partial u(x,\ t)}{\partial t}=\frac{\partial^2 u(x,\ t)}{\partial x^2},\qquad x>0,\qquad t>0, \tag{4.4}$$

$$u(x,\ 0)=0,\qquad x>0, \tag{4.5}$$

$$\frac{\partial u(x,\ t)}{\partial x}\bigg|_{x=0}=\varphi(t),\qquad t>0. \tag{4.6}$$

In this case $P(z) = i \cdot iz = -z$, $p = 1$, $p(p-1)/2 = 0$ and, from (2.17), $Q(z) = -z$. By Theorem 1 there exists a boundary function $\varphi(t)$ having a finite limit as $t \to \infty$ for which the solution $u(x, t)$ of problem (4.4)–(4.6) is not stablized for $t \to \infty$. Suppose, for example, $\varphi(t) \in C^\infty([0, \infty))$, $\varphi(t) \equiv 0$ for $0 \leqslant t \leqslant 1/2$ and $\varphi(t) \equiv 1$ for $t \geqslant 1$. Then the solution of problem (4.4)–(4.6) can be written in the form

$$u(x,\ t)=-\frac{1}{\sqrt{\pi}}\int_0^1\frac{\varphi(\tau)}{\sqrt{t-\tau}}\exp\left[-\frac{x^2}{4(t-\tau)}\right]d\tau-\frac{1}{\sqrt{\pi}}\int_1^t\frac{\exp\left[-\frac{x^2}{4(t-\tau)}\right]}{\sqrt{t-\tau}}\,d\tau. \tag{4.7}$$

We estimate the two summands in the right side of this equality for $t > 1$ and for any $x \geqslant 0$:

$$\frac{1}{\sqrt{\pi}}\left|\int_0^1\frac{\varphi(\tau)}{\sqrt{t-\tau}}\exp\left[-\frac{x^2}{4(t-\tau)}\right]d\tau\right|\leqslant\frac{1}{\sqrt{\pi}}\max_{0\leqslant\tau\leqslant1}|\varphi(\tau)|$$

$$\times\int_0^1\frac{d\tau}{\sqrt{t-\tau}}=\frac{k_0}{\sqrt{t-1}+\sqrt{t}}, \tag{4.8}$$

where $k_0 = (2/\sqrt{\pi}) \max_{0 \leqslant \tau \leqslant 1} |\varphi(\tau)|$,

$$-\frac{1}{\sqrt{\pi}} \int_1^t \frac{1}{\sqrt{t-\tau}} \exp\left[-\frac{x^2}{4(t-\tau)}\right] d\tau = -\frac{x}{\sqrt{\pi}} \int_{\frac{x}{2\sqrt{t-\tau}}}^{\infty} \frac{\exp(-\xi^2)}{\xi^2} d\xi \to -\infty \tag{4.9}$$

as $t \to \infty$ for any $x > 0$. From (4.7), (4.8) and (4.9) it follows that as $t \to \infty$ the solution

$$u(x, t) = -\frac{1}{\sqrt{\pi}} \int_0^t \frac{\varphi(\tau)}{\sqrt{t-\tau}} \exp\left[-\frac{x^2}{4(t-\tau)}\right] d\tau \to -\infty$$

for all $x > 0$.

EXAMPLE 3. Consider the problem

$$\frac{\partial u(x, t)}{\partial t} = -\frac{\partial^4 u(x, t)}{\partial x^4}, \quad x \geqslant 0, \quad t \geqslant 0, \tag{4.10}$$

$$u(x, 0) = 0, \quad x \geqslant 0, \tag{4.11}$$

$$\left.\frac{\partial u(x, t)}{\partial x}\right|_{x=0} = \varphi_1(t), \quad \left.\frac{\partial^4 u(x, t)}{\partial x^4}\right|_{x=0} = \varphi_2(t), \quad t \geqslant 0. \tag{4.12}$$

Here $P(z) = -i\sqrt{2}z^5$, $p = 2$, $p(p-1)/2 = 1$ and hence $Q(z) = -i\sqrt{2}z^4$, $Q(0) = 0$. By Theorem 1 the solution $u(x, t)$ of problem (4.10)–(4.12) is generally not stabilized for arbitrary stabilizing boundary functions $\varphi_1(t)$ and $\varphi_2(t)$. We will show, however, that if the boundary function $\varphi_2(t)$ is integrable on $(0, \infty)$, i.e. if the integral $\int_0^\infty \varphi_2(\tau)\, d\tau = A_2$ exists, then the solution $u(x, t)$ of problem (4.10)–(4.12) will be stabilized for $t \to \infty$ for any stabilizing function $\varphi_1(t)$. If $u(x, t)$ is the solution of problem (4.10)–(4.12), then by virtue of (4.10)

$$\varphi_2(t) = \left.\frac{\partial^4 u(x, t)}{\partial x^4}\right|_{x=0} = -\left.\frac{\partial u}{\partial t}\right|_{x=0},$$

i.e., in addition to conditions (4.12), the function $u(x, t)$ also satisfies the conditions

$$u(x, t)|_{x=0} = -\int_0^t \varphi_2(\tau)\, d\tau,$$

$$\left.\frac{\partial u(x, t)}{\partial x}\right|_{x=0} = \varphi_1(t). \tag{4.12'}$$

According to the result of Example 1, the solution $u(x, t)$ of problem (4.10), (4.11), (4.12′) is stabilized for $t \to \infty$ for any stabilizing boundary functions $\varphi_1(t)$

and $-\int_0^t \varphi_2(\tau)d\tau$, i.e. by virtue of our assumption concerning the function $\varphi_2(t)$ the solution $u(x, t)$ of problem (4.10)–(4.12) is stabilized for any function $\varphi_1(t)$ having a finite limit as $t \to \infty$.

This example shows that, in spite of the fact that the solution $u(x, t)$ of problem (4.10)–(4.12) will not be stabilized for arbitrary stabilizing boundary functions $\varphi_1(t)$ and $\varphi_2(t)$, it is possible to impose conditions on the functions $\varphi_1(t)$ and $\varphi_2(t)$ (e.g. $\varphi_1(t)$ and $\int_0^t \varphi_2(\tau)d\tau$ have finite limits as $t \to \infty$) under which the solution $u(x, t)$ of problem (4.10)–(4.12) will still be stabilized. We note that it is not necessary in this connection to require that the function $\varphi_2(t)$ be stabilizing for $t \to \infty$ but merely to require that the integral $\int_0^\infty \varphi_2(\tau)d\tau$ converge; for example, the solution of problem (4.10)–(4.12) will be stabilized for $t \to \infty$ for any stabilizing function $\varphi_1(t)$ and $\varphi_2(t) = \sin t^2$.

In conclusion, I wish to express my sincere gratitude to V. P. Mihaĭlov and A. K. Guščin for their constant interest in the present work.

BIBLIOGRAPHY

1. I. G. Petrovskiĭ, *Über das Cauchysche Problem für ein System linearer partieller Differentialgleichungen im Gebiete der nichtanalytischen Funktionen,* Bull. Univ. Moscou Sér. Internat. Sect. A Math. Mech. **1** (1937/38), no. 7.

2. E. K. Isakova, *A general boundary value problem for parabolic equations on a plane,* Dokl. Akad. Nauk SSSR **153** (1963), 1249–1252 = Soviet Math. Dokl. **4** (1963), 1804–1808. MR **28** #1407.

3. V. A. Solonnikov, *On boundary value problems for linear parabolic systems of differential equations of general form,* Trudy Mat. Inst. Steklov. **83** (1965) = Proc. Steklov Inst. Math. **83** (1965). MR **35** #1965.

4. A. N. Tihonov, *On boundary conditions containing derivatives of order higher than the order of the equation,* Mat. Sb. **26 (68)** (1950), 35–56; English transl., Amer. Math. Soc. Transl. (1) **4** (1962), 440–466. MR **11**, 440.

5. R. K. Juberg, *A generalization of a result of Tihonov,* Arch. Rational Mech. Anal. **22** (1966), 153–162. MR **33** #429.

Trudy Mat. Inst. Steklov.
126 (1973)

Proc. Steklov Inst. Math.
126 (1973)

A MIXED PROBLEM WITH A DISCONTINUOUS BOUNDARY OPERATOR FOR A HYPERBOLIC EQUATION

UDC 517.947.4

V. I. ČEHLOV

Abstract. In this paper existence and uniqueness theorems are proved for the solution in classes of functions defined by the operator of the equation. The dependence of the smoothness of the solution on the form of the discontinuity manifold of the boundary operator is studied. Cases are indicated when a discontinuity of the boundary operator has no effect on the smoothness of the solution within the domain, as well as cases when a discontinuity is propagated into the domain.

Bibliography: 11 items.

Introduction

In this paper we consider a mixed problem for a hyperbolic equation with constant coefficients

$$\left.\begin{aligned} & A(D_t+ia,\ D_y)\,u(t,\ y) \\ & \equiv \{-(D_t+ia)^2+(\mathbf{B}D_y,\ D_y)\}\,u(t,\ y)=0 \qquad \text{for} \qquad t>0,\ y_1>0, \\ & \text{with initial conditions} \\ & \qquad u(t,\ y)\,|_{t=0}=D_t u(t,\ y)\,|_{t=0}=0,\ y_1>0 \\ & \text{and with boundary conditions} \\ & \qquad u(t,\ y)\,|_{y_1=0}=h^{(0)}(t,\ y') \qquad \text{for} \qquad (M',\ y')>kt, \qquad t>0, \\ & \qquad -iD_\nu u(t,\ y)\,|_{y_1=0}=h^{(1)}(t,\ y') \qquad \text{for} \qquad (M',\ y')<kt, \qquad t>0. \end{aligned}\right\} \quad (1)$$

Here we have adopted the following notation: $y=(y_1, \cdots, y_n)\in R^n$, $y'=(y_2, \cdots, y_n)\in R^{n-1}$; $\mathbf{B}=(b_{i,j})$ is a real constant symmetric positive definite matrix of order $n\times n$; $B_1=(b_{1,1}, \cdots, b_{1,n})$ is the first row of this matrix; $D_t=i\partial/\partial t$ and $D_j=i\partial/\partial y_j$; D_y is the column of elements $D_1, \cdots, D_n$; $-iD_\nu\equiv -i(B_1, D_y)\equiv -i\sum_1^n b_{1,j}D_j$ is the conormal derivative. The vector $M'=(m_2, \cdots, m_n)$ and the

AMS (MOS) subject classifications (1970). Primary 35L05, 35L20, 35M05, 35R05; Secondary 35A05, 35B30.

parameter k in the boundary conditions are real, and $\Sigma_2^n m_j^2 > 0$; the functions $u(t, y)$ and $h^{(l)}(t, y')$, $l = 0, 1$, are also real.

Problem (1) results from the analogous problem for the equation

$$\{-D_t^2 + (\mathbf{B}D_y, D_y)\} v(t, y) = 0$$

upon making the substitution $v(t, y) = e^{at}u(t, y)$, in which a is a fixed positive number.

In order to make the proofs easier to follow, we will successively make three changes in the variables. They simplify the formulas somewhat but are not fundamental changes. It can be assumed without loss of generality that $m_2 \neq 0$, where m_2 is the first component of the vector M' in the boundary conditions. The first change has the form $\tilde{y}_j = y_j$, $j \neq 2$; $\tilde{y}_2 = (M', y')$. We denote the matrix $\mathbf{B}$ after this change by $\tilde{\mathbf{B}} = (\tilde{b}_{ij})$. The second change is $\tilde{\tilde{y}}_1 = \tilde{y}_1$, $\tilde{\tilde{y}}_j = -\tilde{b}_{1,j}\tilde{b}_{1,1}^{-1}\tilde{y}_1 + \tilde{y}_j$, $j \geqslant 2$. Here $\tilde{b}_{1,1} > 0$ by virtue of the positive definiteness of the matrix $\tilde{\mathbf{B}}$. This change transforms $\tilde{\mathbf{B}}$ into a matrix denoted by $\tilde{\tilde{\mathbf{B}}}$. The third change is $x_1 = \tilde{\tilde{b}}_{1,1}^{-1/2}\tilde{\tilde{y}}_1$, $x_2 = \tilde{\tilde{y}}_2$, $x_j = -\tilde{\tilde{b}}_{2,j}\tilde{\tilde{b}}_{2,2}^{-1}\tilde{\tilde{y}}_2 + \tilde{\tilde{y}}_j$, $j \geqslant 3$. Here $\tilde{\tilde{b}}_{1,1}$ and $\tilde{\tilde{b}}_{2,2}$ are positive numbers since $\tilde{\tilde{\mathbf{B}}}$ is a positive definite matrix. This change transforms $\tilde{\tilde{\mathbf{B}}}$ into the matrix

$$\mathbf{A} = \begin{pmatrix} 1 & 0 & 0 \\ 0 & a_{22} & 0 \\ 0 & 0 & \mathbf{A}'' \end{pmatrix}. \tag{2}$$

Here $a_{2,2} = \tilde{\tilde{b}}_{2,2} > 0$ and $A'' = (a_{i,j})$ is a symmetric positive definite matrix in which $a_{i,j} = \tilde{\tilde{b}}_{i,j} - \tilde{\tilde{b}}_{2,i}\tilde{\tilde{b}}_{2,j}\tilde{\tilde{b}}_{2,2}^{-1}$, $3 \leqslant i, j \leqslant n$.

We will also use the notation

$$\mathbf{A}' = \begin{pmatrix} a_{22} & 0 \\ 0 & \mathbf{A}'' \end{pmatrix}.$$

Inasmuch as the matrix $\mathbf{A}$ is symmetric, so that the transposition operation will not be encountered, this notation should not cause any confusion. Under the indicated changes, problem (1) goes over into the form

$$\begin{gathered} \equiv \{-(D_t + ia)^2 + (\mathbf{A}D_x, D_x)\} u(t, x) = 0, \ t > 0, \ x_1 > 0; \\ u(0, x) = D_t u(0, x) = 0, \ x_1 > 0; \\ u(t, x)|_{x_1=0} = h^{(0)}(t, x'), \ x_2 > kt, \ t > 0; \\ -iD_1 u(t, x)|_{x_1=0} = \frac{\partial u}{\partial x_1}(t, 0, x') = h^{(1)}(t, x'), \ x_2 < kt, \ t > 0. \end{gathered} \tag{3}$$

Here $x = (x_1, \cdots, x_n) \in R^n$, $x' = (x_2, \cdots, x_n) \in R^{n-1}$, $D_j = i\partial/\partial x_j$, and D_x is the column of elements $D_1, \cdots, D_n$. We will prove existence and uniqueness theorems for problem (3) and investigate properties characterizing the smoothness of its solution.

The above changes permit us to give a simple qualitative description of the results. The quantity $\sqrt{a_{22}}$ in the equation of (3) denotes the rate of propagation of a surface of constant phase of a plane wave in the direction of the x_2 axis. It turns out that if $|k| < \sqrt{a_{22}}$, the solution $u(t, x)$ will be "smooth" for $x_1 > 0$, its "smoothness" being better the better the "smoothness" of the boundary functions $h^{(0)}$ and $h^{(1)}$, which do not have to satisfy any compatibility conditions on the hyperplane $x_2 = kt$. On the other hand, if $|k| > \sqrt{a_{22}}$, the solution $u(t, x)$ of problem (3) will be "smooth" for $x_1 > 0$ (and right up to the boundary) only if the functions $h^{(0)}$ and $h^{(1)}$ satisfy, in addition to a condition of their own "smoothness," a finite number of compatibility conditions for $x_2 = kt$. These compatibility conditions have a local character for $h^{(0)}$ ($h^{(1)}$) if $k < -\sqrt{a_{22}}$ ($k > \sqrt{a_{22}}$). A precise definition of "smoothness" will be given below.

Problems with a discontinuous boundary operator have been studied for elliptic and parabolic equations by Fichera [1], Peetre [2], Višik and Èskin [3], [4] and other mathematicians. For such equations the solution of a discontinuous boundary value problem is smooth within the domain right up to the boundary except at the manifold of discontinuity of the boundary operator. The solution willl not be smooth on this manifold even for arbitrarily smooth boundary functions.

The above described properties of the solution of problem (3) indicate an essential difference between it and the problem concerning the propagation of a "discontinuity" of the initial functions (see the paper of Courant and Lax [5] and Maslov's monograph [6]). For the Cauchy problem a "discontinuity" of the initial functions creates a "discontinuity" of the solution on the characteristic surfaces. In the problem under consideration the smoothness of the solution within the domain and right up to the boundary essentially depends on the location of the manifold of discontinuity of the boundary operator relative to the characteristic surfaces of the equation.

§1. Auxiliary assertions

1. *Notation.* The elements of all spaces considered, as well as their representatives, will be called *functions.*

We will frequently consider functions of a parameter (denoted by a distinguished argument) with values in a Hilbert space H. We formulate the necessary definitions and assertions pertaining to this case.

$C^k H$ is the set of k times continuously differentiable functions $\varphi(x)$, $x \in R^1$, with values in H, $k = 0, 1, 2, \cdots$; $C^\infty H = \bigcap_0^\infty C^k H$;.

$C_0^\infty H$ is the set of infinitely differentiable functions with compact support in R^1.

SH is the set of infinitely differentiable functions decreasing at infinity together with all of their derivatives more rapidly than any function of the form $|x|^{-m}$, $m = 0, 1, 2, \cdots$.

Finally,

$$L_2H = \left\{ u(x) \,\middle|\, \int_{-\infty}^{+\infty} |u(x)|_H^2\, dx = \|u(x)\|^2 < \infty \right\}.$$

We note that L_2H is a complete space, and $C_0^\infty H$ and SH are dense in L_2H.

The following fact (see [7], page 198) will be used without special mention. If $F(x)$ $(x \in R^1)$ is a (Lebesgue) integrable function with values in $L_2(R^n)$, there exists a (Lebesgue) measurable in R^{n+1} function $f(x; x')$, $x \in R^1$, $x' \in R^n$, uniquely defined everywhere except on a set of measure zero in R^{n+1} and such that $f(x; x') = F(x)$ for almost all $x \in R^1$. Moreover, $f(x; x')$ is integrable with respect to x for almost all x', and $\int_{R^1} f(x; x')dx$ as a function of x' is equal to the element $\int_{R^1} F(x)dx$ of the space $L_2(R^n)$. We will normally use one and the same notation for $F(x)$ and $f(x; x')$.

We mention one additional fact (see [7], page 113). Let φ_0 be a fixed element of a Hilbert space H. The operator $T\psi \equiv (\psi, \varphi_0)$, where $(\cdot, \cdot)$ is the scalar product in H, continuously maps H into R^1. Let $f(x)$ be an integrable function on R^1 with values in H. Then $Tf(x) = (f(x), \varphi_0)$ is integrable on R^1 and

$$\int_a^b Tf(x)\, dx = \int_a^b (f(x), \varphi_0)\, dx = T\int_a^b f(x)\, dx = \left(\int_a^b f(x)\, dx, \varphi_0 \right),$$

where $-\infty \leqslant a < b \leqslant +\infty$.

Suppose $f(x) \in L_2H$ and $\varphi(x) \in L_2(R^1)$. Then $f(x)\varphi(x)$ is integrable and

$$\left| \int_{-\infty}^{+\infty} f(x)\varphi(x)\, dx \right|_H \leqslant \|f(x)\| \, |\varphi(x)|_{L_2(R^1)}.$$

2. *The Fourier transform in* L_2H.

LEMMA 1.1. *If* $u(x) \in L_2H$, *then the following limit exists in* L_2H:

$$\lim_{A \to +\infty} \int_{-A}^{A} e^{ix\xi} u(x)\, dx = u(\xi) \in L_2H,$$

and, in addition,

a) $u(x) = \lim_{A \to +\infty} \int_{-A}^{A} e^{-ix\xi} u(\xi) d\xi$ *in* L_2H;

b) $\|u(\xi)\|^2 = 2\pi \|u(x)\|^2$;

c) *if* $v(x) \in L_2H$ *and* $v(\xi) = \lim_{A \to +\infty} \int_{-A}^{A} e^{ix\xi} v(x)dx$, *then*

$$\int_{-\infty}^{\infty} (u(\xi),\ v(\xi))_H\, d\xi = 2\pi \int_{-\infty}^{\infty} (u(x),\ v(x))_H\, dx.$$

The proof of this lemma repeats with insignificant changes the proofs of analogous assertions in the scalar case (see, for example, [8]).

Further, the Fourier transform operator will be denoted by

$$F[u(x)] = u(\xi) = \int_{-\infty}^{\infty} e^{ix\xi} u(x)\, dx,$$

while the variables x and ξ, as well as t and τ, will always be "dual" in this sense.

COROLLARY 1.1. *If* T *is a bounded linear operator in* H, *then for any* $u(x) \in L_2H$

$$F[Tu(x)] = TF[u(x)].$$

PROOF. Suppose $\varphi(x) \in C_0^\infty H$. Then

$$F[\varphi(x)] = \int_{-\infty}^{\infty} e^{ix\xi} \varphi(x)\, dx,$$

where the limits of integration can be replaced by finite limits. Hence

$$TF[\varphi(x)] = T \int_{-\infty}^{\infty} e^{ix\xi} \varphi(x)\, dx = \int_{-\infty}^{\infty} e^{ix\xi} T\varphi(x)\, dx = F[T\varphi(x)].$$

Suppose $u(x) \in L_2H$, $\varphi_k(x) \in C_0^\infty H$, $\lim_{k\to\infty} \varphi_k(x) = u(x)$ in L_2H. Then $u(\xi) = F[u(x)] \in L_2H$ and we have

$$\begin{aligned}\|TF[u(x)] - F[Tu(x)]\| &\leqslant \|TF[u(x) - \varphi_k(x)]\| + \|TF[\varphi_k(x)] - F[Tu(x)]\| \\ &\leqslant |T| \cdot \|F[u(x) - \varphi_k(x)]\| + \|F[T(\varphi_k(x) - u(x))]\|\end{aligned}$$

$$= 2\pi |T| \cdot \|u(x) - \varphi_k(x)\| + 2\pi \|T(\varphi_k(x) - u(x))\| \leqslant 4\pi |T| \cdot \|u(x) \quad \varphi_k(x)\| \to 0$$

for $k \to \infty$. Q.E.D.

We now consider Fourier transforms of functions with supports on a semiaxis and prove theorems similar to the Paley-Wiener theorems [9].

We first note that if $u(x) \in C_0^\infty H$ and $\operatorname{supp} u(x) \subset R_+^1 = \{x \in R^1 \mid x \geqslant 0\}$, then $u(x)e^{-\eta x} \in C_0^\infty H$ and $F[u(x)e^{-\eta x}] = u(\xi + i\eta) = u(z)$ is an entire function of z.

If $v(x) \in C_0^\infty H$ and $\operatorname{supp} v(x) \subset \overline{R}^1_+$, we have for $\eta > 0$ the estimate

$$|u(z) - v(z)|_H^2 \leqslant \left| \int_0^\infty e^{ix(\xi+i\eta)} (u(x) - v(x))\, dx \right|_H^2$$

$$\leqslant \int_0^\infty e^{-2\eta x} dx \int_0^\infty |u(x) - v(x)|_H^2\, dx = \frac{1}{2\eta} \| u(x) - v(x) \|^2. \tag{1.1}$$

If a function $u(x)$ is defined for $x > 0$, its extension by zero (i.e. by setting it equal to zero where it is not defined) onto all of R^1 will be denoted by $u_+(x)$.

LEMMA 1.2. *If $u_+(x) \in L_2 H$, then $F[e^{-\eta x} u_+(x)] = u_+(\xi + i\eta) = u_+(z) \in L_2 H$ for any $\eta \geqslant 0$, and*

a) $\lim_{\eta \to +\infty} \|u_+(\xi + i\eta)\| = 0$;

b) *$u_+(\xi + i\eta)$ is a uniformly continuous function of η for $\eta \geqslant 0$ with values in $L_2 H$;*

c) *$|u_+(\xi + i\eta)|_H \leqslant c\|u_+\|/\sqrt{\eta}$ as $\eta \longrightarrow +\infty$;*

d) *$u_+(z)$ is an analytic function of z for $\operatorname{Im} z > 0$ with values in H.*

PROOF. a) We have

$$\frac{1}{2\pi} \| u_+(\xi + i\eta) \|^2 = \| e^{-\eta x} u_+(x) \|^2 = \int_0^A e^{-2\eta x} | u_+(x) |_H^2 dx$$

$$+ \int_A^\infty e^{-2\eta x} | u_+(x) |_H^2\, dx \leqslant \int_0^A | u_+(x) |_H^2 dx + e^{-2\eta A} \int_0^\infty | u_+(x) |_H^2 dx < \frac{\varepsilon}{2} + \frac{\varepsilon}{2} = \varepsilon.$$

Here the first (second) summand can be made less than $\epsilon/2$ by choosing a sufficiently small A (large η for fixed A).

b) Suppose $\eta', \eta'' \geqslant 0$. Then we have

$$\frac{1}{2\pi} \| u_+(\xi + i\eta'') - u_+(\xi + i\eta') \|^2 = \int_0^A | u_+(x) |_H^2 | e^{-\eta'' x} - e^{-\eta' x} |^2\, dx$$

$$+ \int_A^\infty | u_+(x) |_H^2 | e^{-\eta'' x} - e^{-\eta' x} |^2\, dx \leqslant A^2 | \eta'' - \eta' |^2 \| u_+(x) \|^2$$

$$+ \int_A^\infty | u_+(x) |_H^2 dx < \frac{\varepsilon}{2} + \frac{\varepsilon}{2} = \varepsilon.$$

Here we have used the obvious inequality

$$| e^{-\eta'' x} - e^{-\eta' x} | \leqslant A | \eta'' - \eta' | \quad \text{for} \quad 0 \leqslant x_1 \leqslant A.$$

The second (first) summand can be made less than $\epsilon/2$ by choosing a sufficiently large A (small difference $|\eta''-\eta'|$ for fixed A).

c), d). Suppose $u_k(x)\in C_0^\infty H$, $\operatorname{supp} u_k(x)\subset\overline{R^1_+}$ and $\lim_{k\to\infty}u_k(x)=u_+(x)$ in L_2H. According to (1.1) we have for $\eta>0$ the inequality

$$|u_k(z)-u_m(z)|_H^2\leqslant\frac{1}{2\eta}\|u_k(x)-u_m(x)\|^2,$$

as well as the inequalities

$$\left|\frac{du_k(z)}{dz}-\frac{du_m(z)}{dz}\right|_H^2\leqslant\int_0^\infty e^{-2\eta x}x^2dx\int_0^\infty|u_k(x)-u_m(x)|_H^2\,dx\leqslant\frac{1}{4\eta^3}\|u_k(x)-u_m(x)\|^2.$$

It follows that the sequence $\{u_k(z)\}$ together with its derivatives converges uniformly in each domain of the form $\operatorname{Im} z=\eta\geqslant\eta_0>0$. Hence the limit function $v(z)=\lim_{k\to\infty}u_k(z)$ is analytic for $\operatorname{Im} z>0$. When $\eta>0$ we can pass to the limit as $k\to\infty$ in the integral $\int_A^B|u_k(\xi+i\eta)-u_+(\xi+i\eta)|_H^2d\xi$ by virtue of the uniform convergence of the $u_k(z)$ for $\operatorname{Im} z>0$ and get

$$\int_A^B|v(\xi+i\eta)-u_+(\xi+i\eta)|_H^2\,d\xi=0$$

for any A and B. Therefore, revising, if necessary, $u_+(z)$, we will get that $v(z)=u_+(z)$ is analytic for $\operatorname{Im} z=\eta>0$.

From estimate (1.1) (when $v\equiv0$) we get

$$|u_k(\xi+i\eta)|_H^2\leqslant\frac{1}{2\eta}\|u_k(x)\|^2,\quad\eta>0.$$

Hence for $k\to\infty$ we have

$$|u_+(\xi+i\eta)|_H^2\leqslant\frac{1}{2\eta}\|u_+(x)\|^2,\quad\eta>0.$$

The proof of the lemma is complete.

We now consider the inverse problem.

LEMMA 1.3. *Suppose* $u(z)=u(\xi+i\eta)\in L_2H$ *for any* $\eta\geqslant0$, *and let the following conditions hold.*

a) $\lim_{\eta\to+\infty}\|u(\xi+i\eta)\|=0$;

b) $u(\xi+i\eta)$ *is a continuous function of* η *for* $\eta\geqslant0$ *with values in* L_2H.

c) $u(z)$ *is an analytic function of* z *for* $\operatorname{Im} z=\eta>0$ *with values in* H.

Then there exists a unique function $u_+(x)\in L_2H$ *such that*

$$u(\xi+i\eta)=F[e^{-\eta x}u_+(x)].$$

PROOF. We make use of a representation formula for a function satisfying the conditions of the lemma,

$$u(z)=\frac{1}{2\pi i}\int_{-\infty+i\eta'}^{+\infty+i\eta'}\frac{u(\zeta)\,d\zeta}{\zeta-z}+\frac{1}{2\pi i}\int_{+\infty+i\eta''}^{-\infty+i\eta''}\frac{u(\zeta)\,d\zeta}{\zeta-z},\qquad 0<\eta'<\operatorname{Im} z<\eta''<+\infty. \tag{1.2}$$

The proof of this formula proceeds in the same way as in the scalar case ([9], page 5) with the use of modified theorems on (i) the integration of functions with values in H and (ii) the completeness of H.

We estimate the norm in H of the second summand in this formula:

$$\left|\int_{+\infty+i\eta''}^{-\infty+i\eta''}\frac{u(\zeta)\,d\zeta}{\zeta-z}\right|_H^2=\lim_{B\to+\infty}\left|\int_{B+i\eta''}^{-B+i\eta''}\frac{u(\zeta)\,d\zeta}{\zeta-z}\right|_H^2$$

$$\leqslant\lim_{B\to+\infty}\left\{\int_{-B+i\eta''}^{B+i\eta''}\frac{|u(\zeta)|_H\,|d\zeta|}{|\zeta-z|}\right\}^2\leqslant\lim_{B\to+\infty}\int_{-B}^{B}\frac{d\xi}{|\xi+i\eta''-z|^2}$$

$$\times\int_{-B}^{B}|u(\xi+i\eta'')|_H^2\,d\xi\leqslant C\|u(\xi+i\eta'')\|^2\to 0$$

as $\eta''\to+\infty$ according to a). Hence (1.2) as $\eta''\to+\infty$ implies

$$u(z)=\frac{1}{2\pi i}\int_{-\infty+i\eta'}^{+\infty+i\eta'}\frac{u(\zeta)\,d\zeta}{\zeta-z}. \tag{1.3}$$

Let

$$u_\eta(x)=F^{-1}[u(\xi+i\eta)]\in L_2H,\quad \eta\geqslant 0.$$

Consider the function $\varphi(x)=e^{-ix\alpha-\beta x}$ for $x>0$, $\varphi(x)=0$ for $x\leqslant 0$, $\beta=\operatorname{Im} z-\eta'$, $\alpha=\operatorname{Re} z$. If $h\in H$, then $\varphi(x)h\in L_2H$ and $F[\varphi(x)h]=hF[\varphi(x)]=h\varphi(\xi)$. According to property c) of Lemma 1.2 we have

$$\int_{-\infty}^{\infty}(u_{\eta'}(x),\ \varphi(x)h)_H\,dx=\frac{1}{2\pi}\int_{-\infty}^{\infty}(u(\xi+i\eta'),\ \varphi(\xi)h)_H\,d\xi,$$

or

$$\int_{-\infty}^{\infty}(\bar\varphi(x)u_{\eta'}(x),\ h)_H\,dx=\frac{1}{2\pi}\int_{-\infty}^{\infty}(\bar\varphi(\xi)u(\xi+i\eta'),\ h)_H\,d\xi.$$

Since $\overline{\varphi}(x) \in L_2(R^1)$, we also have $\overline{\varphi}(\xi) \in L_2(R^1)$, and this means that the functions $\overline{\varphi}(x)u_{\eta'}(x)$ and $\overline{\varphi}(\xi)u(\xi + i\eta')$ are absolutely integrable. We can therefore integrate under the sign of the scalar product:

$$\left(\int_{-\infty}^{\infty} \overline{\varphi}(x)\, u_{\eta'}(x)\, dx,\ h\right)_H = \left(\frac{1}{2\pi}\int_{-\infty}^{\infty} \overline{\varphi}(\xi)\, u(\xi + i\eta')\, d\xi,\ h\right)_H,$$

and this is true for any $h \in H$. Thus

$$\int_{-\infty}^{\infty} \overline{\varphi}(x)\, u_{\eta'}(x)\, dx = \frac{1}{2\pi}\int_{-\infty}^{\infty} \overline{\varphi}(\xi)\, u(\xi + i\eta')\, d\xi. \tag{1.4}$$

By the definition of $\varphi(x)$ we have

$$\int_{-\infty}^{\infty} \overline{\varphi}(x)\, u_{\eta'}(x)\, dx = \int_{0}^{\infty} e^{ix\alpha} e^{-\beta x} u_{\eta'}(x)\, dx.$$

Noting that $\overline{\varphi}(\xi) = -i/(\xi - \alpha - i\beta)$, we now have

$$\frac{1}{2\pi}\int_{-\infty}^{\infty} \frac{-iu(\xi + i\eta')\, d\xi}{\xi - \operatorname{Re} z - i(\operatorname{Im} z - \eta')} = \frac{1}{2\pi i}\int_{-\infty}^{\infty} \frac{u(\xi + i\eta')\, d\xi}{\xi + i\eta' - z} = u(z)$$

(here we have used (1.3)). Comparing the left side with the right side in (1.4), we see that $u(z)$ is the Fourier transform of the function

$$v_+(x) = \exp\left[-(\operatorname{Im} z - \eta')x\right] \cdot u_{\eta'}(x)$$

for $x > 0$, and $v_+(x) = 0$ for $x \leqslant 0$. We prove that as $\eta' \to 0$ the function $v_+(x)$ tends to a function in L_2H that is equal to $\exp(-\operatorname{Im} z \cdot x) \cdot u_0(x)$ for $x > 0$ and to zero for $x \leqslant 0$. We have

$$\int_0^{\infty} \left|\exp\left[-(\operatorname{Im} z - \eta')x\right] u_{\eta'}(x) - \exp(-\operatorname{Im} zx)\, u_0(x)\right|_H^2 dx$$

$$= \int_0^{\infty} \exp\left[-2(\operatorname{Im} z - \eta')x\right] \left|u_{\eta'}(x) - \exp(-\eta' x)\, u_0(x)\right|_H^2 dx$$

$$\leqslant 2\int_0^{\infty} |u_{\eta'}(x) - u_0(x)|_H^2\, dx + 2\int_0^{\infty} |u_0(x)|_H^2 \left[1 - \exp(-\eta' x)\right]^2 dx.$$

We consider the second summand:

$$\int_0^\infty |u_0(x)|_H^2 [1-\exp(-\eta' x)]^2\,dx \leqslant \int_A^\infty |u_0(x)|_H^2\,dx$$

$$+[1-\exp(-\eta' A)]^2 \int_0^\infty |u_0(x)|_H^2\,dx \leqslant \frac{\varepsilon}{2}+\frac{\varepsilon}{2}=\varepsilon.$$

Here the first (second) summand can be made less than $\epsilon/2$ by choosing a sufficiently large A (small η' for fixed A). Hence

$$\lim_{\eta'\to 0}\int_0^\infty |u_0(x)|_H^2 [1-\exp(-\eta' x)]^2\,dx = 0.$$

Inasmuch as

$$\lim_{\eta'\to 0}\int_0^\infty |u_{\eta'}(x)-u_0(x)|_H^2\,dx = 0$$

by condition b), we get

$$\lim_{\eta'\to 0} v_+(x) = v_+(x) = \begin{cases} e^{-\operatorname{Im} z\cdot x}\cdot u_0(x), & x>0 \\ 0\quad, & x\leqslant 0\end{cases}$$

(here we have used the fact that $v_+(x)=F^{-1}[u(\xi+i\eta)]$ does not depend on η'). Thus $u_\eta(x)=F^{-1}[u(\xi+i\eta)]=\exp(-\eta x)\cdot u_0(x)$ for $x>0$, $u_\eta(x)=0$ for $x\leqslant 0$, and for $\eta\to 0$ we get according to condition b) that $u_0(x)=F^{-1}[u(\xi)]=0$ for $x\leqslant 0$. This means that $u_0(x)=u_+(x)$, which completes the proof of the lemma.

3. *The operators* $\Pi_\pm$. Let $R^1_\pm=\{x\in R^1\,|\,x\gtrless 0\}$ (the upper or lower signs are used simultaneously) and $L_2^\pm H=\{u(x)\in L_2H|\ \operatorname{supp} u(x)\subset \overline{R^1_\pm}\}$. Clearly L_2^+H and L_2^-H are orthogonal in the sense of the scalar product

$$((u(x),\ v(x)))=\int_{R^1}(u(x),\ v(x))_H\,dx \ \text{и}\ L_2H=L_2^+H\oplus L_2^-H.$$

We denote the projection operator from L_2H onto $L_2^\pm H$ by $P_\pm$: $P_\pm u(x)=u_\pm(x)=u(x)$ in $R^1_\pm$ and $P_\pm u(x)=u_\pm(x)=0$ in $\overline{R^1_\mp}$. By definition, for $u(x)\in L_2H$

$$\Pi_\pm F[u(x)]=\Pi_\pm u(\xi)=F[P_\pm u(x)]. \tag{1.5}$$

The analytic representation formula for the operators $\Pi_\pm$,

$$\Pi_\pm u(\xi)=\lim_{\delta\to+0}\frac{1}{2\pi i}\int_{\mp\infty}^{\pm\infty}\frac{u(\xi')\,d\xi'}{\xi'-(\xi\pm i\delta)}=\frac{1}{2\pi i}\int_{\mp\infty}^{\pm\infty}\frac{u(\xi')\,d\xi'}{\xi'-(\xi\pm i0)},$$

where the limit is taken in L_2H, is proved in exactly the same way as in the scalar case.

LEMMA 1.4. *If T is a bounded linear operator in H, then for any $u(x) \in L_2H$*

$$\Pi_{\pm} Tu(\xi) = T\Pi_{\pm} u(\xi).$$

PROOF. We have $TP_+u(x) = Tu(x)$ for $x > 0$ and $TP_+u(x) = T0 = 0$ for $x \leqslant 0$. On the other hand, $P_+Tu = Tu(x)$ for $x > 0$ and $P_+Tu(x) = 0$ for $x \leqslant 0$, i.e. $TP_+u(x) = P_+Tu(x)$. Applying the Fourier transform operator to both sides of this equality and using (1.5) and Corollary 1.1, we get

$$F[TP_+u(x)] = TF[P_+u(x)] = T\Pi_+u(\xi) = F[P_+Tu(x)]$$
$$= \Pi_+F[Tu(x)] = \Pi_+TF[u(x)] = \Pi_+Tu(\xi).$$

Q.E.D. An analogous argument can be carried out for Π_-.

As usual, let

$$L_2(R^m) = \left\{u(x'),\ x' \in R^m \,\middle|\, \|u\|^2_{L_2(R^m)} = \int_{R^m} |u(x')|^2 dx' < \infty\right\}.$$

We denote the Fourier transform operator in $L_2(R^m)$ by

$$F_{x' \to \xi'}[u(x')] = u(\xi') = \int_{R^m} (\exp ix'\xi') u(x')\, dx',$$

and the inverse operator by

$$F^{-1}_{x' \to \xi'}[u(\xi')] = u(x') = \frac{1}{(2\pi)^m} \int_{R^m} [\exp(-ix'\xi')] u(\xi')\, d\xi'.$$

If the identity of the variable with respect to which the transform is taken is clear, the notation $x' \to \xi'$ will be dropped.

COROLLARY 1.2. *Suppose $H = L_2(R^m)$, $u(x_1; x') \in L_2H = L_2L_2(R^m)$ and Π_{1+} is the above-introduced operator acting with respect to the variable ξ_1 $(x_1 \to \xi_1)$. Then*

$$\Pi_{1+}F_{x' \to \xi'} = F_{x' \to \xi'}\Pi_{1+}.$$

PROOF. The operator $F_{x' \to \xi'}$ is continuous from $L_2(R^m)$ into $L_2(R^m)$. A reference to Lemma 1.4 completes the proof.

If now $u(x_1, x') = u(x) \in L_2(R^{m+1})$, then $u(x_1, \cdot) = u(x_1, x') \in L_2L_2(R^m)$. For such functions the definition of the operator Π_{1+} is already adequate: $\Pi_{1+}u(\xi_1, x') = F_{x_1 \to \xi_1}[u_+(x_1, x')]$, where $u_+(x_1, x') = u(x)$ for $x_1 > 0$ and

$u_+(x_1, x') = 0$ for $x_1 \leqslant 0$, with

$$F_{x' \to \xi'} \Pi_{1+} u(\xi_1, x') = \Pi_{1+} F_{x' \to \xi'} [u(\xi_1, x')] = \Pi_{1+} u(\xi_1, \xi').$$

In other words, the operator Π_{1+} can be defined in this case by the equality

$$\Pi_{1+} u(\xi_1, \xi') = F_{(x_1, x') \to (\xi_1, \xi')} [u_+(x)],$$

where $u_+(x) = u(x)$ for $x_1 > 0$ and $u_+(x) = 0$ for $x_1 \leqslant 0$. Analogously,

$$\Pi_{1-} u(\xi_1, \xi') = F_{(x_1, x') \to (\xi_1, \xi')} [u_-(x)].$$

REMARK 1.1. In the sequel we will use the same symbol for a space of functions $u(x)$ and the space of Fourier transforms of these functions. The actual meaning of a given symbol will be seen from the notation for the argument of its elements: $u(x)$ as opposed to $u(\xi)$. In particular, if

$$L_2^{\pm}(R^{m+1}) = L_2^{\pm} L_2(R^m) = \{u(x_1, x') \in L_2(R^{m+1}) \mid u(x_1, x') = 0 \quad \text{for} \quad x_1 \gtrless 0\},$$

the corresponding space of Fourier transforms is

$$L_2^{\pm}(R^{m+1}) = L_2^{\pm} L_2(R^m) = \{u(\xi_1, \xi') \in L_2(R^{m+1}) \mid \Pi_{1\pm} u(\xi_1, \xi') = u(\xi_1, \xi')\}.$$

The elements of these spaces will often be denoted by $u_{\pm}(x)$ and $u_{\pm}(\xi)$ respectively, with $x = (x_1, x')$ and $\xi = (\xi_1, \xi')$.

We denote the spaces of test and generalized functions in the usual way: $D(\Omega)$, $D'(\Omega)$, $\Omega \subset R^m$, $S(R^m)$, $S'(R^m)$.

The Sobolev-Slobodeckiĭ spaces are denoted by $H_s(R^m)$, implying that the norm in them is defined with the use of the Fourier transform.

We will also use the notation

$$R_+^{m+1} = \{(x_1, \cdot) \in R^{m+1} \mid x_1 > 0\}, \quad \overline{R_+^{m+1}} = \{(x_1, \cdot) \in R^{m+1} \mid x_1 \geqslant 0\}.$$

LEMMA 1.5. *Let* $u(x) \in L_2^+(R^{m+1})$ *and let* $\alpha(z_1, \xi')$, *where* $z_1 = \xi_1 + i\eta_1$ *and* $\xi' = (\xi_2, \cdots, \xi_{m+1})$ *be for* $\eta_1 \geqslant 0$ *and* $\xi' \in R^m$ (i) *a continuous function of its arguments,* (ii) *analytic with respect to* z_1, (iii) *uniformly (with respect to* $\xi = (\xi_1, \xi') \in R^{m+1}$) *continuous in* η_1, *and* (iv) *such that* $|\alpha(z_1, \xi')| \leqslant C$. *Then* $\alpha(\xi_1, \xi') u(\xi) \in L_2^+(R^{m+1})$.

PROOF. The function

$$u(\xi_1, \xi') = u(\xi_1, \cdot) \in L_2^+(R^{m+1}) = L_2^+ L_2(R^m),$$

i.e.

$$\operatorname{supp} F^{-1}_{x_1\to\xi_1}[u(\xi_1, \cdot)] \subset \overline{R^1_+}.$$

This means that the assertions of Lemma 1.2 hold for $u(\xi_1, \cdot)$. Let us verify that the function $v(z_1, \xi') = \alpha(z_1, \xi')u(z_1, \xi')$ satisfies conditions a)–c) of Lemma 1.3.

a) We have

$$\lim_{\eta_1\to+\infty} \|v(\xi_1+i\eta_1, \cdot)\|^2_{L_2L_2(R^m)} = \lim_{\eta_1\to+\infty} \int_{R^{m+1}} |\alpha(z_1, \xi')u(z_1, \xi')|^2 d\xi_1 d\xi'$$

$$\leqslant C^2 \lim_{\eta_1\to+\infty} \int_{R^{m+1}} |u(z_1, \xi')|^2 d\xi_1 d\xi' = 0.$$

b) Suppose η_1 is a fixed nonnegative number and let $|\Delta| \leqslant \eta_1^{1/2}$ if $\eta_1 > 0$ and $\Delta > 0$ if $\eta_1 = 0$. Then

$$\begin{aligned}
&\|v(z_1+i\Delta, \cdot) - v(z_1, \cdot)\|^2_{L_2L_2(R^m)} \\
&\quad = \|\alpha(z_1+i\Delta, \cdot)u(z_1+i\Delta, \cdot) - \alpha(z_1, \cdot)u(z_1, \cdot)\|^2_{L_2L_2(R^m)} \\
&\quad \leqslant 2\|[\alpha(z_1+i\Delta, \cdot) - \alpha(z_1, \cdot)]u(z_1+i\Delta, \cdot)\|^2_{L_2L_2(R^m)} \\
&\quad + 2\|[u(z_1+i\Delta, \cdot) - u(z_1, \cdot)]\alpha(z_1, \cdot)\|^2_{L_2L_2(R^m)} \\
&\quad \leqslant 2\max_{\xi\in R^{m+1}} |\alpha(z_1+i\Delta, \xi') - \alpha(z_1, \xi')|^2 \|u(z_1+i\Delta, \cdot)\|^2_{L_2L_2(R^m)} \\
&\quad + 2C^2\|u(z_1+i\Delta, \cdot) - u(z_1, \cdot)\|^2_{L_2L_2(R^m)}.
\end{aligned}$$

Inasmuch as $u(\xi_1 + i\eta_1, \cdot)$ is a continuous function of η_1 with values in $L_2L_2(R^m)$, its norm $\|u(\xi_1 + i\eta_1, \cdot)\|_{L_2L_2(R^m)}$ is a continuous nonnegative function of η_1. Recalling that

$$\lim_{\eta_1\to+\infty} \|u(\xi_1+i\eta_1, \cdot)\|_{L_2L_2(R^m)} = 0,$$

we get

$$\|u(\xi_1+i\eta_1, \cdot)\|_{L_2L_2(R^m)} \leqslant M < \infty.$$

It is now clear that both summands of the latter part of the estimate carried out above can be made less than $\epsilon/2$ by choosing a sufficiently small $|\Delta|$.

c) Let

$$\alpha'(z_1, \cdot) = \frac{\partial\alpha(z_1, \xi')}{\partial z_1}, \qquad u'(z_1, \cdot) = \frac{du(z_1, \cdot)}{dz_1}.$$

Suppose $\operatorname{Im} z_1 > 0$ and $\operatorname{Im}(z_1 + \Delta z_1) > 0$. We then have

$$\left\| \frac{1}{\Delta z_1}[\alpha(z_1+\Delta z_1, \cdot)u(z_1+\Delta z_1, \cdot) - \alpha(z_1, \cdot)u(z_1, \cdot)] - \alpha'(z_1, \cdot)u(z_1, \cdot) - \alpha(z_1, \cdot)u'(z_1, \cdot)\right\|_{L_2(R^m)} \leqslant$$

$$\leqslant \left\| \frac{1}{\Delta z_1} [u(z_1+\Delta z_1, \cdot) - u(z_1, \cdot)]\, \alpha(z_1+\Delta z_1, \cdot) - \alpha(z_1, \cdot)\, u'(z_1, \cdot) \right\|$$
$$+ \left\| u(z_1, \cdot) \left\{ \frac{\alpha(z_1+\Delta z_1, \cdot) - \alpha(z_1, \cdot)}{\Delta z_1} - \alpha'(z_1, \cdot) \right\} \right\|_{L_2(R^m)}$$
$$\leqslant \| u'(z_1, \cdot) [\alpha(z_1+\Delta z_1, \cdot) - \alpha(z_1, \cdot)] \|_{L_2(R^m)} +$$
$$+ \left\| \alpha(z_1+\Delta z_1, \cdot) \left[\frac{u(z_1+\Delta z_1, \cdot) - u(z_1, \cdot)}{\Delta z_1} - u'(z_1) \right] \right\|_{L_2(R^m)}$$
$$+ \left\| u(z_1, \cdot) \left[\frac{\alpha(z_1+\Delta z_1, \cdot) - \alpha(z_1, \cdot)}{\Delta z_1} - \alpha'(z_1, \cdot) \right] \right\|_{L_2(R^m)}.$$

Let the summands in the latter expression be denoted in order by K_1, K_2, K_3. We have

$$K_2 \leqslant C \left\| \frac{u(z_1+\Delta z_1, \cdot) - u(z_1, \cdot)}{\Delta z_1} - u'(z_1) \right\|_{L_2(R^m)} \to 0$$

as $\Delta z_1 \to 0$ since $u(z_1, \cdot)$ is analytic. To estimate K_1 we note that $|\alpha'(z_1, \xi')| \leqslant c/r = C_1$ inasmuch as

$$\alpha'(z_1, \xi') = -\frac{1}{2\pi i} \oint_\gamma \frac{\alpha(\zeta_1, \xi')\, d\zeta_1}{(\zeta_1 - z_1)^2},$$

where γ is the circle of radius r in the upper halfplane $\operatorname{Im} z_1 > 0$ with center at z_1. Hence

$$\lim_{\Delta z_1 \to 0} [\alpha(z_1+\Delta z_1, \cdot) - \alpha(z_1, \cdot)] = 0$$

uniformly in $\xi' \in R^m$. This means that $\lim_{\Delta z_1 \to 0} K_1 = 0$. Estimating $\alpha''(z_1, \xi')$ in exactly the same way, we find that $|\alpha''(z_1, \xi')| \leqslant 2c/r^2 = C_2$. Hence

$$\lim_{\Delta z_1 \to 0} \left[\frac{\alpha(z_1+\Delta z_1, \cdot) - \alpha(z_1, \cdot)}{\Delta z_1} - \alpha'(z_1, \cdot) \right] = 0$$

uniformly in $\xi' \in R^m$. This means that $\lim_{\Delta z_1 \to 0} K_3 = 0$. The proof of the lemma is complete.

COROLLARY 1.3. *Let* $\varphi(x) = \varphi(x_1, x') \in S(R^n)$, supp $\varphi(x) \subset \overline{R^{n+1}_+}$, *and for* $\eta_1 \geqslant 0$ *and* $\xi' \in R^n$ *let* $\beta(z_1, \xi')$, *where* $z_1 = \xi_1 + i\eta_1$, *be* (i) *a continuous function of its arguments,* (ii) *analytic with respect to* z_1 *and* (iii) *such that* $|\beta(z_1, \xi')|^2 \leqslant C(|z_1|^2 + |\xi'|^2)^l$, $l > 0$. *Then* $\beta(\xi_1, \xi')\varphi(\xi_1, \xi') \in L_2^+(R^{n+1})$, *i.e.* supp $F^{-1}[\beta(\xi)\varphi(\xi)] \subset \overline{R^{n+1}_+}$.

PROOF. Let l_1 be an integer greater than or equal to $2l$. Then

$$\alpha(z_1, \xi') = \beta(z_1, \xi')\, (z_1 + i\sqrt{|\xi'|^2 + 1})^{-l_1}$$

satisfies the conditions of Lemma 1.5. The function

$$\psi(x_1, x') = F^{-1}[(\xi_1 + i\sqrt{|\xi'|^2+1})^{l_1}\varphi(\xi_1, \xi')] \in S(R^{n+1}),$$

and, clearly, in the same way as for $\varphi(x)$, we have $\operatorname{supp}\psi(x) \subset \overline{R^{n+1}_+}$. The proof is completed by applying Lemma 1.5.

Let $\rho(\xi)$, $\xi \in R^m$, denote a function satisfying the condition

$$\frac{1}{c}(1+|\xi|^2)^{-N} \leqslant |\rho(\xi)|^2 \leqslant c(1+|\xi|^2)^N, \qquad \rho(\xi) \in C^\infty(R^m). \tag{1.6}$$

In the sequel we have to deal with Hilbert spaces of the form

$$H_\rho = \left\{u(\xi) \in S'(R^m) \,\Big|\, |u|^2_{H_\rho} = \int_{R^m} |\rho(\xi)u(\xi)|^2 d\xi < \infty\right\}$$

(here $u(\xi)$ is an ordinary measurable function). If $\rho(\xi)$ satisfies (1.6), it is easily seen that $S(R^m) \subset H_\rho(R^m) \subset S'(R^m)$. We will consider also the spaces $L_2H_{\rho'}(R^m)$, $\rho' = \rho'(\xi')$, $\xi' = (\xi_2, \cdots, \xi_{m+1})$, of functions $u(x_1)$ or $u(\xi_1)$ with values in $H_{\rho'}(R^m)$. As has already been stated, by $u(\xi_1)$ we shall mean $u(\xi_1) = F_{x_1\to\xi_1}[u(x_1)]$ in $L_2H_{\rho'}(R^m)$. Further, we will frequently identify the spaces $L_2H_{\rho'}(R^m)$ and $H_{\rho'}(R^{m+1})$ (see, for example, [7], page 196), as well as the spaces $L_2^+H_{\rho'}(R^m)$ and

$$H^+_{\rho'}(R^{m+1}) = \{u(x_1, \xi') \in H_{\rho'}(R^{m+1}) \,|\, u(x_1, \xi') = 0 \text{ for } x_1 \leqslant 0\},$$

and along with them their Fourier transforms.

COROLLARY 1.4. *Let $u(\xi') \in H_{\rho'}(R^m)$, where $\xi' = (\xi_2, \cdots, \xi_{m+1})$, and for $\eta_1 \geqslant 0$ and $\xi' \in R^m$ let $\alpha(z_1, \xi')$, where $z_1 = \xi_1 + i\eta_1$, be a continuous function of its arguments, analytic with respect to z_1, satisfying the following conditions:*

a) $C \geqslant \int_{-\infty}^{\infty} |\alpha(\xi_1 + i\eta_1, \xi')|^2 d\xi_1 \to 0$ *for* $0 \leqslant \eta_1 \to +\infty$ *uniformly in* ξ' *as* $\eta_1 \to \infty$.

b) $\int_{-\infty}^{\infty} |\alpha(\xi_1 + i(\eta_1 + \Delta), \xi') - \alpha(\xi_1 + i\eta_1, \xi')|^2 d\xi_1 \to 0$ *for* $\Delta > 0, \eta_1 \geqslant 0$, $\eta_1 + \Delta \geqslant 0$ *uniformly in* ξ' *as* $\Delta \to 0, \eta_1 + \Delta \geqslant 0$.

Then $\alpha(\xi_1, \cdot)u(\cdot) \in L_2^+ H_{\rho'}(R^m)$, *i.e.*

$$\operatorname{supp} F^{-1}_{x_1\to\xi_1}[\alpha(\xi_1, \cdot)u(\cdot)] \subset \overline{R^{m+1}_+} = \{(x_1, \xi') \in R^{m+1} \,|\, x_1 \geqslant 0\}.$$

PROOF. We verify that the conditions of Lemma 1.3 are satisfied. Let $v(z_1, \cdot) = \alpha(z_1, \cdot)u(\cdot)$. Then

$$\int_{-\infty}^{\infty} d\xi_1 \int_{R^m} |v(\xi_1 + i\eta_1, \xi')\rho'(\xi')|^2 d\xi' = \int_{R^m} |u(\xi')\rho'(\xi')|^2 d\xi' \int_{-\infty}^{\infty} |\alpha(\xi_1 + i\eta_1, \xi')|^2 d\xi_1 \leqslant C|u|^2_{H_{\rho'}},$$

i.e. $v(\xi_1 + i\eta_1, \cdot) \in L_2 H_{\rho'}(R^m)$ for any $\eta_1 \geqslant 0$. Moreover, from this same equality we have according to condition a) that

$$\lim \|v(\xi_1 + i\eta_1, \cdot)\|_{L_2 H_{\rho'}} = 0$$

for $\eta_1 \to \infty$.

Further, by virtue of condition b)

$$\int_{-\infty}^{\infty} d\xi_1 \int_{R^m} |v(\xi_1 + i\eta_1 + i\Delta, \xi') - v(\xi_1 + i\eta_1, \xi')|^2 |\rho'(\xi')|^2 d\xi'$$
$$= \int_{R^m} |u(\xi')\rho'(\xi')|^2 d\xi' \int_{-\infty}^{\infty} |\alpha(\xi_1 + i\eta_1 + i\Delta, \xi') - \alpha(\xi_1 + i\eta_1, \xi')|^2 d\xi_1 \to 0$$

as $\Delta \to 0$, $\eta_1 \geqslant 0$, $\eta_1 + \Delta \geqslant 0$.

Finally, we prove that $v(z_1, \xi')$ is an analytic function of z_1 for $\operatorname{Im} z_1 > 0$ with values in $H_{\rho'}$. If we make use of the representation (3.11) in [9] and condition a), we will have

$$\alpha(z_1, \xi') = \frac{1}{2\pi i} \int_{-\infty}^{\infty} \frac{\alpha(t_1 + i\varepsilon, \xi')\,dt_1}{t_1 + i\varepsilon - z_1},$$

where $0 < \epsilon < \frac{1}{2} \operatorname{Im} z_1$. By virtue of the uniform convergence of the resultant integrals it is possible to differentiate this equality twice, so as to obtain the estimate

$$\left|\frac{\partial^2 \alpha(z_1, \xi')}{\partial z_1^2}\right| \leqslant \frac{1}{(2\pi)^2} \int_{-\infty}^{\infty} \frac{dt_1}{[(t_1 - \xi_1)^2 + (^1/_2\ \eta_1)^2]^3} \int_{-\infty}^{\infty} |\alpha(t_1 + i\varepsilon, \xi')|^2 dt_1 \leqslant \frac{c_1}{\eta_1^5},$$

where c_1 does not depend on (ξ', η_1). We now have for $\operatorname{Im} z_1 = \eta_1 > 0$

$$\left|u(\xi')\left[\frac{\alpha(z_1 + \Delta z_1, \xi') - \alpha(z_1, \xi')}{\Delta z_1} - \frac{\partial \alpha(z_1, \xi')}{\partial z_1}\right]\right|_{H_{\rho'}}$$
$$= \left|u(\xi')\left[\frac{\partial \alpha}{\partial z_1}(z_1 + \theta\Delta z_1, \xi') - \frac{\partial \alpha}{\partial z_1}(z_1, \xi')\right]\right|_{H_{\rho'}} \to 0$$

as $\Delta z_1 \to 0$ inasmuch as

$$\left|\frac{\partial a}{\partial z_1}(z_1+\theta\Delta z_1,\ \xi')-\frac{\partial a}{\partial z_1}(z_1,\ \xi')\right|\to 0$$

uniformly in ξ' as $\Delta z_1 \to 0$ by virtue of the estimate

$$\left|\frac{\partial^2 a(z_1,\ \xi')}{\partial z_1^2}\right|^2 \leqslant \frac{c}{\eta_1^5} \leqslant \frac{c}{(\eta_1^0)^5},$$

where $0 < \eta_1^0 < \eta_1$ and $0 < \eta_1^0 < \eta_1 + \Delta\eta_1$.

Thus all of the conditions of Lemma 1.3 are satisfied and the proof of the corollary is complete.

We now extend the domain of definition of the operator Π_{1+}.

LEMMA 1.6. *Suppose given a function* $u(\xi_1, \xi') \in H_\rho(R^{n+1})$, $\rho = \rho(\xi)$, *having the representation*

$$u(\xi_1,\ \xi')=u_+(\xi_1,\ \xi')+v(\xi_1,\ \xi'), \tag{1.7}$$

where $u_+(\xi_1, \cdot) \in \underline{L_2^+ H_{\rho'}}(R^n)$, $\rho' = \rho'(\xi')$, $v(\xi_1, \xi') \in H_\gamma(R^{n+1})$, $\gamma = \gamma(\xi)$ *and* $\operatorname{supp} F^{-1}[v(\xi)] \subset \overline{R_-^{n+1}}$, *and suppose each of the three weight functions satisfies with respect to its argument a condition of form* (1.6). *Then the representation* (1.7) *is unique.*

PROOF. Suppose, conversely,

$$u(\xi_1,\ \xi')=u_{1+}(\xi_1,\ \xi')+v_1(\xi_1,\ \xi'), \tag{1.7'}$$

where $u_{1+}(\xi_1, \cdot) \in L_2^+ H_{\rho_1'}(R^n)$, $\rho_1' = \rho_1'(\xi')$, $v_1(\xi_1, \xi') \in H_{\gamma_1}(R^{n+1})$, $\gamma_1 = \gamma_1(\xi)$ and $\operatorname{supp} F^{-1}[v_1(\xi)] \subset \overline{R_-^{n+1}}$.

Let $\varphi(x) \in S(R^{n+1})$, $\operatorname{supp} \varphi(x) \subset \overline{R_+^{n+1}}$. Then $\langle u, \varphi\rangle = \langle u_+, \varphi\rangle = \langle u_{1+}, \varphi\rangle$ by virtue of the properties of the supports of $F^{-1}[v(\xi)]$ and $F^{-1}[v_1(\xi)]$. Hence $\langle u_+ - u_{1+}, \varphi\rangle = 0$ for any $\varphi(x) \in S(R^{n+1})$, $\operatorname{supp} \varphi(x) \subset \overline{R_+^{n+1}}$. Going over to the Fourier transforms, we have

$$\langle u_+(\xi)-u_{1+}(\xi),\ \varphi(\xi)\rangle=\int\limits_{R^{n+1}}(u_+(\xi)-u_{1+}(\xi))\,\bar\varphi(\xi)\,d\xi=0,$$

or, letting $w_+(\xi) = u_+(\xi) - u_{1+}(\xi)$,

$$0=\int\limits_{R^{n+1}} w_+(\xi)\,\bar\varphi(\xi)\,d\xi=\int\limits_{R^{n+1}}\frac{w_+(\xi)}{(1+|\xi'|^2)^M}(1+|\xi'|^2)^M\bar\varphi(\xi)\,d\xi,$$

where $M = \frac{1}{2}\max\{N', N_1'\}$, in which N' and N_1' are the exponents for $\rho'(\xi')$ and $\rho_1'(\xi')$ in condition (1.6). The function

$$W(\xi_1, \xi') = (1 + |\xi'|^2)^{-M} w_+(\xi) \in L_2(R^{n+1})$$

by virtue of the choice of M; hence $W(\xi_1, \cdot) \in L_2 L_2(R^n)$.

Let us prove that $W(\xi_1, \cdot) \subset L_2^+ L_2(R^n)$. It suffices to show, for example, that

$$(1 + |\xi'|^2)^{-M} u_+(\xi_1, \xi') \in L_2^+ L_2(R^n).$$

Let $\psi(x_1, \xi') \in C_0^\infty(R^{n+1})$ and

$$\operatorname{supp} \psi(x_1, \xi') \subset \overline{R_-^{n+1}} = \{(x_1, \xi') \in R^{n+1} \mid x_1 \leqslant 0\}.$$

Then

$$\langle u_+(x_1, \xi'), \psi(x_1, \xi')\rangle = \int_{R^{n+1}} u_+(x_1, \xi') \overline{\psi}(x_1, \xi')\, dx_1 d\xi'$$

$$= \int_{R^{n+1}} \rho'(\xi') u_+(x_1, \xi') \overline{\frac{\psi(x_1, \xi')}{|\rho'(\xi')|^2}} \rho'(\xi')\, dx_1 d\xi' = \left(\left(u_+(x_1, \cdot), \frac{\psi(x_1, \cdot)}{|\rho'(\cdot)|^2}\right)\right)_{L_2 H_{\rho'}} = 0,$$

since

$$\operatorname{supp} \frac{\psi(x_1, \xi')}{|\rho'(\xi')|^2} \subset \overline{R_-^{n+1}}.$$

Hence, in particular,

$$\langle u_+(x_1, \xi'), (1 + |\xi'|^2)^{-M} \psi(x_1, \xi')\rangle = 0,$$

since the function $(1 + |\xi'|^2)^{-M} \psi(x_1, \xi')$ has the same properties as $\psi(x_1, \xi')$. We now have

$$\langle (1 + |\xi'|^2)^{-M} u_+(x_1, \xi'), \psi(x_1, \xi')\rangle$$
$$= \int_{R^{n+1}} (1 + |\xi'|^2)^{-M} u_+(x_1, \xi') \overline{\psi}(x_1, \xi')\, dx_1 d\xi' = 0.$$

The functions $\psi(x_1, \xi')$ of the form here considered are dense in

$$L_2^- L_2(R^n) = L_2^-(R^{n+1}) = \{v(x_1, \xi') \in L_2(R^{n+1}) \mid \operatorname{supp} v(x_1, \xi') \subset \overline{R_-^{n+1}}\},$$

and hence the support of $(1 + |\xi'|^2)^{-M} u_+(x_1, \xi')$ is contained in the halfspace $x_1 \geqslant 0$, i.e.

$$(1 + |\xi'|^2)^{-M} u_+(\xi_1, \xi') = F_{x_1 \to \xi_1}[(1 + |\xi'|^2)^{-M} u_+(x_1, \xi')] \in L_2^+ L_2(R^n).$$

Thus

$$W(\xi_1, \cdot) \in L_2^+ L_2(R^n) \quad \text{and} \quad \int_{R^{n+1}} W(\xi)(1 + |\xi'|^2)^M \bar{\varphi}(\xi)\, d\xi = 0$$

for any $\varphi(\xi) = F[\varphi(x)] \in S(R^{n+1})$, where $\operatorname{supp} \varphi(x) \subset \overline{R_+^{n+1}}$. Inasmuch as $\operatorname{supp} F^{-1}[(1 + |\xi'|^2)^M \varphi(\xi)]$ is also contained in $\overline{R_+^{n+1}}$, while the functions of this form are dense in the subspace

$$\{v(x) \in L_2(R^{n+1}) \mid \operatorname{supp} v(x) \subset \overline{R_+^{n+1}}\},$$

we get $W(x) = F^{-1}[W(\xi)] = 0$. We have thus proved that

$$\int_{R^{n+1}} (1 + |\xi'|^2)^{-2M} |u_+(\xi) - u_{1+}(\xi)|^2\, d\xi = 0.$$

Hence we have

$$\iint_{\Omega_c} |\rho'(\xi')(u_+(\xi) - u_{1+}(\xi))|^2\, d\xi = 0$$

for any bounded measurable set $\Omega_c \subset R^{n+1}$, which implies the equality

$$\int_{\Omega_c} |\rho'(\xi') u_+(\xi)|^2\, d\xi = \int_{\Omega_c} |\rho'(\xi') u_{1+}(\xi)|^2\, d\xi,$$

whose the left side, and hence also the right side, has a limit for $\Omega_c \to R^{n+1}$, so that by virtue of the preceding equality we have

$$\| u_+(\xi_1, \cdot) - u_{1+}(\xi_1, \cdot) \|_{L_2 H_{\rho'}} = 0,$$

as required. From the equality $u_+(\xi) = u_{1+}(\xi)$ it follows that $v(\xi) = v_1(\xi)$, inasmuch as $S(R^{n+1})$ is dense in $H_\gamma(R^{n+1})$. The lemma is proved.

DEFINITION 1.1. Let $u(\xi) \in H_\rho(R^{n+1})$, where $\rho - \rho(\xi)$ satisfies condition (1.6), and suppose $u(\xi)$ is representable in the form (1.7). Then, by definition,

$$\Pi_{1+} u(\xi_1, \xi') = u_+(\xi_1, \xi') \subset L_2^+ H_{\rho'}(R^n)$$

with $\rho' = \rho'(\xi')$ satisfying (1.6).

Lemma 1.6 shows the correctness of this definition. We note that the operator Π_{1+} is thus defined on a dense set in $H_\rho(R^{n+1})$ and, in general, is unbounded.

The operator Π_{1-} is defined analogously in $H_\rho(R^{n+1})$.

LEMMA 1.7. *Let*

$$P_l(\xi_1, \xi') = \sum_{i=0}^{l} a_i(\xi')\xi_1^i, \qquad Q_m(\xi_1, \xi') = \sum_{j=0}^{m} b_j(\xi')\xi_1^j, \qquad b_m = 1,$$

where $b_m = 1$ *and the* $b_j(\xi')$ *are continuous functions, and for some* $\alpha \geqslant 0$

$$(1 + |\xi'|^2)^{-\alpha/2} a_i(\xi') \in L_2(R^n), \quad (1 + |\xi'|^2)^{-\alpha/2} |b_j(\xi')| \leqslant C. \tag{1.8}$$

Suppose the roots $\lambda_j(\xi')$ *of the polynomial* $Q_m(z_1, \xi')$ *in* z_1 *satisfy the condition*

$$\operatorname{Im} \lambda_j(\xi') \leqslant -\delta < 0, \qquad \xi' \in R^n. \tag{1.9}$$

Then the function

$$u(\xi_1, \xi') = \frac{P_l(\xi_1, \xi')}{Q_m(\xi_1, \xi')}$$

belongs to the extended domain of definition of the operator Π_{1+} *and, if* $R_{m-1}(\xi_1, \xi')$ *is the remainder in the division of* $P_l(\xi_1, \xi')$ *by* $Q_m(\xi_1, \xi')$,

$$\Pi_{1+} \frac{P_l(\xi_1, \xi')}{Q_m(\xi_1, \xi')} = \frac{R_{m-1}(\xi_1, \xi')}{Q_m(\xi_1, \xi')}. \tag{1.10}$$

PROOF. We note that $|Q_m(\xi_1, \xi')| \geqslant \delta_1 > 0$, which implies the existence of a function $\rho = \rho(\xi)$ satisfying (1.6) such that $u(\xi_1, \xi') \in H_\rho(R^{n+1})$. It is also clear that, if $S_{l-m}(\xi_1, \xi')$ is the quotient in the division of $P_l(\xi_1, \xi')$ by $Q_m(\xi_1, \xi')$, we can always choose a function $\gamma = \gamma(\xi)$ so that $S_{l-m}(\xi_1, \xi') \in H_\gamma(R^{n+1})$. Let us prove that for some $\rho' = \rho'(\xi')$ satisfying (1.6) we have

$$v(\xi_1, \cdot) = \frac{R_{m-1}(\xi_1, \cdot)}{Q_m(\xi_1, \cdot)} \in L_2 H_{\rho'}(R^n).$$

The coefficients $r_i(\xi')$ of the polynomial $R_{m-1}(\xi_1, \xi')$ satisfy the condition $r_i(\xi') \times (1 + |\xi'|^2)^{-s_1/2} \in L_2(R^n)$ for some $s_1 \geqslant 0$. This follows from the fact that the coefficient of ξ_1^m in the polynomial $Q_m(\xi_1, \xi')$ is equal to 1. Take $\rho' = \rho'(\xi') = (1 + |\xi'|^2)^{-s_2/2}$ and choose s_2 so that $v(\xi_1, \cdot) \in L_2 H_{\rho'}(R^n)$. We have

$$\int_{-\infty}^{\infty} d\xi_1 \int_{R^n} \frac{|R_{m-1}(\xi_1, \xi')|^2 d\xi'}{(1 + |\xi'|^2)^{s_2} |Q_m(\xi_1, \xi')|^2}$$

$$\leqslant \sum_{i=0}^{m-1} \int_{R^n} \frac{|r_i(\xi')|^2 d\xi'}{(1 + |\xi'|^2)^{s_2}} \int_{R^1} \frac{\xi_1^{2(m-1-i)} d\xi_1}{|Q_m(\xi_1, \xi')|^2} \leqslant C \sum_{i=0}^{m-1} \int_{R^n} \frac{|r_i(\xi')|^2 d\xi'}{(1 + |\xi'|^2)^{s_2}}$$

$$\times \max_{\xi_1 \in R^1} \frac{\xi_1^{2(m-1-i)}}{|(\xi_1 - \lambda_{i+1}) \cdots (\xi_1 - \lambda_m)|^2} \int_{R^1} \frac{d\xi_1}{|(\xi_1 - \lambda_1) \cdots (\xi_1 - \lambda_i)|^2}.$$

Let us estimate

$$\beta = \max_{\xi_1 \in R^1} \frac{\xi_1^{2(m-1-i)}}{|(\xi_1 - \lambda_{i+1}) \dots (\xi_1 - \lambda_m)|^2}.$$

We have

$$\beta \leqslant \max_{\xi_1 \in R^1} \frac{\xi_1^2}{|\xi_1 - \lambda_{i+1}|^2} \cdot \ldots \cdot \max_{\xi_1 \in R^1} \frac{\xi_1^2}{|\xi_1 - \lambda_m|^2}$$
$$\leqslant \frac{|\lambda_{i+1}|^2}{(\operatorname{Im} \lambda_{i+1})^2} \cdot \ldots \cdot \frac{|\lambda_m|^2}{(\operatorname{Im} \lambda_m)^2} \leqslant \frac{|b_0(\xi')|^2}{\delta^{2m}},$$

where δ is taken from condition (1.9). It is not difficult to see that

$$\int_{R^1} \frac{d\xi_1}{|(\xi_1 - \lambda_1) \dots (\xi_1 - \lambda_i)|^2} \leqslant C(\delta).$$

By virtue of these estimates we get

$$\int_{-\infty}^{\infty} d\xi_1 \int_{R^n} \frac{|v(\xi_1, \xi')|^2}{(1 + |\xi'|^2)^{s_2}} d\xi' \leqslant C \sum_{i=0}^{m-1} \int_{R^n} \frac{|r_i(\xi') b_0(\xi')|^2 d\xi'}{(1 + |\xi'|^2)^{s_2}}$$
$$\leqslant C \sum_{i=0}^{m-1} \int_{R^n} \frac{|r_i(\xi')|^2 d\xi'}{(1 + |\xi'|^2)^{s_2 - \alpha}} < \infty,$$

if $s_2 - \alpha = s_1$, where α is taken from condition (1.8) and s_1 is such that $r_i(\xi') \times (1 + |\xi'|^2)^{-s_1/2} \in L_2(R^n)$. We thus have

$$u(\xi_1, \xi') = S_{l-m}(\xi_1, \xi') + v(\xi_1, \xi'),$$

where

$$v(\xi_1, \xi') \in L_2 H_{\rho'}(R^n), \quad \operatorname{supp} F^{-1}[S_{l-m}(\xi_1, \xi')] \subset \{(x_1, x') \mid x_1 = 0\},$$

inasmuch as $S_{l-m}(\xi_1, \xi')$ is a polynomial in ξ_1. This means that $u(\xi_1, \xi')$ belongs to the (extended) domain of definition of the operator Π_{1+}, and

$$\Pi_{1+} \frac{P_l(\xi_1, \xi')}{Q_m(\xi_1, \xi')} = \Pi_{1+} \frac{R_{m-1}(\xi_1, \xi')}{Q_m(\xi_1, \xi')}.$$

We now prove (1.10). Let $\varphi(x_1, x') \in C_0^\infty(R^{n+1})$, $\operatorname{supp} \varphi(x) \subset \overline{R_-^{n+1}}$. Then, inasmuch as $u(\xi_1, \xi') \subset H_\rho(R^{n+1}) \subset S'(R^{n+1})$, we have

$$(2\pi)^{n+1} \langle u(x_1, x'), \varphi(x_1, x') \rangle$$
$$= \langle u(\xi_1, \xi'), \varphi(\xi_1, \xi') \rangle = \left\langle R_{m-1}(\xi_1, \xi'), \frac{\varphi(\xi_1, \xi')}{\overline{Q_m}(\xi_1, \xi')} \right\rangle.$$

From the fact that $\varphi(\xi) \subset S(R^{n+1})$, it follows that

$$\psi(x) = F^{-1}[\varphi(\xi)/\overline{Q}_m(\xi)] \in H_s(R^{n+1})$$

for any s. The function $R_{m-1}(\xi_1, \xi')$ is a polynomial in ξ_1, and by virtue of the above-mentioned properties of its coefficients $r_i(\xi')$ we have $F^{-1}[R_{m-1}(\xi_1, \xi')] \in H_{s_3}(R^{n+1})$ for some $s_3 \leqslant 0$. Moreover,

$$\operatorname{supp} F^{-1}_{x \to \xi}[R_{m-1}(\xi_1, \xi')] \subset \{x \mid x_1 = 0\}.$$

The functions $\varphi(\xi_1, \xi')$ and

$$\frac{1}{Q_m(z_1, \xi')} = \frac{1}{(z_1 - \lambda_1) \ldots (z_1 - \lambda_m)}$$

clearly satisfy the conditions of Corollary 1.3 (condition (1.9) is again used here) with the replacement of $\overline{R^{n+1}_+}$ by $\overline{R^{n+1}_-}$, so that $\operatorname{supp} \psi(x) \subset \overline{R^{n+1}_-}$. Taking into account the fact that $\psi(x) \in H_{-s_3}(R^{n+1})$, we get

$$(2\pi)^{n+1} \langle F^{-1}_{x \to \xi}[R_{m-1}(\xi)], \psi(x) \rangle = 0 = \left\langle R_{m-1}(\xi), \frac{\varphi(\xi)}{Q_m(\xi)} \right\rangle.$$

Since this is true for any $\varphi(x) \in C_0^\infty(R^{n+1})$, $\operatorname{supp} \varphi \subset \overline{R^{n+1}_-}$, we have

$$\Pi_{1+} \frac{R_{m-1}(\xi_1, \cdot)}{Q_m(\xi_1, \cdot)} = \frac{R_{m-1}(\xi_1, \cdot)}{Q_m(\xi_1, \cdot)}.$$

Q.E.D.

3. *The null space of the operator* $A(D_t + ia, D_x)$ *in* $L_2(R^{n+1}_+)$. Let $\Omega \subset R^m$. Then, as usual,

$$L_2(\Omega) = \left\{ u(x), \quad x \in \Omega \mid \int_\Omega |u(x)|^2 \, dx < \infty \right\}.$$

With the operator $A(D_t + ia, D_x) = -(D_t + ia)^2 + (\mathbf{A}D_x, D_x)$ we associate the polynomial

$$A(\tau + ia, \xi) = -(\tau + ia)^2 + (\mathbf{A}\xi, \xi) = \xi_1^2 + (\mathbf{A}'\xi', \xi') - (\tau + ia)^2. \quad (1.11)$$

Let

$$\lambda(\tau + ia, \xi') = \sqrt{(\tau + ia)^2 - (\mathbf{A}'\xi', \xi')} \quad (1.12)$$

with the condition $\lambda(ia, \xi') = i\sqrt{a^2 + (\mathbf{A}'\xi', \xi')}$. Here, as before, $x = (x_1, \cdots, x_n)$, $x' = (x_2, \cdots, x_n)$, $\xi' = (\xi_2, \cdots, \xi_n)$ and $\xi = (\xi_1, \cdots, \xi_n)$. It follows from this choice of root branch that $\lambda(s + ia, \xi')$ is an analytic function of $s = \tau + i\sigma$, $\sigma > -a$, for which it can be directly established that

$$\operatorname{Im}\lambda(s+ia,\ \xi')\geqslant a+\sigma. \tag{1.13}$$

The polynomial $A(\tau+ia,\ \xi)$ admits the factorization

$$A(\tau+ia,\ \xi)=(\xi_1+\lambda)(\xi_1-\lambda). \tag{1.14}$$

REMARK 1.2. In the sequel the arguments (either $(\tau+ia,\ \xi)$ or $(\tau+ia,\ \xi')$) of the symbols $A(\cdot,\ \cdot)$, $\lambda(\cdot,\ \cdot)$ and similar ones will often not be indicated for the sake of brevity.

In describing the null space of the operator $A(D_t+ia,\ D_x)$ in $L_2(R_+^{n+1})$ we will find it convenient for what follows to consider $u(t,\ x)\in L_2(R_+^{n+1})$ as an element of $L_2(R^{n+1})$ that is equal to zero for $x_1\leqslant 0$, i.e. we will give a description of the null space of the operator $A(D_t+ia,\ D_x)$ in $L_2^+(R^{n+1})$. We note that $A(D_t+ia,\ D_x)u(t,\ x)\in D'(R^{n+1})$ (more precisely, $S'(R^{n+1})$) for $u(t,\ x)\in L_2^+(R^{n+1})$, and that $A(D_t+ia,\ D_x)u(t,\ x)=0$ for $x_1>0$ is understood, as usual, in the sense that $\langle A(D_t+ia,\ D_x)u(t,\ x),\ \varphi(t,\ x)\rangle=0$ for any $\varphi(t,\ x)\in D(R^{n+1})$, $\varphi(t,\ x)=0$ for $x_1\leqslant 0$.

LEMMA 1.8. *If $u(t,\ x)\subset L_2(R^{n+1})$, $u(t,\ x)=0$ for $x_1\leqslant 0$, and $A(D_t+ia,\ D_x)u(t,\ x)=0$ for $x_1>0$, then*

$$u(\tau,\ \xi)=F[u(t,\ x)]=\frac{c(\tau,\ \xi')}{\xi_1+\lambda(\tau+ia,\ \xi')}, \tag{1.15}$$

in which at least $(\tau^2+|\xi'|^2+1)^{-1}c(\tau,\ \xi')\in L_2(R^n)$.

The proof is a modification of the proof of a theorem on functions with a point support (see, for example, [10], §9.4). Suppose $u(t,\ x)$ satisfies the conditions of the lemma. Then supp $A(D_t+ia,\ D_x)u(t,\ x)$ is concentrated on the hyperplane $x_1=0$. If $\varphi(t,\ x)\in S(R^{n+1})$, then $\langle Au,\ \varphi\rangle=\langle u,\ A^*\varphi\rangle$, where $A^*=A(D_t-ia,\ D_x)$ is the formal adjoint of the differential expression $A(D_t+ia,\ D_x)$. If, in addition, $\varphi(t,\ x)=0$ for $|x_1|<\delta$, then $\langle Au,\ \varphi\rangle=0$.

Let us prove that $\langle Au,\ \varphi\rangle=0$ for any function $\varphi(t,\ s)\in S(R^{n+1})$ that together with its derivatives up to order 2 vanishes for $x_1=0$. To this end we construct a sequence $\varphi_\nu(t,\ x)\in S(R^{n+1})$, $\nu=1,\ 2,\ \cdots$, such that $\lim_{\nu\to\infty}\varphi_\nu(t,\ x)=\varphi(t,\ x)$ uniformly together with derivatives up to order 2 and $\varphi_\nu(t,\ x)=0$ in a neighborhood of the hyperplane $x_1=0$. With the use of a function $\rho(x_1)$ such that $\rho(x_1)=0$ for $|x_1|>1$, $\rho(x_1)=1$ for $|x_1|<1/2$, $\rho(x_1)\in C_0^\infty(R^1)$, we construct the sequence $\varphi_\nu(t,\ x)=\varphi(t,\ x)(1-\rho(\nu x_1))$ and prove that it is the desired one, i.e. $\lim_{\nu\to\infty}\psi_\nu(t,\ x)=0$ uniformly together with derivatives up to order 2, where $\psi_\nu(t,\ x)=\varphi(t,\ x)-\varphi_\nu(t,\ x)=\varphi(t,\ x)\rho(\nu x_1)$. We have

$$\frac{d^j}{dx^j}\rho(\nu x_1)=\rho^{(j)}(\nu x_1)\nu^j,$$

and the estimate $|\rho^{(j)}(\nu x_1)| \leqslant R$ for $0 \leqslant j \leqslant 2$. Therefore

$$\left|\frac{d^j}{dx_1^j}\rho(\nu x_1)\right|\leqslant R\nu^j.$$

We now estimate the derivatives of $\psi_\nu(t, x)$. Let $k = (k_0, k_2, \cdots, k_n)$ denote an integral vector, let $|k| = k_0 + k_2 + \cdots + k_n$ and let

$$D_k=\frac{\partial^{|k|}}{\partial t^{k_0}\partial x_2^{k_2}\ldots\partial x_n^{k_n}},\quad k_j\geqslant 0,\ 0\leqslant k_1\leqslant 2,\ k_1 \text{ an integer},\ 0\leqslant k_1+|k|\leqslant 2.$$

We then have

$$\begin{aligned}\frac{\partial^{k_1}}{\partial x_1^{k_1}}D_k\psi_\nu(t,x)&=\frac{\partial^{k_1}}{\partial x_1^{k_1}}\{\rho(\nu x_1)D_k\varphi_\nu(t,x)\}\\&=\sum_{j=0}^{k_1}C_{k_1}^j\left(\frac{\partial^{k_1-j}}{\partial x_1^{k_1-j}}D_k\varphi\right)\frac{\partial^j}{\partial x_1^j}\rho(\nu x_1).\end{aligned}\tag{1.16}$$

According to the mean value theorem we have

$$\begin{aligned}\left|\frac{\partial^{2-|k|}}{\partial x_1^{2-|k|}}D_k\varphi(t,x)\right|&=\left|x_1\frac{\partial^{2-|k|+1}}{\partial x_1^{2-|k|+1}}D_k\varphi(t,\theta,x')\right|\\&\leqslant\max_{|x_1|<1/\nu}\left|\frac{\partial^{2-|k|+1}}{\partial x_1^{2-|k|+1}}D_k\varphi\right|\frac{1}{\nu}=C_k\frac{1}{\nu},\quad 0<|\theta|<|x_1|.\end{aligned}$$

Here we have used the facts that the summands in (1.16) vanish except when $|x_1| < 1/\nu$, that $\varphi(t, x) \in S(R^{n+1})$ and that

$$\frac{\partial^{2-|k|}}{\partial x_1^{2-|k|}}D_k\varphi(t,0,x')=0.$$

Extending these estimates for the lowest derivatives, we find

$$\left|\frac{\partial^{2-|k|-j}}{\partial x_1^{2-|k|-j}}D_k\varphi(t,x)\right|\leqslant C_{k,j}\frac{1}{\nu^{j+1}}.$$

Hence we have

$$\left|\frac{\partial^{k_1-j}}{\partial x_1^{k_1-j}}D_k\varphi(t,x)\right|\leqslant C(k,k_1)\frac{1}{\nu^{2-|k|-k_1+j+1}}\quad\text{for } |x_1|\leqslant\frac{1}{\nu}.$$

Returning to the sum in (1.16), we get

$$\left|\frac{\partial^{k_1}}{\partial x_1^{k_1}} D_k\psi_\nu(t, x)\right| \leqslant \sum_{j=0}^{k_1} \frac{C(k, k_1)|}{\nu^{2-|k|-k_1+j+1}} R\nu^j \leqslant C_1\frac{1}{\nu},$$

inasmuch as $|k| + k_1 \leqslant 2$. Hence

$$\frac{\partial^{k_1}}{\partial x_1^{k_1}} D_k\psi_\nu(t, x) \to 0$$

uniformly for $\nu \to \infty$ and $|k| + k_1 \leqslant 2$.

We now note that $\langle Au, \varphi_\nu\rangle = 0$ for any $\varphi_\nu(t, x)$ and $A^*\varphi_\nu(t, x) \to A^*\varphi(t, x)$ uniformly. Hence we obtain the chain of equalities

$$0 = \lim_{\nu\to\infty}\langle Au, \varphi_\nu\rangle = \lim_{\nu\to\infty}\langle u, A^*\varphi_\nu\rangle = \lim_{\nu\to\infty}\int_{R^{n+1}} u(t, x)\, A^*\varphi_\nu(t, x)\, dt\, dx$$

$$= \int_{R^{n+1}} u(t, x)\, A^*\varphi(t, x)\, dt\, dx = \langle u, A^*\varphi\rangle = \langle Au, \varphi\rangle.$$

This proves that $\langle Au, \varphi\rangle = 0$ for any function $\varphi(t, x) \in S(R^{n+1})$ that together with its derivatives up to order 2 vanishes for $x_1 = 0$.

We will find a representation of the generalized function $A(D_t + ia, D_x)u(t, x) \in S'(R^{n+1})$. Let $\varphi(t, x) \in S(R^{n+1})$. According to Taylor's theorem we have

$$\varphi(t, x) = \sum_{j=0}^{2} \frac{1}{j!} \frac{\partial^j\varphi(t, 0, x')}{\partial x_1^j} x_1^j + r_2(t, x). \tag{1.17}$$

We note that if $\rho(x_1)$ is the function introduced above, then $\langle Au, \varphi\rangle = \langle Au, \rho\varphi\rangle$, since $\operatorname{supp} Au = \{(t, 0, x')\}$. We multiply (1.17) by $\rho(x_1)$:

$$\rho(x_1)\varphi(t, x) = \sum_{j=0}^{2} \frac{1}{j!} \frac{\partial^j\varphi(t, 0, x')}{\partial x_1^j} \rho(x_1) x_1^j + \rho(x_1) r_2(t, x).$$

We have

$$\frac{\partial^j\varphi(t, 0, x')}{\partial x_1^j} \rho(x_1) x_1^j \in S(R^{n+1}) \qquad 0 \leqslant j \leqslant 2, \qquad \rho(x_1)\varphi(t, x) \in S(R^{n+1}),$$

and hence $\rho(x_1) r_2(t, x) \in S(R^{n+1})$. Therefore

$$\langle Au, \varphi\rangle = \left\langle Au, \sum_{j=0}^{2} \frac{1}{j!} \frac{\partial^j\varphi(t, 0, x')}{\partial x_1^j} \rho(x_1) x_1^j\right\rangle + \langle Au, \rho(x_1) r_2\rangle.$$

Here $\rho(x_1)r_2(t, x)$ together with its derivatives up to order 2 vanishes for $x_1 = 0$. According to what was proved above $\langle Au, \rho(x_1)r_2(t, x)\rangle = 0$. Thus

$$\langle Au, \varphi\rangle = \left\langle Au, \sum_{j=0}^{2} \frac{1}{j!} \frac{\partial^j \varphi(t, 0, x')}{\partial x_1^j} \rho(x_1) x_1^j \right\rangle$$

$$= \sum_{j=0}^{2} \frac{1}{j!} \left\langle Au, \frac{\partial^j \varphi(t, 0, x')}{\partial x_1^j} \rho(x_1) x_1^j \right\rangle = \sum_{j=0}^{2} \frac{1}{j!} \left\langle u, A^* \left\{ \frac{\partial^j \varphi(t, 0, x')}{\partial x_1^j} \rho(x_1) x_1^j \right\} \right\rangle.$$

As was noted above,

$$A^* = -(D_t - ia)^2 + D_1^2 + (\mathbf{A}' D_{x'}, D_{x'}),$$

which implies

$$A^* \left\{ \frac{\partial^j \varphi(t, 0, x')}{\partial x_1^j} \rho(x_1) x_1^j \right\}$$

$$= \left\{ [-(D_t - ia)^2 + (\mathbf{A}' D_{x'}, D_{x'})] \frac{\partial^j \varphi(t, 0, x')}{\partial x_1^j} \right\} \rho(x_1) x_1^j$$

$$+ \frac{\partial^j \varphi(t, 0, x')}{\partial x_1^j} \{D_1^2 \rho(x_1) x_1^j\} = \alpha_{0_j}(x_1) A'^* \frac{\partial^j \varphi(t, 0, x')}{\partial x_1^j} + \alpha_{1_j}(x_1) \frac{\partial^j \varphi(t, 0, x')}{\partial x_1^j}.$$

Here $A' = -(D_t + ia)^2 + (\mathbf{A}' D_{x'}, D_{x'})$, A'^* is the formal adjoint of A', and the meaning of $\alpha_{0_j}(x_1)$ and $\alpha_{1_j}(x_1)$, which are real-valued functions of x_1, is clear from the formula.

Using the above expression, we find that

$$\langle Au, \varphi\rangle = \sum_{j=0}^{2} \frac{1}{j!} \left\{ \left\langle u, \alpha_{0_j}(x_1) A'^* \frac{\partial^j \varphi(t, 0, x')}{\partial x_1^j} \right\rangle \right.$$

$$\left. + \left\langle u, \alpha_{1_j}(x_1) \frac{\partial^j \varphi(t, 0, x')}{\partial x_1^j} \right\rangle \right\} = \sum_{j=0}^{2} \frac{1}{j!} \left\{ \int_{R^n} dt dx' \int_{-\infty}^{\infty} dx_1 u(t, x) \alpha_{0_j}(x_1) \right.$$

$$\left. \times A'^* \frac{\partial^j \varphi(t, 0, x')}{\partial x_1^j} + \int_{R^n} dt dx' \int_{-\infty}^{\infty} dx_1 u(t, x) \alpha_{1_j}(x_1) \frac{\partial^j \varphi(t, 0, x')}{\partial x_1^j} \right\}.$$

Let

$$\beta_{k,j}(t, x') = \int_{-\infty}^{\infty} u(t, x) \alpha_{k,j}(x_1) dx_1, \quad (\operatorname{supp} \alpha_{k,j}(x_1) \subset [-1, 1]).$$

We have $\beta_{k,j}(t, x') \in L_2(R^n)$, since

$$\int_{R^n} dtdx'\,|\beta_{k,j}(t,x')|^2 \leqslant \int_{-1}^{1} |\alpha_{k,j}(x_1)|^2 dx_1 \int_{R^n} dtdx' \times \int_{-1}^{1} |u(t,x)|^2 dx_1 \leqslant C\,|u|^2_{L_2(R^{n+1})}.$$

In this notation we have

$$\langle \mathsf{A}u, \varphi\rangle = \sum_{j=0}^{2} \frac{1}{j!}\left\{\left\langle \beta_{0j}(t,x'),\ \mathsf{A}'^{*}\frac{\partial^j\varphi(t,0,x')}{\partial x_1^j}\right\rangle + \left\langle \beta_{1,j}(t,x'),\ \frac{\partial^j\varphi(t,0,x')}{\partial x_1^j}\right\rangle\right\} = \sum_{j=0}^{2}\frac{1}{j!}\left\langle \mathsf{A}'\beta_{0,j}(t,x') + \beta_{1,j}(t,x'),\ \frac{\partial^j\varphi(t,0,x')}{\partial x_1^j}\right\rangle.$$

The resultant functionals can be represented in the form of the direct product $(\mathsf{A}'\beta_{0,j} + \beta_{1,j}) \times \delta^{(j)}(x_1)$. Letting

$$\gamma_j(t,x') = \frac{1}{j!}\{\mathsf{A}'\beta_{0,j}(t,x') + \beta_{1,j}(t,x')\}, \tag{1.18}$$

we get

$$\langle \mathsf{A}u, \varphi\rangle = \sum_{j=0}^{2} \langle \gamma_j(t,x') \times \delta^{(j)}(x_1),\ \varphi(t,x)\rangle,$$

which implies

$$\mathsf{A}(D_t + ia, D_x)\,u(t,x) = \sum_{j=0}^{2} \gamma_j(t,x') \times \delta^{(j)}(x_1),$$

or, in terms of Fourier transforms,

$$\mathsf{A}(\tau + ia, \xi)\,u(\tau,\xi) = \sum_{j=0}^{2} c_j(\tau,\xi')\,\xi_1^j,$$

where the $c_j(\tau,\xi')$ are measurable functions that can differ only by numerical factors from the $\gamma_j(\tau,\xi')$. Recalling that $u(\tau,\xi) \subset L_2(R^{n+1})$, we see that this is possible only if $c_2(\tau,\xi') = 0$ for almost all $(\tau,\xi') \in R^n$, i.e.

$$\mathsf{A}(\tau + ia, \xi)\,u(\tau,\xi) = c_0(\tau,\xi') + c_1(\tau,\xi')\,\xi_1,$$

or

$$u(\tau,\xi) = \frac{c_1(\tau,\xi')\,\xi_1 + c_0(\tau,\xi')}{\mathsf{A}(\tau + ia, \xi)}. \tag{1.19}$$

We now again make use of the fact that $u(t, x) = 0$ for $x_1 \leqslant 0$. According to Lemma 1.2, this means that the function $u(\cdot, \xi_1, \cdot)$ of the parameter $\xi_1 \in R^1$ with values in $H = L_2(R^n)$ can be analytically continued into the halfplane $\operatorname{Im} z_1 = \operatorname{Im}(\xi_1 + i\eta_1) = \eta_1 > 0$. Hence the right side of (1.19) must also be analytically continuable into this domain. Inasmuch as

$$A(\tau + ia, \xi) = (\xi_1 + \lambda)(\xi_1 - \lambda), \qquad \operatorname{Im} \lambda \geqslant a > 0,$$

the numerator $c_1(\tau, \xi')\xi_1 + c_0(\tau, \xi')$ must be divisible by $\xi_1 - \lambda$, i.e. $c_0(\tau, \xi') = -\lambda c_1(\tau, \xi')$, and we obtain (1.15).

Finally, we note that $c(\tau, \xi')$ differs from $\gamma_1(\tau, \xi')$ only by a numerical factor while, according to (1.18), $[\tau^2 + |\xi'|^2 + 1]^{-1}\gamma_1(\tau, \xi') \in L_2(R^n)$. Hence this is also true for $c(\tau, \xi')$. Of course, in order for $u(\tau, \xi) \in L_2(R^{n+1})$ the function $c(\tau, \xi')$ must satisfy a stronger condition, which we do not require. The proof of Lemma 1.8 is complete.

REMARK 1.3. Lemma 1.8 can obviously also be proved for a homogeneous differential operator $A(D_t, D_1, D_{x'})$ of order m for which the polynomial $A(\tau, \xi_1, \xi')$ has "separated" roots in ξ_1 [11].

LEMMA 1.9. *Let* $\lambda(\tau + ia, \xi')$ *be the symbol defined by* (1.12). *Then*

$$\lim_{A \to +\infty} \frac{1}{2\pi} \int_{-A}^{A} \frac{e^{-ix_1\xi_1} d\xi_1}{\xi_1 + \lambda} = -ie^{ix_1\lambda} \tag{1.20}$$

uniformly on any set of the form $[\epsilon, +\infty) \times \Omega_c$, *where* Ω_c *is a bounded set in* R^n: $\tau^2 + |\xi'|^2 \leqslant c$, $\epsilon > 0$.

The proof is carried out by means of an integration in the complex plane and consists in estimating the following integral over a semicircle:

$$\frac{1}{2\pi} \int_{-\pi}^{0} \frac{iA \exp i\varphi \cdot \exp(-ix_1 A \exp i\varphi)\, d\varphi}{A \exp i\varphi + \lambda}.$$

This integral is estimated in the obvious way by "excising the ends".

§2. Function spaces (the case $|k| < \sqrt{a_{22}}$)

1. *Auxiliary formulas.* As before, let $\lambda(\tau + ia, \xi') = \sqrt{(\tau + ia)^2 - (A'\xi', \xi')}$. Also, let

$$\begin{aligned} \tilde{\lambda}(\tau' + ia, \xi') &= \lambda(\tau' - k\xi_2 + ia, \xi') \\ &= \sqrt{(\tau' - k\xi_2 + ia)^2 - a_{22}\xi_2^2 - (A''\xi'', \xi'')}, \qquad \xi'' = (\xi_3, \ldots, \xi_n), \end{aligned} \tag{2.1}$$

with the condition $\tilde{\lambda}(k\xi_2 + ia, \xi') = \lambda(ia, \xi') = i\sqrt{a^2 + (\mathbf{A}'\xi', \xi')}$;

$$\varkappa(\tau' + ia, \xi'') = \sqrt{(\tau' + ia)^2 - (1 - \theta^2)(\mathbf{A}''\xi'', \xi'')} \tag{2.2}$$

with the condition $\kappa(ia, \xi'') = i\sqrt{a^2 + (1 - \theta^2)(\mathbf{A}''\xi'', \xi'')}$, $\theta = k/\sqrt{a_{22}}$; and

$$\eta_{\pm}(\tau' + ia, \xi') = \sqrt{\frac{\varkappa(\tau' + ia, \xi'') \pm \theta(\tau' + ia)}{1 \pm \theta} \pm \sqrt{a_{22}}(1 \mp \theta)\xi_2} \tag{2.3}$$

with the conditions

$$\eta_-(\tau' + ia, \xi') \to i\sqrt{\sqrt{a_{22}}(1 + \theta)\xi_2}, \quad \text{and} \quad \eta_+(\tau' + ia, \xi') \to \sqrt{\sqrt{a_{22}}(1 - \theta)\xi_2}$$

as $\xi_2 \to +\infty$. Under this choice of branches, from (2.1)–(2.3) we have

$$\tilde{\lambda}(\tau' + ia, \xi') = \eta_+(\tau' + ia, \xi')\,\eta_-(\tau' + ia, \xi'). \tag{2.4}$$

If we make the substitution $\tau' - k\xi_2 = \tau$, we get

$$\lambda(\tau + ia, \xi') = \eta_+(\tau + k\xi_2 + ia, \xi')\,\eta_-(\tau + k\xi_2 + ia, \xi'). \tag{2.5}$$

The functions $\eta_{\pm}(\tau' + ia, z_2, \xi'')$ are analytic with respect to $z_2 = \xi_2 + i\eta_2$ in the upper and lower halfplanes respectively, more precisely for $\eta_2 > -a/\sqrt{a_{22}}(1 - \theta)$ and $\eta_2 < a/\sqrt{a_{22}}(1 + \theta)$ respectively, and are not equal to zero in these domains. In fact, taking the square of both sides of (2.2), we have

$$\varkappa_2^2 - a^2 = \varkappa_1^2 - \tau'^2 + (1 - \theta^2)(\mathbf{A}''\xi'', \xi''), \qquad \varkappa_1\varkappa_2 = \tau' a,$$

where $\kappa_1 = \operatorname{Re}\kappa$ and $\kappa_2 = \operatorname{Im}\kappa$. Hence

$$\begin{aligned}\varkappa_2^2(\varkappa_2^2 - a^2) &= \varkappa_2^2\varkappa_1^2 - \varkappa_2^2\tau'^2 + \varkappa_2^2(1 - \theta^2)(\mathbf{A}''\xi'', \xi'')\\ &= \tau'^2(a^2 - \varkappa_2^2) + \varkappa_2^2(1 - \theta^2)(\mathbf{A}''\xi'', \xi'')\end{aligned}$$

and, finally,

$$\varkappa_2^2 - a^2 = \frac{\varkappa_2^2}{\tau'^2 + \varkappa_2^2}(1 - \theta^2)(\mathbf{A}''\xi'', \xi'').$$

It follows that $\kappa_2 \geqslant a$, with equality being possible only when $\xi'' = 0$. Hence $\operatorname{Im}[\kappa + \theta(\tau' + ia)] \geqslant (1 + \theta)a$. It also follows that the expression under the radical sign in the right side of (2.3) does not vanish in the domains indicated above.

We note in addition that the functions $\lambda(s+ia,\xi')$, $\kappa(s+k\xi_2+ia,\xi'')$ and $\eta_\pm(s+k\xi_2+ia,\xi')$ are analytic with respect to $s=\tau+i\sigma$ and do not vanish in the halfplane $\sigma>-a$. For example, $\kappa(s+k\xi_2+ia,\xi'')$ is analytic by virtue of the fact that the expression under the radical sign in (2.2) does not vanish for $\sigma>-a$.

Let the symbol $q(\tau+ia,\xi')$ be defined by the formula

$$|q(\tau+ia,\xi')|^2=\frac{1}{|\eta_+(\tau+k\xi_2+ia,\xi')|^2\operatorname{Im}\lambda(\tau+ia,\xi')} \tag{2.6}$$

with the following conditions:

I) The function $q(s,\xi')$, $s=\tau+i\sigma$, $\xi'\in R^{n-1}$, is analytic with respect to s for $\operatorname{Im}s>0$ and infinitely differentiable with respect to ξ'.

II) $|q(\tau+ia,\xi')A(\tau+ia,\xi)|\geqslant c(a)>0$, $\xi\in R^n$.

III) $|q(\tau+ia,\xi')|\leqslant c_1(a)$, $(\tau,\xi')\in R^n$, $a>0$.

Let us prove that these conditions can be satisfied. To this end we note that $\operatorname{Im}\lambda(\tau+ia,\xi)$ can be represented in the form of a product:

$$\operatorname{Im}\lambda(\tau+ia,\xi')=\mu_+(\tau,\xi')\mu_-(\tau,\xi'), \tag{2.7}$$

where $\mu_\pm(s,\xi')$ are analytic with respect to s and do not vanish for $\operatorname{Im}s>-a$ and for $\operatorname{Im}s<a$ respectively and satisfy the condition

$$|\mu_+(\tau,\xi')|=|\mu_-(\tau,\xi')|=[\operatorname{Im}\lambda(\tau+ia,\xi')]^{1/2}.$$

when $s=\tau$. This is proved by an ordinary factorization of $\operatorname{Im}\lambda(\tau+ia,\xi')$ with the use of the fact that $\operatorname{Im}\lambda(\tau+ia,\xi')\geqslant a>0$, $\operatorname{Im}\lambda(\tau+ia,\xi')\to a$ for $|\tau|\to\infty$. The proof can be found in [11].

The functions $\mu_\pm(\tau,\xi')$ in (2.7) will be infinitely differentiable with respect to ξ'. On the basis of the factorization (2.7) we can assert that the function

$$q(s,\xi')=\frac{1}{\eta_+(s+k\xi_2,\xi')\mu_+(s,\xi')} \tag{2.8}$$

satisfies I). The fulfilment of III) follows from the fact that

$$\frac{1}{\operatorname{Im}\lambda(\tau+ia,\xi')}\leqslant\frac{1}{a},$$

while

$$|\eta_+(\tau+k\xi_2+ia,\xi')|^2\geqslant\frac{\operatorname{Im}\varkappa(\tau+k\xi_2+ia,\xi'')+\theta a}{1+\theta}\geqslant a.$$

To verify II) we note that

$$\min_{\xi_1}|A(\tau+ia,\xi)|^2\geqslant c|\lambda(\tau+ia,\xi')|^2\operatorname{Im}\lambda(\tau+ia,\xi'),$$

which is established by a direct calculation. Hence

$$|q(\tau+ia,\ \xi')\ A(\tau+ia,\ \xi)|^2 \geqslant c\,|\eta_-(\tau+k\xi_2+ia,\ \xi')|^2\operatorname{Im}\lambda(\tau+ia,\ \xi')\geqslant caa=ca^2.$$

The symbol $q(\tau+ia,\ \xi')$ defined by (2.6) and (2.8) will be used in the definitions below of function spaces.

We cite in addition two formulas that will be needed below:

$$\int_{-\infty}^{\infty}\frac{d\xi_1}{|A(\tau+ia,\ \xi)|^2}=\frac{\pi}{2\,|\lambda(\tau+ia,\ \xi')|^2\operatorname{Im}\lambda(\tau+ia,\ \xi')}, \tag{2.9}$$

$$\int_{-\infty}^{\infty}\frac{\xi_1^2 d\xi_1}{|A(\tau+ia,\ \xi)|^2}=\frac{\pi}{2\operatorname{Im}\lambda(\tau+ia,\ \xi')}. \tag{2.10}$$

These integrals are calculated by going over into the complex plane.

2. *The spaces* V_q *and* V_q^+. Let the symbol $q(s,\ \xi')$ be defined by (2.6) and (2.8).

DEFINITION 2.1.

$$V_q\equiv V_q(R^{n+1}) =\{u(t,\ x)\in L_2(R^{n+1})\,|\,q(\tau+ia,\ \xi')\ A(\tau+ia,\ \xi)\,u(\tau,\ \xi)\in L_2(R^{n+1})\},$$

with norm

$$|u|_q=|q\,Au(\tau,\ \xi)|_{L_2(R^{n+1})}.$$

Here $u(\tau,\ \xi)=F[u(t,\ x)]$.

DISCUSSION OF DEFINITION 2.1. The functions of the class $S(R^n)$ are dense in V_q with respect to the norm $|\cdot|_q$. For suppose $u(t,\ x)\in V_q$. Then $v(\tau,\ \xi)=q\ A\ u(\tau,\ \xi)\in L_2(R^{n+1})$. Hence there exists a sequence $\{v_i(t,\ x)\}$ such that $v_i(t,\ x)\in C_0^\infty(R^{n+1})$ and $\lim_{i\to\infty}v_i=v$ in $L_2(R^{n+1})$. According to II),

$$u_i(\tau,\ \xi)=\frac{v_i(\tau,\ \xi)}{q\ A}\in L_2(R^{n+1})$$

and hence $u_i(t,\ x)\in V_q$ and $|u_i-u|_q=|v_i-v|_{L_2(R^{n+1})}\to 0$ as $i\to\infty$. The functions $v_i(\tau,\ \xi)\in S(R^{n+1})$, and hence by virtue of I) and II) we have $u_i(\tau,\ \xi)\in S(R^{n+1})$, i.e. $u_i(t,\ x)\in S(R^{n+1})$.

V_q is a complete space. For if the sequence $\{u_i\}$ is fundamental in V_q, then $\{v_i(\tau,\ \xi)=q\ A\ u_i(\tau,\ \xi)\}$ is fundamental in $L_2(R^{n+1})$, and if $v(\tau,\ \xi)=\lim_{i\to\infty}v_i(\tau,\ \xi)$, then

$$u(\tau,\ \xi)=\frac{v(\tau,\ \xi)}{q\,A}\in L_2(R^{n+1})$$

by virtue of II). But then $u(t, x) = F^{-1}[u(\tau, \xi)] \in V_q$ and $\lim_{i\to\infty}u_i = u$.

Let $R^{n+1}_{\pm} = \{(t, x) \in R^{n+1} | x_1 \gtrless 0\}$, the upper or lower signs being taken simultaneously.

DEFINITION 2.2. V_q^+ is the set of functions $u(t, x) \in L_2(R^{n+1}_{\pm})$, for each of which there exists a function $\hat{u}(t, x) \in V_q$ such that $\hat{u}(t, x) = u(t, x)$ for $x_1 > 0$. The norm in V_q^+ is

$$|u|_{V_q^+}=|u|_q^+=\inf|\hat{u}|_q, \tag{2.11}$$

where the infimum is taken over all of the extensions of $u(t, x)$ onto R^{n+1} in $V_q(R^{n+1})$.

DISCUSSION OF DEFINITION 2.2. V_q^+ is a complete space. For if

$$\overset{\circ}{V}{}_q^-(R^{n+1})=\{u(t,\ x)\in V_q\,|\,\operatorname{supp} u(t,\ x)\subset\overline{R^{n+1}_-}\},$$

then $\overset{\circ}{V}{}_q^-(R^{n+1})$ is a closed subspace in V_q, and clearly V_q^+ is isomorphic to the factor space $V_q/\overset{\circ}{V}{}_q^-(R^{n+1})$. Hence V_q^+ is complete inasmuch as V_q is complete.

We introduce in V_q^+ an equivalent norm that is more convenient for what follows. Let $u(t, x) \in V_q^+$, $u_+(t, x) = u(t, x)$ for $x_1 > 0$, $u_+(t, x) = 0$ for $x_1 \leqslant 0$, $\hat{u}(t, x) \in V_q$ and $\hat{u}(t, x) = u(t, x)$ for $x_1 > 0$. Let

$$A_-(\tau+ia,\ \xi)=(\xi_1-\lambda)(\xi_1+\bar{\lambda}). \tag{2.12}$$

Since $|A_-(\tau + ia, \xi)| = |A(\tau + ia, \xi)|$, we have $qA_-\hat{u}(\tau, \xi) \in L_2(R^{n+1})$ and $\Pi_+ qA_-\hat{u}(\tau, \xi) \in L_2^+(R^{n+1})$. If $v_-(\tau, \xi) = \hat{u}(\tau, \xi) - u_+(\tau, \xi)$, then $v_-(\tau, x) \in L_2(R^{n+1})$ and $\operatorname{supp} v_-(t, x) \subset \overline{R^{n+1}_-}$, i.e. $v_-(\tau, \xi) \in L_2^-(R^{n+1})$. Clearly $F^{-1}[qA_-v_-(\tau, \xi)] \in H_{-2}(R^{n+1})$. We will prove that $\operatorname{supp} F^{-1}[qA_-v_-(\tau, \xi)] \in \overline{R^{n+1}_-}$. Let $\varphi(t, x) \in C_0^\infty(R^{n+1})$ and $\operatorname{supp}\varphi(t, x) \subset \overline{R^{n+1}_+}$. Then

$$\langle qA_-v_-(\tau,\xi),\varphi(\tau,\xi)\rangle=\langle v_-(\tau,\ \xi);\ \overline{qA_-\varphi(\tau,\ \xi)}\rangle.$$

We prove that $\overline{qA_-\varphi(\tau, \xi)} \in L_2^+(R^{n+1})$, i.e. that $\operatorname{supp} F^{-1}[qA_-\varphi(\tau, \xi)] \subset \overline{R^{n+1}_+}$. To this end it suffices to note that $\bar{A}_-(\tau + ia, \xi_1, \xi')$ is a polynomial in ξ_1, i.e. the function $\bar{q}(\tau + ia, \xi')\bar{A}_-(\tau + ia, z_1, \xi')$ is analytic with respect to z_1 for $\operatorname{Im} z_1 = \eta_1 \geqslant 0$. By definition this function is continuous with respect to τ and ξ, and

$$|\bar{q}(\tau+ia,\ \xi')\ \bar{A}_-(\tau+ia,\ \xi)|\leqslant c(|z_1|^2+\tau^2+|\xi'|^2).$$

It follows by Corollary 1.3 that $\bar{q}\ \bar{A}_-\ \varphi(\tau,\ \xi) \in L_2^+(R^{n+1})$, and hence

$$\langle v_-(\tau,\ \xi),\ \bar{q}\ \bar{A}_-\varphi(\tau,\ \xi)\rangle = (v_-(\tau,\ \xi),\ \bar{q}\ \bar{A}_-\varphi(\tau,\ \xi))_{L_2(R^{n+1})} = 0,$$

i.e. $\operatorname{supp} F^{-1}[q A_-v_-(\tau,\ \xi)] \in \overline{R_-^{n+1}}$. According to what has been proved,

$$u_+(\tau,\ \xi) = \hat{u}(\tau,\ \xi) - v_-(\tau,\ \xi),$$

$$q A_-u_+(\tau,\ \xi) = q A_-\hat{u}(\tau,\ \xi) - q A_-v_-(\tau,\ \xi) = \Pi_{1+}q A_-\hat{u}(\tau,\ \xi) + w_-(\tau,\ \xi),$$

where

$$w_-(\tau,\ \xi) = \Pi_{1-}q A_-\hat{u}(\tau,\ \xi) - q A_-v_-(\tau,\ \xi),$$

$$w_-(t,\ x) \in H_{-2}(R^{n+1}), \qquad \operatorname{supp} w_-(t,\ x) \in \overline{R_-^{n+1}}.$$

Consequently $q A_-u_+(\tau,\ \xi)$ belongs to the extended domain of definition of the operator Π_{1+}, and

$$\Pi_{1+}q A_-u_+(\tau,\ \xi) = \Pi_{1+}q A_-\hat{u}(\tau,\ \xi) \in L_2^+(R^{n+1}). \tag{2.13}$$

Thus for any $u(t,\ x) \in V_q^+$ we can put

$$|u|'_{V_q^+} = |\Pi_{1+}q A_-u_+(\tau,\ \xi)|_{L_2(R^{n+1})}, \tag{2.14}$$

where $u_+(t,\ x) \in L_2^+(R^{n+1})$, as before.

We will prove that $|u|'_{V_q^+} = |u|_q^+$. Let $\hat{u}(t,\ x)$ be an extension of $u(t,\ x)$ in V_q. Then

$$\hat{u}_0(\tau,\ \xi) = \frac{1}{q A_-}\Pi_{1+}q A_-u_+(\tau,\ \xi) \in V_q$$

according to what has been proved. On the basis of (2.13) we have

$$\hat{u}_0(\tau,\ \xi) = \frac{1}{q A_-}\Pi_{1+}q A_-\hat{u}(\tau,\ \xi) = \hat{u} \quad \frac{1}{q A_-}\Pi_{1-}q A_-\hat{u}(\tau,\ \xi).$$

We will prove that

$$\frac{1}{q A_-}\Pi_{1-}q A_-\hat{u}(\tau,\ \xi) \in L_2^-(R^{n+1}),$$

for which we revert to Lemma 1.5. The function $[q(\tau + ia,\ \xi')\ A_-(\tau + ia,\ z_1,\ \xi')]^{-1}$ is analytic with respect to z_1 for

$\operatorname{Im} z_1 \leqslant 0$. We prove that it is bounded for $\operatorname{Im} z_1 \leqslant 0$ and continuous in $\eta_1 = \operatorname{Im} z_1$, uniformly with respect to $(\tau, \xi) \in R^{n+1}$. We have

$$|A_-(\tau + ia, z_1, \xi')| = |(\xi_1 - (\lambda - i\eta_1))(\xi_1 + \bar{\lambda} + i\eta_1))| = |(\xi_1 - \beta)(\xi_1 + \beta)|,$$

where $\beta = \lambda - i\eta_1$. Let $\beta_1 = \operatorname{Re}\beta = \operatorname{Re}\lambda$, $\beta_2 = \operatorname{Im}\beta = \operatorname{Im}\lambda - \eta_1 \geqslant a + |\eta_1|$, $\eta_1 \leqslant 0$. Then

$$|A_-(\tau + ia, z_1, \xi')|^2 = |\xi_1^2 - \beta^2|^2 = (\xi_1^2 - \beta_1^2 + \beta_2^2)^2 + 4\beta_1^2\beta_2^2 = (\xi_1^2 - \beta_1^2)^2 + 2\beta_2^2\xi_1^2 + 2\beta_1^2\beta_2^2 + \beta_2^4 \geqslant \beta_2^2(\beta_1^2 + \beta_2^2) = \beta_2^2|\beta|^2.$$

We note in addition that $\operatorname{Im}\beta \geqslant \operatorname{Im}\lambda$, $|\beta| \geqslant |\lambda|$ for $\eta_1 \leqslant 0$. From all that has been said we get

$$\left|\frac{1}{q(\tau + ia, \xi')\, A_-(\tau + ia, z_1, \xi')}\right|^2 \leqslant \frac{|\eta_+|^2 \operatorname{Im}\lambda}{(\operatorname{Im}\beta)^2 |\beta|^2} \leqslant \frac{|\eta_+|^2}{\operatorname{Im}\beta\, |\lambda|^2} \leqslant \frac{1}{\operatorname{Im}\lambda \cdot |\eta_-|^2} \leqslant \frac{1}{a^2}.$$

To prove the uniform continuity in η_1, we estimate the derivative

$$\frac{\partial}{\partial\eta_1}[q(\tau + ia, \xi')\, A_-(\tau + ia, z_1, \xi')]^{-1},$$

noting that $\partial/\partial\eta_1 = i\partial/\partial z_1$. We have

$$\left|\frac{d}{dz_1}\frac{1}{qA_-}\right| = \left|\frac{1}{qA_-^2}\frac{dA_-}{dz_1}\right| = \left|\frac{1}{qA_-}\right|\left|\frac{2(z_1 - i\operatorname{Im}\lambda)}{A_-}\right|.$$

Now let $\beta = z_1 - i\operatorname{Im}\lambda$, $\beta_1 = \operatorname{Re}\beta = \xi_1$ and $\beta_2 = \operatorname{Im}\beta = \eta_1 - \operatorname{Im}\lambda$. We then have

$$\left|\frac{z_1 - i\operatorname{Im}\lambda}{A_-}\right|^2 = \left|\frac{\beta}{(\beta - \lambda_1)(\beta + \lambda_1)}\right|^2.$$

As before, we have

$$|(\beta - \lambda_1)(\beta + \lambda_1)|^2 = |\lambda_1^2 - \beta^2|^2 \geqslant \beta_2^2|\beta|^2,$$

so that

$$\left|\frac{z_1 - i\operatorname{Im}\lambda}{A_-}\right|^2 \leqslant \frac{1}{\beta_2^2} \leqslant \frac{1}{(\operatorname{Im}\lambda)^2} \leqslant \frac{1}{a^2} \quad \text{for } \eta_1 \leqslant 0.$$

Since also $|qA_-|^{-1} \leqslant a^{-1}$, we get

$$\left|\frac{d}{dz_1}\frac{1}{qA_-}\right| \leqslant \frac{2}{a^2}.$$

This proves that the function $[q(\tau + ia, \xi')A_-(\tau + ia, z_1, \xi)]^{-1}$ is continuous in η_1 for $\eta_1 \leqslant 0$, uniformly with respect to $(\tau, \xi) \in R^{n+1}$. It can now be seen that Lemma 1.5 applies to the function $[qA_-]^{-1}\Pi_{1-}qA_-\hat{u}(\tau, \xi)$ and hence that this function belongs to $L_2^-(R^{n+1})$. Consequently $\Pi_{1+}\hat{u}_0(\tau, \xi) = \Pi_{1+}\hat{u}(\tau, \xi)$, i.e. $\hat{u}_0(t, x)$ is an extension of $u(t, x)$ onto R^{n+1} in V_q, with

$$|\hat{u}_0|_q = |\Pi_{1+} q A_- u_+|_{L_2(R^{n+1})} = |\Pi_{1+} q A_- \hat{u}|_{L_2(R^{n+1})} \leqslant |q A_- \hat{u}|_{L_2(R^{n+1})} = |\hat{u}|_q.$$

This is true for any $\hat{u} \in V_q$ such that $\hat{u}(t, x) = u(t, x)$ for $x_1 > 0$. Hence

$$|u|'_{V_q^+} = |\hat{u}_0|_q = \inf |\hat{u}|_q = |u|_q^+,$$

which proves the coincidence of the norms (2.11) and (2.14).

3. *The spaces* $X_q^i(R^n)$. *The traces of functions in* V_q *on the hyperplanes* $x_1 =$ const.

DEFINITION 2.3. $X_q^1(R^n)$ is the closure of $S(R^n)$ in the norm

$$|\varphi(t, x')|^2_{X_q^1} = \int_{R^n} |\eta_-(\tau + k\xi_2 + ia, \xi')\varphi(\tau, \xi')|^2 \, d\tau d\xi'. \tag{2.15}$$

DEFINITION 2.4. $X_q^0(R^n)$ is the closure of $S(R^n)$ in the norm

$$|\varphi(t, x')|^2_{X_q^0} = \int_{R^n} \frac{|\varphi(\tau, \xi')|^2 \, d\tau d\xi'}{|\eta_+(\tau + k\xi_2 + ia, \xi')|^2}. \tag{2.16}$$

Suppose $\varphi(t, x) \in S(R^{n+1})$, and let

$$\varphi_{x_1}^{(k)}(\tau, \xi') = \frac{\partial^k}{\partial x_1^k} \varphi(\tau, x_1, \xi') = \frac{1}{2\pi} \int_{-\infty}^{\infty} e^{-ix_1\xi_1}(-i\xi_1)^k \varphi(\tau, \xi) \, d\xi_1.$$

From Cauchy's inequality it follows at once that

$$|\varphi_{x_1}^{(k)}(\tau, \xi')|^2 \leqslant \frac{1}{4\pi^2} \int_{-\infty}^{\infty} \frac{\xi_1^{2k} d\xi_1}{|q A|^2} \int_{-\infty}^{\infty} |q A \varphi(\tau, \xi)|^2 \, d\xi_1,$$

which implies by virtue of (2.9), (2.10) and (2.6) the following estimate for $k = 0, 1$.

$$|\varphi_{x_1}^{(k)}|^2_{X_q^{1-k}} \leqslant C |\varphi|_q^2. \tag{2.17}$$

DEFINITION 2.5. If $u(t, x) \in V_q$, then by the trace of $\partial^k u(t, x)/\partial x_1^k, k = 0, 1$, for $x_1 =$ const is meant the function $u_{x_1}^{(k)}(t, x') \in X_q^{1-k}(R^n)$ such that $u_{x_1}^{(k)} = \lim_{i\to\infty} \varphi_{ix_1}^{(k)}$ in $X_q^{1-k}(R^n)$, where $\lim_{i\to\infty}\varphi_i = u$ in V_q for $\varphi_i(t, x) \in S(R^{n+1})$.

If $\{\varphi_i(t, x)\}$ is a sequence from Definition 2.5, it follows from estimate (2.17) that the sequence $\{\varphi_{ix_1}^{(k)}\}$ converges in $X_q^{1-k}(R^n)$, and hence the traces

$u_{x_1}^{(k)}(t, x')$, $k = 0, 1$, are defined for any $u(t, x) \in V_q$. It is also clear from estimate (2.17) that Definition 2.5 does not depend on the choice of the sequence $\{\varphi_i(t, x)\}$.

Inasmuch as estimate (2.17) also holds for $u(t, x) \in V_q$ and $u_{x_1}^{(k)}(t, x') \in X_q^{1-k}(R^n)$, we have

LEMMA 2.1. *The mapping* $u(t, x) \to \{u_{x_1}^{(0)}(t, x'), u_{x_1}^{(1)}(t, x')\}$ *is uniformly (with respect to* x_1*) continuous from* V_q *into* $X_q^1(R^n) \times X_q^0(R^n)$.

LEMMA 2.2. *If* $u(t, x) \in V_q$, $k = 0, 1$, *then* $u_{x_1}^{(k)}(t, x') \in CX_q^{1-k}(R^n)$ *as a function of* x_1.

PROOF. According to Lemma 2.1, it suffices to carry out the proof for a function $\varphi(t, x) \in S(R^{n+1})$. We have

$$\varphi_{x_1+\Delta x_1}^{(k)}(\tau, \xi') - \varphi_{x_1}^{(k)}(\tau, \xi') = \frac{1}{2\pi}\int_{-\infty}^{\infty} (-i\xi_1)^k \varphi(\tau, \xi)\{\exp[-i\xi_1(x_1 + \Delta x_1)]$$

$$-\exp(-i\xi_1 x_1)\}\, d\xi_1 = \frac{1}{2\pi}\int_{-M}^{M} [A\varphi(\tau, \xi)] \frac{\{\exp[-i\xi_1(x_1+\Delta x_1)] - \exp(-i\xi_1 x_1)\}(-i\xi_1)^k}{A}\, d\xi_1$$

$$+\frac{1}{2\pi}\int_{|\xi_1|>M} [A\varphi(\tau, \xi)] \frac{\{\exp[-i\xi_1(x_1+\Delta x_1) - \exp(-ix_1\xi_1)\}(-i\xi_1)^k}{A}\, d\xi_1.$$

We denote the summands in order by R_1 and R_2. For R_2, on the basis of Cauchy's inequality, (2.9) and (2.10) we have

$$|R_2|^2 \leqslant \frac{c_k}{|\lambda|^{2(1-k)} \operatorname{Im} \lambda} \int_{|\xi_1|>M} |A\varphi(\tau, \xi)|^2\, d\xi_1.$$

The estimate $|\exp(-i\xi_1 \Delta x_1) - 1|^2 \leqslant M^2 (\Delta x_1)^2$ for $|\xi_1| \leqslant M$ is obvious. Hence, using Cauchy's inequality, (2.9) and (2.10), we have for R_1 the estimate

$$|R_1|^2 \leqslant \frac{c_k M^2 (\Delta x_1)^2}{|\lambda|^{2(1-k)} \operatorname{Im} \lambda} \int_{-\infty}^{\infty} |A\varphi(\tau, \xi)|^2\, d\xi_1.$$

Hence

$$|\varphi_{x_1+\Delta x_1}^{(k)}(t, x') - \varphi_{x_1}^{(k)}(t, x')|^2_{X_q^{1-k}}$$
$$\leqslant c_k \left\{M^2 (\Delta x_1)^2 |\varphi|_q^2 + \int_{R^n} d\tau d\xi' \int_{|\xi_1|\geqslant M} |q A\varphi(\tau, \xi)|^2\, d\xi_1\right\} < \varepsilon, \qquad k = 0, 1,$$

if M is sufficiently large and $|\Delta x_1|$ is sufficiently small. The proof is complete. We note in addition to Lemma 2.2 that $u_{x_1}^{(0)}(t, x') \in C^1 X_q^0(R^n)$ as a function of x_1 [11].

We now prove an inequality needed for the subsequent investigation of the traces of functions in V_q:

$$\max_{\xi_1} \frac{\xi_1^{2k}}{|\mathscr{A}(\tau+ia,\ \xi)|^2} \leqslant \frac{C_k}{|\lambda|^{2(1-k)}\operatorname{Im}\lambda},\ k=0,1. \tag{2.18}$$

Let $\lambda_1 = \operatorname{Re}\lambda$ and $\lambda_2 = \operatorname{Im}\lambda$. We have

$$|A(\tau+ia,\ \xi)|^2 = |\xi_1^2-\lambda^2|^2 = \xi_1^2-\lambda_1^2)^2+2(\xi_1^2+\lambda_1^2)\lambda_2^2+\lambda_2^4.$$

Hence $|A(\tau+ia,\ \xi)|^2 \geqslant \lambda_2^2|\lambda|^2$, and also $\xi_1^{-2}|A(\tau+ia,\ \xi)|^2 \geqslant 2\lambda_2^2$, which implies the validity of (2.18).

LEMMA 2.3. *If $u(t,x)\in V_q$, then*

$$u_{x_1}^{(k)}(\tau,\ \xi') = F_{x_1\to\xi_1}^{-1}[(-i\xi_1)^k u(\cdot,\ \xi_1,\ \cdot)]$$
$$= \frac{1}{2\pi}\int_{-\infty}^{\infty} e^{-ix_1\xi_1}(-i\xi_1)^k u(\cdot,\ \xi_1,\ \cdot)\,d\xi_1, \qquad k-0,\ 1,$$

where the Fourier transform is taken in $L_2X_q^{1-k}(R^n)$.

PROOF. We first note that $(-i\xi_1)^k u(\cdot,\xi_1,\cdot)\in L_2X_q^{1-k}(R^n)$. In fact, using (2.18), we have

$$|(-i\xi_1)^k u(\cdot,\ \xi_1,\ \cdot)|^2 = \frac{\xi_1^{2k}}{|q A|^2}|q\,A u(\tau,\ \xi)|^2$$
$$\leqslant \frac{c}{|q|^2|\lambda|^{2(1-k)}(\operatorname{Im}\lambda)^2}|q\,A u(\tau,\ \xi)|^2.$$

Hence when $k = 1$, for example, we have

$$\|(-i\xi_1)^k u(\cdot,\ \xi_1,\ \cdot)\|^2_{L_2X_q^0(R^n)} = \int_{-\infty}^{\infty} d\xi_1 \int_{R^n} \frac{|-i\xi_1 u(\tau,\ \xi)|^2\,d\tau d\xi'}{|\eta_+(\tau+k\xi_2+ia,\ \xi')|^2}$$
$$\leqslant c\int_{\infty}^{\infty} d\xi_1\int_{R^n}\frac{1}{\operatorname{Im}\lambda}|q\,A u(\tau,\ \xi)|^2\,d\tau d\xi' \leqslant c_1|u|_q^2.$$

The equality for $k = 0$ is obtained in exactly the same way. This means that $(-i\xi_1)^k u(\cdot,\xi_1,\cdot)\in L_2X_q^{1-k}(R^n)$. Lemma 1.1 therefore implies the existence of the function

$$u^{(k)}(\cdot,\ x_1,\ \cdot) = F_{x_1\to\xi_1}^{-1}[(-i\xi_1)^k u(\cdot,\ \xi_1,\ \cdot)] \in L_2X_q^{1-k}(R^n),$$

which according to what has been proved satisfies the estimate

$$\| u^{(k)}(\cdot,\ x_1,\ \cdot)\|^2_{L_2X_q^{1-k}} = \frac{1}{2\pi}\|(-i\xi_1)^k u(\cdot,\ \xi_1,\ \cdot)\|_{L_2X_q^{1-k}} \leqslant c\,|u|^2_q. \qquad (2.19)$$

It remains to prove that $u^{(k)}(\cdot, x_1, \cdot) = u^{(k)}_{x_1}(\tau, \xi')$. Let $\varphi_p(t, x) \in S(R^{n+1})$, $\lim_{p\to\infty}\varphi_p = u$ in V_q. It then follows from (2.19) that

$$\lim_{p\to\infty} \varphi_p(\tau,\ \xi)\cdot(-i\xi_1)^k = (-i\xi_1)^k u(\tau,\ \xi)$$

in $L_2X_q^{1-k}(R^n)$, and hence

$$\lim_{p\to\infty} F^{-1}_{x_1\to\xi_1}[(-i\xi_1)^k \varphi_p(\tau,\ \xi)] = F^{-1}_{x_1\to\xi_1}[(-i\xi_1)^k u(\cdot,\ \xi_1,\ \cdot)]$$

in $L_2X_q^{1-k}(R^n)$. But

$$F^{-1}_{x_1\to\xi_1}[(-i\xi_1)^k \varphi_p(\tau,\ \xi)] = \varphi^{(k)}_{p,\,x_1}(\tau,\ \xi'),$$

the function $\varphi^{(k)}_{p,\,x_1}(t, x')$ converging to $u^{(k)}_{x_1}(t, x')$ in the norm of $X_q^{1-k}(R^n)$ uniformly with respect to x_1. Hence after an appropriate modification we have

$$u^{(k)}(\cdot,\ x_1,\ \cdot) = F^{-1}_{x_1\to\xi_1}[(-i\xi_1)^k u(\cdot,\ \xi_1,\ \cdot)] \underset{L_2X_q^{1-k}}{=} \lim F^{-1}_{x_1\to\xi_1}[(-i\xi_1)^k \varphi(\cdot,\ \xi_1,\ \cdot)]$$
$$= \lim \varphi^{(k)}_{p,\,x_1}(\tau,\ \xi') \underset{CX_q^{1-k}}{=} u^{(k)}_{x_1}(\tau,\ \xi')$$

(CX_q^{1-k} is with the uniform metric). The proof of Lemma 2.3 is complete.

Finally, we note that each function in $X_q^k(R^n)$ can be extended onto R^{n+1} as an element of $V_q(R^{n+1})$ [11].

4. *The spaces* $X_q^i(R^n_{k\pm})$. Let $R^n_{k\pm} = \{(t, x') \in R^n | x_2 \gtrless kt\}$, the upper or lower signs being taken simultaneously.

DEFINITION 2.6. $X_q^1(R^n_{k+})$ is the set of functions $h(t, x') \in L_2(R^n_{k+})$ for each of which there exists an element $\hat{h}(t, x') \in X_q^1$ such that $\hat{h}(t, x') = h(t, x')$ in R^n_{k+}; the norm in $X_q^1(R^n_{k+})$ is

$$|h|_{X_q^1(R^n_{k+})} = \inf |\hat{h}|_{X_q^1},$$

where the infimum is taken over all of the extensions of $h(t, x)$ in $X_q^0(R^n)$.

If

$$\mathring{X}_q^1(\overline{R^n_{k-}}) = \{h(t,\ x') \in X_q^1(R^n) \,|\, \operatorname{supp} h(t,\ x') \subset \overline{R^n_{k-}}\},$$

then $\mathring{X}_q^1(\overline{R^n_{k-}})$ is a closed subspace in $X_q^1(R^n)$ and $X_q^1(R^n_{k+})$ is isomorphic to the quotient space $X_q^1(R^n)/\mathring{X}_q^1(\overline{R^n_{k-}})$.

DEFINITION 2.7. $X_q^0(R^n_{k-})$ is the set of functions $h(t, x') \in D'(R^n_{k-})$ such that for each of them there exists an element $\hat{h}(t, x') \in X_q^0(R^n)$ with $\hat{h}(t, x') = h(t, x')$

in R^n_{k-}; the norm in $X^0_q(R^n_{k-})$ is

$$|h|_{X^0_q(R^n_{k-})}=\inf|\hat{h}|_{X^0_q},$$

where the infimum is taken over all extensions $\hat{h}(t,x')$ in $X^0_q(R^n)$.

As in the preceding case, $X^0_q(R^n_{k-})$ is isomorphic to $X^0_q(R^n)/\overset{\circ}{X}{}^0_q(\overline{R^n_{k+}})$.

§3. Existence and uniqueness theorem ($|k|<\sqrt{a_{22}}$)

1. *The null space of the operator* $A(D_t+ia,D_x)\equiv-(D_t+ia)^2+(AD_x,D_x)$ *in* V^+_q. Suppose $u(t,x)\in V^+_q$ and $A(D_t+ia,D_x)u(t,x)=0$ in $D'(R^{n+1}_+)$. Then $u(t,x)\in L_2(R^{n+1}_+)$, and hence the representation (1.15) holds for $u_+(t,x)$. We sharpen the conditions which the function $c(\tau,\xi')$ must satisfy in this case.

LEMMA 3.1. *Suppose* $u(t,x)\in V^+_q$, $A(D_t+ia,D_x)u=0$. *Then*

$$c(\tau,\xi')=iu^{(0)}_0(\tau,\xi')=\frac{u^{(1)}_0(\tau,\xi')}{\lambda(\tau+ia,\xi')},\tag{3.1}$$

where

$$u^{(0)}_0(t,x')=\lim_{x_1\to+0}u^{(0)}_{x_1}(t,x')\quad in\quad X^1_q(R^n),$$

$$u^{(1)}_0(t,x')=\lim_{x_1\to+0}u^{(1)}_{x_1}(t,x')\quad in\quad X^0_q(R^n).$$

PROOF. Let $\hat{u}(t,x)$ be a fixed extension of $u(t,x)$ onto R^{n+1} in V_q. We first note that the limit boundary traces $u^{(0)}_0(t,x')$ and $u^{(1)}_0(t,x')$ are meaningful. In fact, $u^{(k)}_{x_1}(t,x')=\hat{u}^{(k)}_{x_1}(t,x')$, $k=0,1$, for $x_1>0$, the $\hat{u}^{(k)}_{x_1}(t,x')$ being continuous functions of x_1, $x_1\in R^1$, i.e. the limit

$$\lim_{x_1\to+0}\hat{u}^{(k)}_{x_1}(t,x')=\lim_{x_1\to+0}u^{(k)}_{x_1}(t,x')$$

exists in $X^{1-k}_q(R^n)$.

We prove that the condition $u(t,x)\in V^+_q$ necessarily implies that

$$(t,x')=F^{-1}[c(\tau,\xi')]\in X^1_q(R^n).$$

On the basis of (2.14) we find that

$$q(\tau+ia,\xi')A_-(\tau+ia,\xi)u_+(\tau,\xi)=qc(\tau,\xi')[\xi_1-\lambda-2i\operatorname{Im}\lambda]+\frac{4iq\lambda\operatorname{Im}\lambda\cdot c(\tau,\xi')}{\xi_1+\lambda}.$$

Inasmuch as $(\tau^2+|\xi'|^2+1)^{-1}c(\tau,\xi')\in L_2(R^n)$, we can apply Lemma 1.7:

$$\Pi_{1+}qA_-u_+(\tau,\xi)-\frac{4iq\lambda\operatorname{Im}\lambda\cdot c(\tau,\xi')}{\xi_1+\lambda}\in L_2(R^{n+1})$$

(the latter being true by assumption and (2.14)). Hence, by Fubini's theorem,

$$\infty > \int_{R^{n+1}} \frac{|q\lambda \operatorname{Im}\lambda \cdot c(\tau, \xi')|^2}{|\xi_1+\lambda|^2} d\tau d\xi' d\xi_1$$

$$= \pi \int_{R^n} |\eta_-(\tau + k\xi_2 + ia, \xi')\, c(\tau, \xi')|^2 \, d\tau d\xi' = \pi |c|^1_{X^1_q}.$$

We will now prove (3.1). First, as we noted in the proof of Lemma 2.3,

$$(-i\xi_1)^k \hat{u}(\tau, \xi_1, \xi') \in L_2 X_q^{1-k}(R^n),$$

where $\hat{u}(t, x)$ is an extension of $u(t, x)$ onto R^{n+1} in V_q and $k = 0, 1$. Second,

$$(-i\xi_1)^k u_+(\tau, \xi) = (-i\xi_1)^k \hat{u}(\tau, \xi) + (-i\xi_1)^k v(\tau, \xi).$$

Here $v(\tau, \xi) \in L_2(R^{n+1})$, and supp $v(t, x)$, and so also supp $F^{-1}[(-i\xi_1)^k v(\tau, \xi)]$, belong to $\overline{R^{n+1}_-}$. Hence

$$(-i\xi_1)^k u_+(\tau, \xi) = \Pi_{1+}(-i\xi_1)^k \hat{u}(\tau, \xi) + w_-(\tau, \xi).$$

Here

$$w_-(\tau, \xi) = \Pi_{1-}(-i\xi_1)^k \hat{u}(\tau, \xi) + (-i\xi_1)^k v(\tau, \xi),$$

$$\Pi_{1+}(-i\xi_1)^k \hat{u}(\tau, \xi) \in L_2^+ X_q^{1-k}(R^n), \text{ a } \operatorname{supp} w(t, x) \subset \overline{R^{n+1}_-}.$$

Hence $(-i\xi_1)^k u_+(\tau, \xi)$ belongs to the domain of definition of the operator Π_{1+}, and

$$\Pi_{1+}(-i\xi_1)^k u_+(\tau, \xi) = \Pi_{1+}(-i\xi_1)^k \hat{u}(\tau, \xi).$$

When $k = 0$ we have

$$\Pi_{1+} u_+(\tau, \xi) = \frac{c(\tau, \xi')}{\xi_1 + \lambda}.$$

Hence when $x_1 > 0$ we get

$$u^{(0)}_{x_1}(\tau, \xi') = F^{-1}_{x_1 \to \xi_1}[\hat{u}(\tau, \xi_1, \xi')]$$

$$= F^{-1}_{x_1 \to \xi}[\Pi_{1+}\hat{u}(\tau, \xi)] = F^{-1}_{x_1 \to \xi_1}\left[\frac{c(\tau, \xi')}{\xi_1 + \lambda}\right] \text{ in } L_2 X_q^1(R^n).$$

Temporarily letting $\eta_- = \eta_-(\tau + k\xi_2 + ia, \xi')$ (formula (2.3)), we note that the preceding implies

$$\int_0^\infty dx_1 \int_{R^n} d\tau d\xi' \left| \eta_- \left[u_{x_1}^{(0)}(\tau, \xi') - \frac{1}{2\pi} \int_{-A}^{A} \frac{e^{-ix_1\xi_1} c(\tau, \xi')\, d\xi_1}{\xi_1 + \lambda} \right] \right|^2 \to 0$$

as $A \to +\infty$. In particular, if $\epsilon > 0$ and Ω_C is a ball of radius C in R^n, then

$$\int_\varepsilon^\infty dx_1 \int_{\Omega_C} d\tau d\xi' \left| \eta_- \left[u_{x_1}^{(0)}(\tau, \xi') - \frac{1}{2\pi} \int_{-A}^{A} \frac{e^{-ix_1\xi_1} c(\tau, \xi')\, d\xi_1}{\xi_1 + \lambda} \right] \right|^2 \to 0$$

as $A \to +\infty$.

According to Lemma 1.9, we can pass to the limit under the integral sign, as a result of which we get

$$\int_\varepsilon^\infty dx_1 \int_{\Omega_C} d\tau d\xi' \, | \eta_- [u_{x_1}^{(0)}(\tau, \xi') - (-i)\, c(\tau, \xi')\, e^{ix_1\lambda}] |^2 \to 0$$

for any $\epsilon > 0$ and any $\Omega_C \in R^n$. This means that $u_{x_1}^{(0)}(\tau, \xi') = -ic(\tau, \xi')e^{ix_1\lambda}$ in $L_2X_q^1$ for $x_1 > 0$. Both sides of this equality are continuous functions of x_1 for $x_1 > 0$ with values in $X_q^1(R^n)$, so that the indicated equality is satisfied for any $x_1 > 0$ in $X_q^1(R^n)$. As $x_1 \to +0$ the left (and hence the right) side of this equality has a limit in $X_q^1(R^n)$ equal to $u_0^{(0)}(\tau, \xi')$.

We determine more precisely the limit of the right side. For $x_1 \to +0$ we have $\lim e^{ix_1\lambda} = 1$ uniformly on any bounded set in R^n. Hence we at once get

$$c(\tau, \xi') = \lim_{x_1 \to +0} c(\tau, \xi')\, e^{ix_1\lambda} \text{ in } X_q^1(R^n).$$

Therefore

$$u_{x_1}^{(0)}(\tau, \xi') = -ic(\tau, \xi')\, e^{ix_1\lambda}, \quad x_1 \geqslant 0, \tag{3.2}$$

in $X_q^1(R^n)$, this equality being understood in the limit sense when $x_1 = 0$. This proves the first equality in (3.1). Now suppose $k = 1$. Then

$$\Pi_{1+}(-i\xi_1)\, u_1(\tau, \xi) = \Pi_{1+} \frac{(-i\xi_1)\, c(\tau, \tau')}{\xi_1 + \lambda} = \Pi_{1+}\left\{ \;ic(\tau, \xi') + \frac{i\lambda c(\tau, \zeta')}{\xi_1 + \lambda} \right\} = \frac{i\lambda c(\tau, \xi')}{\xi_1 + \lambda}$$

(the latter equality follows from Lemma 1.7). Hence

$$u_{x_1}^{(1)}(\tau, \xi') = F_{x_1 \to \xi_1}^{-1}[(-i\xi_1)\, \hat{u}(\tau, \xi)] = F_{x_1 \to \xi_1}^{-1}[(-i\xi_1)\, u_+(\tau, \xi)] - F_{x_1 \to \xi_1}^{-1}\left[\frac{i\lambda c(\tau, \xi')}{\xi_1 + \lambda}\right]$$

in $L_2X_q^0(R^n)$ for $x_1 > 0$. Letting

$$v_{x_1}(\tau, \xi') = [i\lambda]^{-1}\, u_{x_1}^{(1)}(\tau, \xi') \in L_2X_q^1(R^n),$$

we arrive at a situation analogous to the preceding, and hence

$$v_{x_1}(\tau, \xi') = -ic(\tau, \xi')\, e^{ix_1\lambda} \quad \text{for} \quad x_1 \geqslant 0$$

in $X_q^1(R^n)$, i.e.

$$u_{x_1}^{(1)}(\tau, \xi') = \lambda c(\tau, \xi')\, e^{ix_1\lambda} \qquad x_1 \geqslant 0, \tag{3.3}$$

in $X_q^0(R^n)$. When $x_1 = 0$ this equality is understood in the limit sense.

Equality (3.3) implies the second equality in (3.1), while from (3.2) and (3.3) we have the useful relation

$$u_{x_1}^{(1)}(\tau, \xi') = i\lambda u_{x_1}^{(0)}(\tau, \xi'), \; x_1 \geqslant 0, \tag{3.4}$$

which is understood in the limit sense when $x_1 = 0$. The proof of Lemma 3.1 is complete.

DEFINITION 3.1.

$$\mathfrak{A}_q = \{u(t, x) \in V_q^+ \mid A(D_t + ia, D_x)\, u(t, x) = 0\}$$

is the null space of the operator $A(D_t + ia, D_x)$ in V_q^+.

THEOREM 3.1. *The assertion* $u(t, x) \in \mathfrak{A}_q$ *is equivalent to the fact that*

$$u_+(\tau, \xi) = \frac{iu_0^{(0)}(\tau, \xi')}{\xi_1 + \lambda(\tau + ia, \xi')}, \tag{3.5}$$

where $u_0^{(0)}(t, x') = \lim_{x_1 \to +0} u_{x_1}^{(0)}(t, x')$ *in* $X_q^1(R^n)$.

PROOF. The necessity of (3.5) was proved in Lemma 3.1. We must prove the sufficiency. Suppose $u_+(\tau, \xi)$ is determined by (3.5), where

$$u_0^{(0)}(t, x) = F^{-1}[u_0^{(0)}(\tau, \xi')] \in X_q^1(R^n).$$

We will verify that the function $u(t, x) = F^{-1}[u_+(\tau, \xi)]$ for $x_1 > 0$ belongs to V_q^+. If ξ_1 is replaced by $z_1 = \xi_1 + i\eta_1$, $\eta_1 \geqslant 0$, in (3.5), the resultant function $u_+(\tau, z_1, \xi')$ satisfies, as is easily verified, the conditions of Lemma 1.3 with $H = X_q^1(R^n)$. Applying this lemma, we get

$$u_+(t, x_1, x') = F^{-1}[u_+(\tau, \xi_1, \xi')] = 0 \quad \text{for} \quad x_1 < 0,$$

and hence $u_+(t, x_1, x')$ is the extension of $u(t, x)$ by zero onto R^{n+1} (in $L_2X_q^1(R^n)$). According to (2.14) we now find, as in Lemma 3.1, that

$$\Pi_{1+}q\, A_{-}u_{+}(\tau, \xi) = \frac{-4q\lambda \operatorname{Im}\lambda \cdot u_0^{(0)}(\tau, \xi')}{\xi_1 + \lambda} \in L_2(R^{n+1}).$$

The inclusion here follows from the fact that

$$\int_{-\infty}^{+\infty} \frac{d\xi_1}{|\xi_1 + \lambda|^2} = \frac{\pi}{\operatorname{Im}\lambda}, \quad \text{while} \quad u_0^{(0)}(t, x') \in X_q^1(R^n)$$

(recall that $q = q(\tau + ia, \xi')$ is defined by formula (2.6)).

Thus $u(t, x) \in V_q^+$ and

$$|u|^2_{V_q^+} = |\Pi_{1+}q\, A_{-}u_{+}(\tau, \xi)|^2_{L_2(R^{n+1})} \leqslant C|u_0^{(0)}|^2_{X_q^1}. \tag{3.6}$$

Further, we have

$$A(\tau + ia, \xi)u_{+}(\tau, \xi) = iu_0^{(0)}(\tau, \xi')(\xi_1 - \lambda).$$

Inasmuch as $u_0^{(0)}(t, x') \in X_q^1(R^n)$, the function $A(\tau + ia, \xi)u_+(\tau, \xi)$ belongs to the domain of definition of the operator Π_{1+} and $\Pi_{1+}Au_+(\tau, \xi) = 0$, i.e.

$$\langle A(D_t + ia, D_x)u_+(t, x), \varphi(t, x)\rangle = 0$$

for any $\varphi(t, x) \in C_0^\infty(R^{n+1})$, supp $\varphi(t, x) \in R_+^{n+1}$. Consequently

$$A(D_t + ia, D_x)u(t, x) = 0.$$

The proof of the theorem is complete.

2. *Statement of the problem. Existence and uniqueness theorems.* We formulate problem (3) in terms of the spaces introduced above.

0) Let $h^{(0)}(t, x') \in X_q^1(R_{k+}^n)$, $h^{(1)}(t, x') \in X_q^0(R_{k-}^n)$, $h^{(0)}(t, x') = 0$ and $h^{(1)}(t, x') = 0$ for $t < 0$. Find functions $u(t, x)$ and $v^{(i)}(t, x')$, $i = 0, 1$, satisfying the following conditions:

1) $u(t, x) \in \mathfrak{A}_q$, $v^{(i)}(t, x') \in X_q^{1-i}(R^n)$, $i = 0, 1$.

2) supp $v^{(0)}(t, x') \subset R_{k-}^n$, supp $v^{(1)}(t, x') \subset R_{k+}^n$, $u(t, x) = 0$ for $t < x_1$.

3) $\lim_{x_1 \to +0} u_{x_1}^{(0)}(t, x') = u_0^{(0)}(t, x') = \hat{h}^{(0)}(t, x') + v^{(0)}(t, x')$, where the limit is taken in $X_q^1(R^n)$.

4) $\lim_{x_1 \to +0} \partial u(t, x)/\partial x_1 = u_0^{(1)}(t, x') = \hat{h}^{(1)}(t, x') + v^{(1)}(t, x')$, where the limit is taken in $X_q^0(R^n)$.

Here $\hat{h}^{(0)}$ and $\hat{h}^{(1)}$ are fixed extensions of $h^{(0)}$ and $h^{(1)}$ in $X_q^1(R^n)$ and $X_q^0(R^n)$ respectively.

THEOREM 3.2. *For fixed extensions* $\hat{h}^{(0)}(t, x')$ *and* $\hat{h}^{(1)}(t, x')$ *there exists a unique solution* $u(t, x)$ *of problem* 0)–4) *that does not depend on the choice of the extensions of* $h^{(0)}$ *and* $h^{(1)}$ *and satisfies the estimate*

$$|u|_{V_q^+} \leqslant C\{|h^{(0)}|_{X_q^1(R_{k+}^n)} + |h^{(1)}|_{X_q^0(R_{k-}^n)}\}. \tag{3.7}$$

REMARK 3.1. For the sake of brevity we do not present here a proof of the fact that $u(t, x) = 0$ for $t < x_1$ inasmuch as it is not needed for the subsequent assertions on smoothness. Such a proof can be found in [11]. It is carried out with the use of a representation formula for the solution that will be obtained here on the basis of Lemmas 1.2 and 1.3.

PROOF. Lemma 3.1 and Theorem 3.1 establish a one-to-one and bicontinuous correspondence between $\mathfrak{A}_q$ and $X_q^1(R^n)$. Therefore the matter reduces to a determination of the function $u_0^{(0)}(\tau, \xi')$ in the representation (3.5).

Conditions 3) and 4) are equivalent to the conditions (we use (3.1))

3′) $u_0^{(0)}(\tau, \xi') = \hat{h}^{(0)}(\tau, \xi') + v^{(0)}(\tau, \xi')$;

4′) $i\lambda(\tau + ia, \xi')u_0^{(0)}(\tau, \xi') = \hat{h}^{(1)}(\tau, \xi') + v^{(1)}(\tau, \xi')$.

Using the factorization (2.5), we rewrite this system in the form

$$\eta_-(\tau + k\xi_2 + ia, \xi')u_0^{(0)}(\tau, \xi') = \eta_-(\tau + k\xi_2 + ia, \xi')\hat{h}^{(0)}(\tau, \xi') + \eta_-(\tau + k\xi_2 + ia, \xi')v^{(0)}(\tau, \xi'),$$

$$\eta_-(\tau + k\xi_2 + ia, \xi')u_0^{(0)}(\tau, \xi') = [i\eta_+(\tau + k\xi_2 + ia, \xi')]^{-1}\hat{h}^{(1)}(\tau, \xi') + [i\eta_+(\tau + k\xi_2 + ia, \xi')]^{-1}v^{(1)}(\tau, \xi').$$

After making the substitution $\tau + k\xi_2 = \tau'$ and letting

$$\begin{aligned} u_0^{(0)}(\tau' - k\xi, \xi') &= U^{(0)}(\tau', \xi'), \\ \hat{h}^{(i)}(\tau' - k\xi_2, \xi') &= H^{(i)}(\tau', \xi'), \\ v^{(i)}(\tau' - k\xi_2, \xi') &= V^{(i)}(\tau', \xi'), \qquad i = 0, 1, \end{aligned} \tag{3.8}$$

we get

3″) $\eta_-(\tau' + ia, \xi')U^{(0)}(\tau', \xi') = \eta_-(\tau' + ia, \xi')H^{(0)}(\tau, \xi') + \eta_-(\tau' + ia, \xi') \times V^{(0)}(\tau', \xi')$;

4″) $\eta_-(\tau' + ia, \xi')U^{(0)}(\tau', \xi') = [i\eta_+(\tau' + ia, \xi')]^{-1}H^{(1)}(\tau', \xi') + [i\eta_+(\tau' + ia, \xi')]^{-1}V^{(1)}(\tau', \xi')$.

If the functions $v^{(i)}(t, x')$, $i = 0, 1$, satisfy 1), then, clearly,

$$1') \begin{cases} \eta_-(\tau' + ia, \xi')V^{(0)}(\tau', \xi') \in L_2(R^n); \\ [\eta_+(\tau' + ia, \xi')]^{-1}V^{(1)}(\tau', \xi') \in L_2(R^n). \end{cases}$$

If the functions $v^{(i)}(t, x')$ satisfy 2), then

$$2')\begin{cases} \operatorname{supp} V^{(0)}(t, x') \subset \overline{R^n_-} = \{(t, x') \in R^n \mid x_2 \leqslant 0\}; \\ \operatorname{supp} V^{(1)}(t, x') \subset \overline{R^n_+} = \{(t, x') \in R^n \mid x_2 \geqslant 0\}. \end{cases}$$

Let us prove the second assertion in 2'). We have

$$V^{(1)}(t, x') = F^{-1}_{(t, x') \to (\tau', x')}[v^{(1)}(\tau' - k\xi_2, \xi')] = v^{(1)}(t, x_2 + kt, x'')$$

and hence $V^{(1)}(t, x_2, x'') = 0$ for $x_2 < 0$, since then $x_2 + kt < kt$ and $v^{(1)}(t, x_2 + kt, x'') = 0$. Here, as before, $x'' = (x_3, \cdots, x_n)$. The first assertion in 2') is verified analogously.

Thus 1), 2) implies 1'), 2'). In regard to the converse assertion we note the following: the fact that 1') implies 1) is verified by the simple substitution $\tau' = \tau + k\xi_2$, while 2) is obtained from 2') by inverting the proof of the fact that 2) implies 2'). All of this is done with (3.8) being taken into account. Consequently conditions 1), 2) and 1'), 2') for $v^{(0)}$ and $v^{(1)}$ are equivalent.

We now prove that

$$\begin{gathered} \Pi_{2+}\eta_-(\tau' + ia, \xi')V^{(0)}(\tau', \xi') = 0, \\ \Pi_{2-}[\eta_+(\tau' + ia, \xi')]^{-1}V^{(1)}(\tau', \xi') = 0. \end{gathered} \tag{3.9}$$

First, the application of Π_{2+} and Π_{2-} is valid according to 1'). Further, let

$$\varphi(t, x') \in C_0^\infty(R^n), \quad \operatorname{supp}\varphi(t, x') \subset R^n_+, \quad \varphi(\tau', \xi') = F[\varphi(t, x')].$$

Then

$$\varphi(\tau', \xi') \in L_2^+L_2(R^{n-1}) = L_2^+(R^n)$$

and (setting $\eta_- = \eta_-(\tau' + ia, \xi')$)

$$\begin{aligned} &(\eta_-V^{(0)}(\tau', \xi'), \varphi(\tau', \xi'))_{L_2(R^n)} \\ &\quad = \langle \eta_-V^{(0)}(\tau', \xi'), \varphi(\tau', \xi')\rangle = \langle V^{(0)}(\tau', \xi'), \bar{\eta}_-\varphi(\tau', \xi')\rangle. \end{aligned}$$

According to (2.3) the function $\eta_-(\tau' + ia, z_2, \xi'')$ is analytic with respect to $z_2 = \xi_2 + i\eta_2$ for $\eta_2 < a/\sqrt{a_{22}}(1 + \theta)$ and is not equal to zero in this domain. This means that the complex conjugate $\bar{\eta}_-(\tau' + ia, \xi_2, \xi'')$ of $\eta_-(\tau' + ia, \xi_2, \xi'')$ can be analytically continued into the domain $z_2 = \xi_2 + i\eta_2$, $\eta_2 > a/\sqrt{a_{22}}(1 + \theta)$ of the complex z_2 $(= \xi_2 + i\eta_2)$ plane. By choice, $\bar{\eta}_-(\tau' + ia, z_2, \xi'')$ and $\varphi(\tau', \xi')$ admit an application of Corollary 1.3, so that

$$\operatorname{supp} F^{-1}[\bar{\eta}_-(\tau'+ia,\ \xi')\varphi(\tau',\ \xi')] \subset \overline{R^n_+}$$

and, in addition,

$$\bar{\eta}_-(\tau'+ia,\ \xi')\varphi(\tau',\ \xi') \in S(R^n).$$

It follows according to 2′) that

$$(\eta_-(\tau'+ia,\ \xi')V^{(0)}(\tau',\ \xi'),\ \varphi(\tau',\ \xi'))_{L_2(R^n)}=0$$

for any $\varphi(t, x') \in C_0^\infty(R^n)$, $\operatorname{supp}\varphi \subset R^n_+$. The set of Fourier transforms of such functions is dense in $L_2^+(R^n)$, so that

$$\eta_-(\tau'+ia,\ \xi')V^{(0)}(\tau',\ \xi') \in L_2^-(R^n),$$

as required. It is proved in exactly the same way that

$$\Pi_{2-}[\eta_+(\tau'+ia,\ \xi')]^{-1}V^{(1)}(\tau',\ \xi')=0.$$

Assuming that the solution of system 3″), 4″) exists in the indicated classes and using (3.9), we now find upon applying the operators Π_{2+} and Π_{2-} to each of the equations 3″) and 4″) that

$$\Pi_{2+}\eta_-U^{(0)}(\tau',\ \xi')=\Pi_{2+}\eta_-H^{(0)}(\tau',\ \xi'),$$
$$\Pi_{2-}\eta_-U^{(0)}(\tau',\ \xi')=\Pi_{2-}\frac{1}{i\eta_+}H^{(1)}(\tau',\ \xi'),$$
$$\Pi_{2-}\eta_-U^{(0)}(\tau',\ \xi')=\Pi_{2-}\eta_-H^{(0)}(\tau',\ \xi')+\eta_-V^{(0)}(\tau',\ \xi'),$$
$$\Pi_{2+}\eta_-U^{(0)}(\tau',\ \xi')=\Pi_{2+}\frac{1}{i\eta_+}H^{(1)}(\tau',\ \xi')+\frac{1}{i\eta_+}V^{(1)}(\tau',\ \xi').$$

Here again we have let $\eta_\pm = \eta_\pm(\tau' + ia,\ \xi')$.

Taking the sum of the first two equations, we get

$$U^{(0)}(\tau',\xi')=\frac{1}{i\eta_-}\left[i\Pi_{2+}\eta_-H^{(0)}(\tau',\ \xi')+\Pi_{2-}\frac{H^{(1)}(\tau',\ \xi')}{\eta^+}\right]. \tag{3.10}$$

Further, we find that

$$V^{(0)}(\tau',\ \xi')=\frac{1}{i\eta_-}\Pi_{2-}\left[\frac{H^{(1)}(\tau',\ \xi')}{\eta_+}-i\eta_-H^{(0)}(\tau',\ \xi')\right], \tag{3.11}$$

$$V^{(1)}(\tau',\ \xi')=\eta_+\Pi_{2+}\left[i\eta_-H^{(0)}(\tau',\ \xi')-\frac{H^{(1)}(\tau',\ \xi')}{\eta_+}\right]. \tag{3.12}$$

Suppose now $U^{(0)}(\tau',\ \xi')$, $V^{(i)}(\tau',\ \xi')$, $i = 0, 1$, are determined by (3.10)–(3.12). Then

$$\left.\begin{aligned} u_0^{(0)}(\tau, \xi') &= U^{(0)}(\tau + k\xi_2, \xi'), \\ v^{(i)}(\tau, \xi') &= V^{(i)}(\tau + k\xi_2, \xi'), \end{aligned}\right\} \tag{3.13}$$

$$u_0^{(0)}(t, x') = F^{-1}[u_0^{(0)}(\tau, \xi')], \qquad v^{(i)}(t, x') = F^{-1}[v^{(i)}(\tau, \xi')].$$

The membership of these functions in the corresponding function classes is obvious. To be able to assert that (3.10)–(3.13) give the solution of system 3′), 4′) or, equivalently, that (3.10)–(3.12) give the solution of system 3″), 4″) it suffices to verify the fulfilment of 2′). This verification is the converse of the proof of (3.9) and will therefore be omitted.

Thus formulas (3.10)–(3.13) and (3.5) determine a unique solution of problem 0)–4) for fixed $\hat{h}^{(0)}(t, x')$ and $\hat{h}^{(1)}(t, x')$.

To prove that $u(t, x)$ is independent of the choice of extensions of $h^{(0)}(t, x')$ and $h^{(1)}(t, x')$, we take other extensions $\hat{h}^{(0)}(t, x')$ and $\hat{h}^{(1)}(t, x')$. Let $H_1^{(i)}(\tau', \xi')$, $i = 0, 1$, denote the corresponding functions in (3.8) and let $U_1^{(0)}(\tau', \xi')$ and $V_1^{(i)}(\tau', \xi')$ denote the corresponding functions in (3.10)–(3.12). If we prove that $U^{(0)}(\tau', \xi') = U_1^{(0)}(\tau', \xi')$, the indicated independence of $u(t, x)$ will follow. We have

$$U^{(0)}(\tau', \xi') - U_1^{(0)}(\tau', \xi') = \frac{1}{i\eta_-}\Big\{i\Pi_{2+}\eta_-[H^{(0)}(\tau', \xi') - H_1^{(0)}(\tau', \xi')] + \Pi_{2-}\frac{1}{\eta_+}[H^{(1)}(\tau', \xi') - H_1^{(1)}(\tau', \xi')]\Big\}.$$

Here again $\eta_\pm = \eta_\pm(\tau' + ia, \xi')$. The function $g^{(0)}(t, x') = \hat{h}^{(0)}(t, x') - \hat{h}_1^{(0)}(t, x')$ has its support in $\overline{R^n_{k-}}$ and belongs to $X_q^1(R^n)$; therefore, in exactly the same way as in the proof of (3.9), we establish that the function

$$G^{(0)}(\tau', \xi') = g^{(0)}(\tau' - k\xi_2, \xi') = F_{(t, x') \to (\tau', \xi')}[g^{(0)}(t, x_2 + kt, x'')]$$

is such that

$$\Pi_{2+}[\eta_- G^{(0)}(\tau', \xi')] = \Pi_{2+}\eta_-[H^{(0)}(\tau', \xi') - H_1^{(0)}(\tau', \xi')] = 0.$$

Analogously,

$$\Pi_{2-}\frac{1}{\eta_+}[H^{(1)}(\tau', \xi') - H_1^{(1)}(\tau', \xi')] = 0,$$

i.e. $U^{(0)}(\tau', \xi') = U_1^{(0)}(\tau', \xi')$, as required.

Let us prove estimate (3.7). By virtue of the substitution $\tau + k\xi_2 = \tau'$, (3.8) and (3.10) we have

$$|u_0^{(0)}|^2_{X_q^1} = \int_{R^n} |\eta_-(\tau' + ia, \xi') U^{(0)}(\tau', \xi')|^2 d\tau' d\xi'$$

$$\leqslant 2|\Pi_{2+}\eta_- H^{(0)}(\tau', \xi')|^2_{L_2(R^n)} + 2|\Pi_{2-}[\eta_+]^{-1} H^{(1)}(\tau', \xi')|^2_{L_2(R^n)} \leqslant$$

$$\leqslant 2|\eta_- H^{(0)}(\tau',\ \xi')|^2_{L_2(R^n)} + 2|[\eta_+]^{-1} H^{(1)}(\tau',\ \xi')|^2_{L_2(R^n)}$$
$$= 2|\hat{h}^{(0)}|^2_{X^1_q} + 2|\hat{h}^{(1)}|^2_{X^0_q}.$$

Here we have used the notation $\eta_\pm = \eta_\pm(\tau' + ia,\ \xi')$ and the fact that the operators $\Pi_{2\pm}$ are orthogonal projections in $L_2(R^n)$. Inasmuch as $u_0^{(0)}(t,\ x')$ does not depend on the choice of extensions $\hat{h}^{(0)}$ and $\hat{h}^{(1)}$, we can take the infimum over all possible extensions in the right side of the obtained inequality and get

$$|u_0^{(0)}|^2_{X^1_q} \leqslant C\{|h^{(0)}|^2_{X^1_q(R^n_{k+})} + |h^{(1)}|^2_{X^0_q(R^n_{k-})}\}.$$

This result together with (3.6) implies (3.7).

The proof that $u(t,\ x) = 0$ for $t < x_1$ is omitted (see Remark 3.1). The proof of Theorem 3.2 is complete.

§4. Theorem on the smoothness of the solution ($|k| < \sqrt{a_{22}}$)

1. *The spaces* V_s. *Intrinsic smoothness of a function.* Let $\chi(x_1) \in C_0^\infty(R^1)$, supp $\chi(x_1) \subset \{x_1 \in R^1 | x_1 \geqslant \delta > 0\}$, δ being an arbitrary positive number, and $\chi(z_1) = \chi(\xi_1 + i\eta_1) = F[\chi(x_1)\exp(-\eta_1 x_1)]$. Then the following estimates hold:

$$|z_1^m \chi(z_1)| \leqslant C_m \exp(-\delta \operatorname{Im} z_1), \qquad \operatorname{Im} z_1 \geqslant 0, \tag{4.1}$$

$$\int_{-\infty}^{+\infty} |z_1^m \chi(z_1)|^2\, d\xi_1 \leqslant C_m \exp(-2\delta \operatorname{Im} z_1), \qquad \operatorname{Im} z_1 > 0. \tag{4.2}$$

The first of them immediately results from estimating the Fourier integral by taking the maximum of the modulus of the integrand, while the second is implied by Parseval's formula.

The intrinsic smoothness question will be studied in a scale of spaces generated by powers of the operator

$$\mathsf{A}(D_t + ia,\ D_x) = -(D_t + ia)^2 + (\mathbf{A}D_x,\ D_x).$$

DEFINITION 4.1.

$$V_s \equiv V_s(R^{n+1}) = \{u(t,\ x) \in L_2(R^{n+1}) | [-(D_t + ia)^2 + (\mathbf{A}D_x,\ D_x)]^s u(t,\ x) \in L_2(R^{n+1})\},$$

$$|u|^2_{V_s} = |\mathsf{A}^s(D_t + ia,\ D_x)\, u(t,\ x)|^2_{L_2(R^{n+1})};$$

s is a positive integer.

It is easily seen that $S(R^{n+1})$ is dense in V_s and that V_s is a complete space.

LEMMA 4.1. *Let* $u(t,\ x) \in \mathfrak{A}_q$ *and let* $u_+(t,\ x)$ *denote its extension by zero to* R^{n+1}. *In order that* $u_+(t,\ x)\chi(x_1) \in V_s$ *for any* $\chi(x_1) \in C_0^\infty(R^1)$, supp $\chi(x_1) \subset \{x_1 | x_1 > 0\} = R^1_+$, *it is sufficient that*

$$\int_{R^n} |u_0^{(0)}(\tau, \xi')|^2 |\lambda|^{2s} e^{-\delta \operatorname{Im} \lambda} d\tau d\xi' < \infty \tag{4.3}$$

for any $\delta > 0$.

REMARK 4.1. Condition (4.3) is also necessary [11].

PROOF. Suppose $u_+(\tau, \xi)$ is given by (3.5) ($u(t, x) \in \mathfrak{A}_q$) and $u_0^{(0)}(\tau, \xi')$ satisfies (4.3). We will prove that

$$F[u_+(\tau, \xi)\chi(x_1)] = u_0^{(0)}(\tau, \xi')\chi(\xi_1 + \lambda). \tag{4.4}$$

For $\varphi(t, x) \in C_0^\infty(R^{n+1})$ we have

$$\begin{aligned}(2\pi)^{n+1}\langle u_+(t, x)\chi(x_1), \varphi(t, x)\rangle &= (2\pi)^{n+1}\langle u_+(t, x), \chi(x_1)\varphi(t, x)\rangle \\ &= \langle u_+(\tau, \xi), F[\chi(x_1)\varphi(t, x)]\rangle = \int_{R^{n+1}} u_+(\tau, \xi) \\ &\times \frac{1}{2\pi}\int_{-\infty}^{\infty} \overline{\chi(\xi_1 - \alpha_1)}\,\overline{\varphi(\tau, \alpha_1, \xi')}\, d\alpha_1 d\tau d\xi.\end{aligned}$$

Here we have taken into account that $u_+(\tau, \xi) \in L_2(R^{n+1})$. The function $|u_+(\tau, \xi)\,\overline{\chi}(\xi_1 - \alpha_1)\overline{\varphi}(\tau, \alpha_1, \xi')|$ is integrable over R^{n+2}. In fact, $\chi(\xi_1) \in S(R^1)$, and $\varphi(\tau, \alpha_1, \xi') \in S(R^{n+1})$, so that for any $k \geqslant 0$ we have

$$|\overline{\chi}(\xi_1 - \alpha_1)| \leqslant C_k[1 + |\xi_1 - \alpha_1|^2]^{-k}, \quad |\overline{\varphi}(\tau, \alpha_1, \xi')| \leqslant b_k[1 + \tau^2 + |\xi'|^2 + \alpha_1^2]^{-2k}.$$

Using the inequality

$$[1 + |\xi_1 - \alpha_1|^2][1 + \tau^2 + |\xi'|^2 + \alpha_1^2]^2 \geqslant \frac{1}{8}[1 + \tau^2 + |\xi'|^2 + \xi_1^2][1 + \alpha_1^2], \tag{$*$}$$

which we will prove later on, we get

$$|\overline{\chi}(\xi_1 - \alpha_1)\overline{\varphi}(\tau, \alpha_1, \xi')|^2 \leqslant d_k(1 + \tau^2 + |\xi|^2)^{-k}(1 + \alpha_1^2)^{-k}.$$

Inasmuch as $u_+(\tau, \xi) \in L_2(R^{n+1})$, this implies the integrability of the function $|u_+(\tau, \xi)\,\overline{\chi}(\xi_1 - \alpha_1)\overline{\varphi}(\tau, \alpha_1, \xi')|$. Taking into account the fact that $\chi(x_1)$ is a real function, so that $\overline{\chi}(\xi_1 - \alpha_1) = \chi(\xi_1 - \alpha_1)$, and using Fubini's theorem, we find that

$$\begin{aligned}&\langle u_+(t, x)\chi(x_1), \varphi(t, x)\rangle \\ &= \frac{1}{(2\pi)^{n+1}}\int_{R^{n+1}} d\tau d\xi' d\alpha_1 \overline{\varphi}(\tau, \alpha_1, \xi')\frac{1}{2\pi}\int_{-\infty}^{\infty} u_+(\tau, \xi)\chi(\alpha_1 - \xi_1)\, d\xi_1 =\end{aligned}$$

$$= \frac{1}{(2\pi)^{n+1}} \left\langle \frac{1}{2\pi} \int_{-\infty}^{\infty} u_+(\tau, \alpha_1, \xi') \chi(\xi_1 - \alpha_1) \, d\alpha_1, \ \varphi(\tau, \xi) \right\rangle.$$

Hence

$$F[u_+(t, x) \chi(x_1)] = \frac{1}{2\pi} \int_{-\infty}^{\infty} u_+(\tau, \alpha_1, \xi') \chi(\xi_1 - \alpha_1) \, d\alpha_1.$$

Using (3.5), we get

$$F[u_+(t, x) \chi(x_1)] = \frac{u_0^{(0)}(\tau, \xi')}{2\pi i} \int_{-\infty}^{\infty} \frac{\chi(\xi_1 - \alpha_1) \, d\alpha_1}{\alpha_1 + \lambda}.$$

Closing now the contour of integration in the lower halfplane and using (4.1), we find that

$$F[u_+(t, x) \chi(x_1)] = u_0^{(0)}(\tau, \xi') \chi(\xi_1 + \lambda),$$

as required. We note that in proving (4.4) we used the representation (3.5) of $u_+(\tau, \xi)$ and the fact that $u_+(\tau, \xi) \in L_2(R^{n+1})$.

Let us prove the inequality (*) used above. It is obvious when $\xi_1 = 0$. If $\xi_1 \neq 0$, we put $\alpha_1 = t\xi_1$ and use the notation $(1 + \tau^2 + |\xi'|^2) = y^2 \geqslant 1$. Then the inequality takes the form

$$(1 + \xi_1^2 (t-1)^2)(y^2 + t^2\xi_1^2)^2 \geqslant \frac{1}{8} (y^2 + \xi_1^2)(1 + t^2\xi_1^2).$$

If $t^2 \geqslant 1/4$, this inequality is obviously satisfied. If $t^2 \leqslant 1/4$, then $1/4 \leqslant (t-1)^2 \leqslant 9/4$. Inasmuch as

$$y^2 + t^2\xi_1^2 \geqslant 1 + t^2\xi_1^2,$$

$$(1 + \xi_1^2(t-1)^2)(y^2 + t^2\xi_1^2)^2 \geqslant {}^1/_2 (1 + y^2 + \xi_1^2(t-1)^2 + t^2\xi_1^2)$$
$$\geqslant {}^1/_2 (y^2 + 1/4\xi_1^2) \geqslant {}^1/_8 (y^2 + \xi_1^2),$$

the inequality is also satisfied.

We now turn our attention directly to verifying condition (4.3). According to Definition 4.1, we must verify that

$$[-(D_t + ia)^2 + (\mathbf{A}D_x, D_x)]^s u_+(t, x) \chi(x_1) \in L_2(R^{n+1}).$$

We have

$$\begin{aligned} F[\{-(D_t + ia)^2 &+ (\mathbf{A}D_x, D_x)\}^s u_+(t, x) \chi(x_1)] \\ &= (\xi_1 - \lambda)^s (\xi_1 + \lambda)^s u_0^{(0)}(\tau, \xi') \chi(\xi_1 + \lambda) \\ &= u_0^{(0)}(\tau, \xi') \sum_{k=0}^{s} (\xi_1 + \lambda)^{2s-k} (-2\lambda)^k \chi(\xi_1 + \lambda) C_s^k. \end{aligned}$$

Using (4.2), we get

$$\int\limits_{R^{n+1}} | A^s(\tau + ia, \xi) F[u_+(t, x)\chi(x_1)]|^2 \, d\tau d\xi$$
$$\leqslant C(s) \int\limits_{R^n} |u_0^{(0)}(\tau, \xi')|^2 |\lambda|^{2s} \exp(-2\delta \operatorname{Im} \lambda) \, d\tau d\xi' < \infty.$$

The proof of the lemma is complete.

DEFINITION 4.2. A function $u(t, x) \in V_q^+$ has a *smoothness of the type of the space* V_s (or simply, of type V_s) *in* R_+^{n+1} if $u_+(t, x)\chi(x_1) \in V_s$ for any $\chi(x_1) \in C_0^\infty(R^1)$, supp $\chi(x_1) \subset R_+^1$.

We note that it is not necessary that $u(t, x) \in V_s^+$.

The definition of V_s^+ is analogous to the definition of V_q^+.

COROLLARY 4.1. *Let* $u(t, x) \in \mathfrak{A}_q$. *In order for* $u(t, x)$ *to have a smoothness of type* V_s *in* R_+^{n+1} *it is sufficient that the condition*

$$\int\limits_{R^n} |u_0^{(0)}(\tau, \xi')|^2 |\lambda|^{2s} \exp(-\delta \operatorname{Im} \lambda) \, d\tau d\xi' < \infty \tag{4.5}$$

be satisfied for any $\delta > 0$.

The proof consists of a reference to Lemma 4.1.

We note that condition (4.5) is also necessary [11].

COROLLARY 4.2. *Condition* (4.5) *is equivalent to the following condition*: *for any* $\delta > 0$

$$\int\limits_{R^n} |u_0^{(0)}(\tau, \xi')|^2 (\tau^2 + 1)^s \exp(-\delta \operatorname{Im} \lambda) \, d\tau d\xi' < \infty. \tag{4.6}$$

PROOF. By the definition of $\lambda(\tau + ia, \xi')$ (formula (1.12)) we have $\operatorname{Re} \lambda \operatorname{Im} \lambda = \tau a$, which implies $(\operatorname{Re} \lambda)^2 = \tau^2 a^2/(\operatorname{Im} \lambda)^2 \leqslant \tau^2$, since $\operatorname{Im} \lambda \geqslant a$. Let $\epsilon > 0$ and choose C_ϵ^2 so that $(\operatorname{Im} \lambda)^{2s} e^{-\epsilon \operatorname{Im} \lambda} \leqslant C_\epsilon^2$. Then

$$|\lambda|^{2s} e^{-\delta \operatorname{Im} \lambda} \leqslant (\tau^2 + (\operatorname{Im} \lambda)^2)^s e^{-\delta \operatorname{Im} \lambda} \leqslant C_1 (\tau^2 + 1)^s (\operatorname{Im} \lambda)^{2s} e^{-\delta \operatorname{Im} \lambda}$$
$$\leqslant C_1 (\tau^2 + 1)^s e^{-(\delta - \varepsilon) \operatorname{Im} \lambda}) (\operatorname{Im} \lambda)^{2s} e^{-\varepsilon \operatorname{Im} \lambda} \leqslant C_1 C_\varepsilon^2 (\tau^2 + 1)^{2s} e^{-\delta' \operatorname{Im} \lambda},$$

where $\delta' = \delta - \epsilon$. For each $\delta > 0$ we can take $\epsilon = \delta/2$, so that $\delta' = \delta/2 > 0$. Thus condition (4.6) implies (4.5).

Conversely, suppose (4.5) is satisfied. For any $\epsilon > 0$ we can choose C_ϵ^2 so that $e^{\epsilon \operatorname{Im} \lambda} [\operatorname{Im} \lambda]^{-2s} \geqslant C_\epsilon^2$. Then

$$|\lambda|^{2s} e^{-\delta \operatorname{Im} \lambda} = (\tau^2 a^2 + (\operatorname{Im} \lambda)^4)^{2s} e^{-\delta \operatorname{Im} \lambda} (\operatorname{Im} \lambda)^{-2s}$$
$$\geqslant (\tau^2 a^2 + a^4)^{2s} e^{-\delta' \operatorname{Im} \lambda} e^{\varepsilon \operatorname{Im} \lambda} (\operatorname{Im} \lambda)^{-2s} \geqslant C_1(a) C_\varepsilon^2 [\tau^2 + 1]^{2s} e^{-\delta' \operatorname{Im} \lambda},$$

where $\delta' = \delta + \epsilon$. For any $\delta' > 0$ we can take $\epsilon = \delta'/2$, so that $\delta = \delta' - \epsilon = \delta'/2 > 0$. Hence (4.5) implies (4.6). The proof of Corollary 4.2 is complete.

2. *Theorem on the intrinsic smoothness in* R_+^{n+1} *of the solution of problem* 0) – 4) *in* §3.2.

THEOREM 4.1. *Suppose the functions* $h^{(0)}(t, x')$ *and* $h^{(1)}(t, x')$ *in problem* 0) – 4) *have extensions* $\hat{h}^{(0)}(t, x')$ *and* $\hat{h}^{(1)}(t, x')$ *to all of* R^n *satisfying the conditions*

$$\int_{R^n} |\tau + k\xi_2 + ia|^{2s} |\eta_-(\tau + k\xi_2 + ia)|^2 |\hat{h}^{(0)}(\tau, \xi')|^2 d\tau d\xi' < \infty, \tag{4.7}$$

$$\int_{R^n} |\tau + k\xi_2 + ia|^{2s} |\eta_+(\tau + k\xi_2 + ia)|^{-2} |\hat{h}^{(1)}(\tau, \xi')|^2 d\tau d\xi' < \infty. \tag{4.8}$$

Then the solution $u(t, x)$ *of problem* 0) – 4) *has a smoothness of type* V_s *in* R_+^{n+1}.

REMARK 4.2. Conditions (4.7) and (4.8) imply a smoothness of the functions $h^{(0)}(t, x')$ and $h^{(1)}(t, x')$ in the direction of the tangent to the manifold of discontinuity $x_2 = kt$, since $-i(\tau + k\xi_2)$ is the Fourier transform of the operator $\partial/\partial t + k\partial/\partial x_2$, which is the operator of differentiation along the manifold $x_2 = kt$.

PROOF. Suppose $u(t, x)$ is the solution of problem 0)–4) and $U^{(0)}(\tau', \xi')$ and $H^{(i)}(\tau', \xi')$, $i = 0, 1$, are as in the proof of Theorem 3.2. Conditions (4.7) and (4.8) mean that

$$\int_{R^n} |\tau' + ia|^{2s} |\eta_-(\tau' + ia, \xi') H^{(0)}(\tau', \xi')|^2 d\tau' d\xi' < \infty, \tag{4.7'}$$

$$\int_{R^n} |\tau' + ia|^{2s} \left| \frac{H^{(1)}(\tau', \xi')}{\eta_+(\tau' + ia, \xi')} \right|^2 d\tau d\xi' < \infty. \tag{4.8'}$$

From (3.10) we see that $U^{(0)}(\tau', \xi')$ also satisfies (4.7'), which means that

$$\int_{R^n} |\tau + k\xi_2 + ia|^{2s} |\eta_-(\tau + k\xi_2 + ia, \xi')|^2 |u_0^{(0)}(\tau, \xi')|^2 d\tau d\xi' < \infty. \tag{4.9}$$

We will prove that (4.9) implies (4.6). It suffices to prove the inequality

$$C(\delta)(\tau^2 + 1)^s e^{-\delta \operatorname{Im} \lambda} \leqslant |\tau + k\xi_2 + ia|^{2s} |\eta_-(\tau + k\xi_2 + ia, \xi')|^2 \tag{4.10}$$

for any $\delta > 0$. According to (2.3) we have

$$|\eta_-(\tau' + ia, \xi')|^2 \geqslant \left| \frac{\operatorname{Im} \varkappa - \theta a}{1 - \theta} \right| \geqslant a,$$

since $\operatorname{Im} \kappa = \operatorname{Im} \kappa(\tau' + ia, \xi'') \geqslant a$. Hence the right side of (4.10) is bounded from below by a^{2s+1}.

Inasmuch as $e^{-\delta \operatorname{Im} \lambda} \leqslant 1$, inequality (4.10) will follow if we establish that

$$\frac{\tau^{2s} e^{-\delta \operatorname{Im} \lambda}}{[(\tau + k\xi_2)^2 + a^2]^s} \leqslant C(\delta),$$

and this follows from the inequality

$$\frac{\tau^2 e^{-\delta \operatorname{Im} \lambda}}{(\tau + k\xi_2)^2 + a^2} \leqslant C(\delta) \quad \text{for any} \quad \delta > 0. \tag{4.11}$$

We prove the latter. If $k = 0$, it is obvious. Suppose $0 < |k| < \sqrt{a_{22}}$ and ϵ is such that $0 < \epsilon < \min\{k^2, a_{22} - k^2\}$. We divide the (τ, ξ_2) plane into three parts:

$$\Omega_1 = \{(\tau, \xi_2) \mid (k^2 + \varepsilon)\xi_2^2 \leqslant \tau^2\},$$
$$\Omega_2 = \{(\tau, \xi_2) \mid \tau^2 \leqslant (k^2 - \varepsilon)\xi_2^2\},$$
$$\Omega_3 = \{(\tau, \xi_2) \mid (k^2 - \varepsilon)\xi_2^2 \leqslant \tau^2 \leqslant (k^2 + \varepsilon)\xi_2^2\}.$$

If $(\tau, \xi_2) \in \Omega_1 \cup \Omega_2$, it is elementary to prove that $\tau^2/[(\tau + k\xi_2)^2 + a^2] \leqslant C_1$, and hence (4.11) is true. Suppose $(\tau, \xi_2) \in \Omega_3$. To estimate $\operatorname{Im} \lambda(\tau + ia, \xi')$ we note that

$$2(\operatorname{Im}\lambda)^2 = |\lambda|^2 - \operatorname{Re}\lambda^2 = \{(\tau^2 - (\mathbf{A}'\xi', \xi') - a^2)^2 + 4a^2\tau^2\}^{1/2}$$
$$- (\tau^2 - (\mathbf{A}'\xi', \xi') - a^2) \geqslant a_{22}\xi_2^2 - \tau^2 + (\mathbf{A}''\xi'', \xi'') + a^2,$$

since $(\mathbf{A}'\xi', \xi') = a_{22}\xi_2^2 + (\mathbf{A}''\xi'', \xi'')$, while in Ω_3 we have $\tau^2 < a_{22}\xi_2^2$. Hence

$$2(\operatorname{Im}\lambda)^2 \geqslant a_{22}\xi_2^2 - (k^2 + \varepsilon)\xi_2^2 = \xi_2^2(a_{22} - (k^2 + \varepsilon)) = \varepsilon_2^2\xi_2^2, \quad -\delta \operatorname{Im}\lambda \leqslant -\varepsilon_3|\xi_2|,$$

where $\epsilon_3 = \epsilon_3(\delta) > 0$ and $e^{-\delta \operatorname{Im} \lambda} \leqslant e^{-\epsilon_3 |\xi_2|}$. Finally,

$$\frac{\tau^2 e^{-\delta \operatorname{Im} \lambda}}{(\tau + k\xi_2)^2 + a^2} \leqslant \frac{(k^2 + \varepsilon)|\xi_2|^2 e^{-\varepsilon_3|\xi_2|}}{a^2} \leqslant C_2(\delta),$$

which proves (4.11), and hence (4.10) and (4.6), for $u_0^{(0)}(\tau, \xi')$. An application of Corollary 4.2 completes the proof of Theorem 4.1.

REMARK 4.3. Inasmuch as

$$|A(\tau + ia, \xi)|^2 = (\tau^2 - (\mathbf{A}\xi, \xi))^2 + 2a^2(\tau^2 + (\mathbf{A}\xi, \xi)) + a^4$$
$$\geqslant a^2(\tau^2 + (\mathbf{A}\xi, \xi) + a^2),$$

we have $V_s \supset H_s(R^{n+1})$. It therefore follows from Theorem 4.1 that one can obtain a solution of problem 0) – 4) of any smoothness in R_+^{n+1} by requiring a sufficient smoothness of the boundary functions $h^{(0)}(t, x')$ and $h^{(1)}(t, x')$ (i.e. by taking s sufficiently large in (4.7) and (4.8)) and by not requiring that they satisfy any compatibility conditions on the manifold of discontinuity $x_2 = kt$ of the boundary operator.

REMARK 4.4. In order for the solution of problem 0) – 4) to have a smoothness right up to the boundary better than that guaranteed by Theorem 3.2 it is necessary that the boundary functions $h^{(0)}(t, x')$ and $h^{(1)}(t, x')$ satisfy certain compatibility conditions.

We identify the notion of "localness" of conditions for the compatibility of $h^{(0)}$ and $h^{(1)}$ on the manifold $x_2 = kt$ with the possibility of varying the functions $h^{(0)}$ and $h^{(1)}$ for $x_2 > kt + \epsilon$ and $x_2 < kt - \epsilon$, $\epsilon > 0$, respectively without violating these compatibility conditions. The variation occurs, of course, in fixed function spaces of the type, for example, of the considered spaces $H_\rho(R^n)$, $\rho = \rho(\tau, \xi')$. It can then be proved that no "local" (in this sense) compatibility conditions can give a solution of problem 0) – 4) having a smoothness of type V_1 in R_+^{n+1} right up to the boundary, i.e. belonging to V_1^+, for any smoothness of $h^{(0)}(t, x')$ and $h^{(1)}(t, x')$ in their domains of definition [11].

§5. Function spaces ($|k| > \sqrt{a_{22}}$)

1. We will distinguish the cases a) $k > \sqrt{a_{22}}$ and b) $k < -\sqrt{a_{22}}$. Where it is possible, the presentation is conducted for both cases a) and b) simultaneously.

Let $\overset{\vee}{\lambda}(\tau' + ia, \xi')$ be as in (1.12). This function can now be analytically continued with respect to the complex variable $z_2 = \xi_2 + i\eta_2$ into the halfplane where $k\eta_2 \leqslant 0$, i.e. where a) $\eta_2 \leqslant 0$ for $k > \sqrt{a_{22}}$ and b) $\eta_2 \geqslant 0$ for $k < -\sqrt{a_{22}}$. This can easily be verified by defining z_2 as a root of the equation

$$[\overset{\vee}{\lambda}(\tau' + ia, z_2, \xi'')]^2 = (\tau' - kz_2 + ia)^2 - a_{22}z_2^2 - (\mathbf{A}''\xi'', \xi'') = 0.$$

We also note that $\overset{\vee}{\lambda}(\tau' + ia, z_2, \xi'')$ does not vanish for $\eta_2 \leqslant 0$ ($\eta_2 \geqslant 0$) in case a) (case b)).

2. *The spaces* R_s, Y_s^i *and* Z^i. Let s be a nonnegative integer.

DEFINITION 5.1. Let $m = 1, 0$ in cases a) and b) respectively. Then

$$R_s \equiv R_s(R^{n+1}) = \Big\{ u(t,\ x) \in L_2(R^{n+1}) \,\Big|\, |u|^2_{R_s} \tag{5.1}$$
$$= \int_{R^{n+1}} (\xi_1^{2(2s+m)} + \tau^{4s} + |\xi'|^{4s} + a^{4s}) \, |u(\tau,\ \xi)|^2 \, d\tau d\xi < \infty \Big\};$$

$$Y_s^1 \equiv Y_s^1(R^n) = \Big\{ h(t,\ x') \in L_2(R^n) \,\Big|\, |h|^2_{Y_s^1}$$
$$= \int_{R^n} |\lambda(\tau + ia,\ \xi')|^{2m} (\tau^2 + |\xi'|^2 + a^2)^{2s} \, |h(\tau,\ \xi')|^2 \, d\tau d\xi' < \infty \Big\};$$

$$Y_s^0 \equiv Y_s^0(R^n) = \Big\{ h(t,\ x') \in H_{m-1}(R^n) \,\Big|\, |h|^2_{Y_s^0}$$
$$= \int_{R^n} |\lambda(\tau + ia,\ \xi')|^{2(m-1)} (\tau^2 + |\xi'|^2 + a^2)^{2s} \, |h(\tau,\ \xi')|^2 \, d\tau d\xi' < \infty \Big\};$$

$$Z^i \equiv Z^i(R^n) = \Big\{ r(t,\ x') \in H_{-3/2}(R^n) \,\Big|\, |r|^2_{Z^i}$$
$$= \int_{R^n} \frac{|\lambda(\tau + ia,\ \xi')|^{2(m+i-1)}}{\operatorname{Im} \lambda(\tau + ia,\ \xi')} \, |r(\tau,\ \xi')|^2 \, d\tau d\xi' < \infty \Big\}, \qquad i = 0,\ 1.$$

DEFINITION 5.2. Let $m = 1, 0$ in cases a) and b) respectively. $R_s^+ = R_s(R_+^{n+1})$ is the set of functions $u(t, x) \in L_2(R_+^{n+1})$ for each of which there exists a function $\hat{u}(t, x) \in R_s$ such that $\hat{u}(t, x) = u(t, x)$ on R_+^{n+1}. The norm in R_s^+ is

$$|u|_{R_s^+} = \inf |\hat{u}|_{R_s}, \tag{5.2}$$

where the infimum is taken over all of the extensions of $u(t, x)$ onto R_+^{n+1} in R_s.

It is shown in exactly the same way as before that (i) the spaces R_s, Y_s^i, Z^i and R_s^+, $i = 0, 1$, are complete, (ii) the functions from $C_0^\infty(R^{n+1})$ or $C_0^\infty(R^n)$ are dense in R_s, Y_s^i and Z^i, and (iii) the norm (5.2) is equivalent to the following norm, which will be denoted by the same symbol:

$$|u|_{R_s^+} = \left| \Pi_{1+} \left[\left[\xi_1 - i(\tau^{4s} + |\xi'|^{4s} + a^{4s})^{\frac{1}{2(2s+m)}} \right]^{2(2s+m)} u_+(\tau,\ \xi) \right] \right|_{L_2(R^{m+1})}, \tag{5.3}$$

where, as usual, $u_+(t, x)$ is the extension by zero of $u(t, x)$ onto R^{n+1} and $2s + m \neq 0$. If, on the other hand, $2s + m = 0$ (case b)), we have $R_0 = L_2(R^{n+1})$ and $R_0^+ = L_2(R_+^{n+1})$.

DEFINITION 5.3.

$$\mathfrak{A}_s = \{ u(t,\ x) \in R_s^+ \mid A(D_t + ia,\ D_x) u(t,\ x)$$
$$\equiv [-(D_t + ia)^2 + (\mathbf{A} D_x,\ D_x)] \, u(t,\ x) = 0 \}$$

is the null space of the operator $A(D_t + ia, D_x)$ in R_s^+.

The equality $A(D_t + ia, D_x)u(t, x) = 0$ in Definition 5.3 is understood in $D'(R_+^{n+1})$. We assert that $\mathfrak{A}_s$ is a closed subspace of R_s^+.

For suppose $u \in R_s^+$, $\hat{u}$ is an extension of u onto R^{n+1} in R_s and $\psi(t, x) \in D(R_+^{n+1})$. We extend $\psi(t, x)$ by zero onto R^{n+1}. Preserving the notation, we then have $\psi(t, x) \in D(R^{n+1})$ and, by the definition of $\hat{u}$, $\langle u, \psi\rangle = \langle \hat{u}, \psi\rangle$. It readily follows that $|\langle u, \psi\rangle| = |\langle \hat{u}, \psi\rangle| \leqslant C(\psi)|\hat{u}|_{R_s}$. Taking the infimum over all extensions, we now get $|\langle u, \psi\rangle| \leqslant C(\psi)|u|_{R_s^+}$, i.e. convergence in R_s^+ implies convergence in $D'(R_+^{n+1})$.

Suppose now $\lim_{i\to\infty} u_i = u$ in R_s^+. Then $\lim_{i\to\infty} u_i = u$ in $D'(R_+^{n+1})$, and hence

$$\lim_{i\to\infty} A(D_t + ia, D_x)\, u_i(t, x) = A(D_t + ia, D_x)\, u(t, x)$$

in $D'(R_+^{n+1})$. This proves that $\mathfrak{A}_s$ is closed.

LEMMA 5.1. *The assertion*

$$u(t, x) \in \mathfrak{A}_0$$

and the assertion

$$u_+(\tau, \xi) = \frac{ir(\tau, \xi')}{\xi_1 + \lambda(\tau + ia, \xi')}, \tag{5.4}$$

where $r(t, x') \in Z^1$, are equivalent, and the resultant one-to-one mapping of $\mathfrak{A}_0$ onto Z^1 is bicontinuous.

PROOF. Since $\mathfrak{A}_0 \subset L_2(R_+^{n+1})$, it follows from Lemma 1.8 that the representation (5.4) holds at least with $r(t, x') \in H_{-2}(R^n)$. Further, according to (5.3) for $m = 1$ we have $\Pi_{1+}(\xi_1 - i)u_+(\tau, \xi) \in L_2(R^{n+1})$. But

$$(\xi_1 - i)\, u_+(\tau, \xi) = \frac{-i(\lambda + i)\, r(\tau, \xi')}{\xi_1 + \lambda} + ir(\tau, \xi'),$$

which according to Lemma 1.7 implies

$$\Pi_{1+}(\xi_1 - i)\, u_+(\tau, \xi) = \frac{-i(\lambda + i)\, r(\tau, \xi')}{\xi_1 + \lambda} \in L_2(R^{n+1}).$$

This means that $r(t, x') \in Z^1$. When $m = 0$ (case b)) we have

$$\Pi_{1+} u_+(\tau, \xi) = u_+(\tau, \xi) = \frac{ir(\tau, \xi')}{\xi_1 + \lambda(\tau + ia, \xi')} \in L_2(R^{n+1}).$$

It follows that $r(\tau, \xi') \in Z^1$. The formulas obtained for $m = 1, 0$ clearly imply that $|r|_{Z^1} \leqslant C|u|_{R_0^+}$.

Now supppse (5.4) holds with $r(t, x') \in Z^1$. Here again we have $(m = 1, 0)$

$$\Pi_{1+} (\xi_1^m - i) u_+ (\tau, \xi) = \frac{i[(-\lambda)^m - i] r(\tau, \xi')}{\xi_1 + \lambda}.$$

A direct integration shows that

$$|\Pi_{1+} (\xi_1^m - i) u_+ (\tau, \xi)|_{L_2(R^{n+1})} \leqslant C |r|_{Z^1}.$$

This means that $u(t, x) \in R_0^+$. Let us verify that $u(t, x) \in \mathfrak{A}_0$. We have

$$A(\tau + ia, \xi) u_+ (\tau, \xi) = [\xi_1 - \lambda] ir(\tau, \xi').$$

Inasmuch as $r(\tau, \xi') \in Z^1$, it follows that $A(\tau + ia, \xi) u_+(\tau, \xi)$ belongs to the domain of definition of the operator Π_{1+} and

$$\Pi_{1+} A(\tau + ia, \xi) u_+(\tau, \xi) = 0,$$

i.e.

$$\langle A(D_t + ia, D_x) u_+(t, x), \varphi(t, x)\rangle = 0$$

for any $\varphi(t, x) \in C_0^\infty (R^{n+1})$, supp $\varphi(t, x) \subset R_+^{n+1}$. The required result follows. The proof of the lemma is complete.

3. *The traces of elements of* $\mathfrak{A}_s$ *on the hyperplanes* $x_1 = \text{const}$. In general, a function $u(t, x) \in R_0$ for $m = 0$ (case b)) does not have traces on the hyperplanes $x_1 = \text{const}$, nor does its derivative $\partial u/\partial x_1$ for $m = 1$. But if it is also an element of $\mathfrak{A}_0 \subset R_0^+$, then the traces of the function itself and its derivative $\partial u/\partial x_1$ can be defined.

Suppose $r(t, x') \subset S(R^n)$, $u_+(\tau, \xi)$ is defined by (5.4), and $u(t, x) = F^{-1}[u_+(\tau, \xi)]$ for $x_1 > 0$, $\chi(x_1) \in C_0^\infty(R^1)$, supp $\chi(x_1) \subset R_+^1$. In the proof of Lemma 4.1 it was also proved (formula (4.4)) that if $u_+(\tau, \xi) \in L_2(R^{n+1})$ and has the form (5.4), then $F[u_+(t, x)\chi(x_1)] = r(\tau, \xi')\chi(\xi_1 + \lambda)$. Here $\chi(z_1)$ satisfies (4.1) and (4.2). If $r(\tau, \xi') \in S(R^n)$, it follows that $F[u_+(t, x)\chi(x_1)] \in S(R^{n+1})$, i.e. $u_+(t, x)\chi(x_1) \in S(R^{n+1})$. We can then say that

$$u_+(\tau, x_1, \xi')\chi(x_1) = F^{-1}_{x_1 \to \xi_1}[r(\tau, \xi')\chi(\xi_1 + \lambda)] = r(\tau, \xi') e^{ix_1\lambda}\chi(x_1).$$

If we take $\chi(x_1)$ equal to unity in a neighborhood of a fixed $x_1^0 > 0$, then

$$u_+(\tau, x_1^0, \xi') = r(\tau, \xi') \exp ix_1^0\lambda$$

and

$$\frac{\partial u_+(\tau, x_1^0, \xi')}{\partial x_1} = i\lambda r(\tau, \xi') \exp ix_1^0\lambda$$

are the Fourier transforms of the restrictions of $u_+(t, x)$ and $\partial u_+(t, x)/\partial x_1$ to the hyperplane $x_1 = x_1^0$. These restrictions are naturally called the traces of $u_+(t, x)$ and $\partial u_+(t, x)/\partial x_1$ for $x_1 = x_1^0$ and are denoted by $u_{x_1^0}^{(0)}(t, x')$ and $u_{x_1^0}^{(1)}(t, x')$ respectively. The functions $u(t, x)$ of the form just considered are dense in $\mathfrak{A}_0$ inasmuch as the functions of class $S(R^n)$ are dense in Z^1 and Lemma 5.1 holds. This permits us to correctly give

DEFINITION 5.4. Suppose $u(t, x) \in \mathfrak{A}_0$, $u_+(\tau, \xi)$ is as in (5.4), $r_k(t, x') \in S(R^n)$ and $\lim_{k\to\infty} r_k = r$ in Z^1. By the *traces* of the functions $u(t, x)$ and $\partial u(t, x)/\partial x_1$ on the hyperplanes $x_1 = \text{const} > 0$ are respectively meant functions $u_{x_1}^{(0)}(t, x')$ and $u_{x_1}^{(1)}(t, x')$ such that

$$u_{x_1}^{(j)}(t, x') = \lim_{k\to\infty} u_{k, x_1}^{(j)}(t, x') = \lim_{k\to\infty} F^{-1}_{(t, x')\to(\tau, \xi')} [(i\lambda)^j r_k(\tau, \xi') \exp ix_1\lambda]$$

in Z^{1-j}, $j = 0, 1$. Here $\{u_k(t, x)\}$ is the sequence in $\mathfrak{A}_0$ constructed from the functions r_k in accordance with (5.4).

LEMMA 5.2. *If $u(t, x) \in \mathfrak{A}_0$, then the following assertions are true*:

$$1)\ u_{x_1}^{(j)}(\tau, \xi') = F[u_{x_1}^{(j)}(t, x')] = (i\lambda)^j r(\tau, \xi') \exp ix_1\lambda, \qquad x_1 > 0, \tag{5.5}$$

where $r(\tau, \xi')$ is from (5.4).

2) *$u_{x_1}^{(j)}(t, x')$ is a continuous function of x_1, $x_1 > 0$, with values in Z^{1-j} such that*

$$\lim_{x_1\to+0} u_{x_1}^{(j)}(t, x') = F^{-1}[(i\lambda)^j r(\tau, \xi')] \tag{5.6}$$

in Z^{1-j}.

3) *$u_{x_1}^{(j)}(t, x')$ is also a continuous function of x_1, $x_1 > 0$, with values in Y^{1-j}.*

PROOF. 1) If $r(t, x') \in Z^1$, then, inasmuch as $|\exp ix_1\lambda| \leqslant \exp(-x_1 \operatorname{Im} \lambda)$, $\operatorname{Im} \lambda \geqslant a > 0$,

$$u_j(t, x') = F^{-1}_{(t, x')\to(\tau, \xi')} [(i\lambda)^j r(\tau, \xi') \exp ix_1\lambda] \in Z^{1-j}.$$

Moreover, if $r_k(t, x')$ is the sequence in Definition 5.4, then

$$|u_j(t, x') - F^{-1}[(i\lambda)^j r_k(\tau, \xi') e^{ix_1\lambda}]|^2_{Z^{1-j}}$$

$$= \int_{R^n} \frac{|\lambda|^{2m}}{\operatorname{Im}\lambda} |(r(\tau, \xi') - r_k(\tau, \xi')) \exp ix_1\lambda]|^2 \, d\tau d\xi^2$$

$$\leqslant C|r(t, x') - r_k(t, x')|^2_{Z^1} \to 0 \quad \text{as} \quad k\to\infty.$$

This proves 1).

2) Let $x_1 > 0$ and $|\Delta| < x_1/2$. Then for any ball $\Omega_C \subset R^n$ we have

$$|\exp i(x_1+\Delta)\lambda - \exp ix_1\lambda|^2 \leqslant \exp(-2x_1 \operatorname{Im}\lambda)\cdot|\exp i\Delta\lambda - 1|^2 \leqslant C(\Omega_C)\cdot|\Delta|^2 \exp(-2x_1 \operatorname{Im}\lambda);$$

while, if $(\tau, \xi') \in R^n \backslash \Omega_C$, then

$$|\exp i(x_1+\Delta)\lambda - \exp ix_1\lambda|^2 \leqslant \exp[-2(x_1-|\Delta|)\operatorname{Im}\lambda]\,|1-\exp i|\Delta|\lambda|^2 \leqslant 4\exp(-x_1\operatorname{Im}\lambda).$$

Using these estimates and (5.5), we find that

$$|u^{(j)}_{x_1+\Delta}(t, x') - u^{(j)}_{x_1}(t, x')|^2_{Z^{1-j}} = \int_{R^n} \frac{|\lambda|^{2m}|r(\tau, \xi')|^2}{\operatorname{Im}\lambda}|\exp i(x_1+\Delta)\lambda - \exp ix_1\lambda|^2 d\tau d\xi'$$

$$\leqslant C_1(\Omega_C)|\Delta|^2 \int_{\Omega_C} \frac{|\lambda|^{2m}|r(\tau, \xi')|^2}{\operatorname{Im}\lambda}\exp(-2x_1\operatorname{Im}\lambda)\,d\tau d\xi'$$

$$+4\int_{R^n\backslash\Omega_C} \frac{|\lambda|^{2m}|r(\tau, \xi')|^2}{\operatorname{Im}\lambda}\exp(-x_1\operatorname{Im}\lambda)\,d\tau d\xi' < \frac{\varepsilon}{2}+\frac{\varepsilon}{2}=\varepsilon.$$

Here the second (first) summand can be made less then $\epsilon/2$ by choosing a sufficiently large Ω_C (small $|\Delta|$ for fixed Ω_C). This proves the continuity of $u^{(j)}_{x_1}(t, x')$ for $x_1 > 0$. Suppose now $x_1 \to +0$. Then (5.6) is established by means of exactly the same estimate as the one just carried out (by putting $x_1 = 0$ and $\Delta > 0$).

3) If the same estimate is carried out for $|u^{(j)}_{x_1+\Delta} - u^{(j)}_{x_1}|_{Y_0^{1-j}}$, the only difference will be the absence of the denominator $\operatorname{Im}\lambda$. But the presence of the exponential factors (in which x_1 is a fixed positive number) saves the situation and the estimate is carried out in exactly the same way. The proof of Lemma 5.2 is complete.

COROLLARY 5.1. *If $u(t, x) \in \mathfrak{A}_0$ and the limit $\lim_{x_1\to+0} u^{(0)}_{x_1}(t, x') = u_0(t, x')$ exists in Y_0^1, then $u_0(t, x') = r(t, x')$ and hence $r(t, x') \in Y_0^1$.*

PROOF. From the definitions of the spaces Y_0^1 and Z^1 it is clear that $u_0(t, x') \in Z^1$ and $\lim_{x_1\to+0} u^{(0)}_{x_1} = u_0$ in Z^1. But then $u_0(t, x') = r(t, x')$ according to assertion 2) of Lemma 5.2, and hence $r(t, x') \in Y_0^1$.

LEMMA 5.3. *If $r(t, x') \in Y_s^1$, then the function $u(t, x)$ defined by* (5.4) *belongs to $\mathfrak{A}_s$ and*

$$|u|_{R_s^+} \leqslant C|r|_{Y_s^1}, \tag{5.7}$$

$$\lim_{x_1 \to +0} u_{x_1}^{(0)}(t, x_1) = r(t, x') \quad in \quad Y_s^1,$$

$$\lim_{x_1 \to +0} u_{x_1}^{(1)}(t, x') = F^{-1}_{(t, x') \to (\tau, \xi')}[i\lambda r(\tau, \xi')] \quad in\ Y_s^0.$$

PROOF. The assertion $A(D_t + ia, D_x)u(t, x) = 0$ is verified in exactly the same way as in the proof of Lemma 5.1.

Let us prove (5.7). According to (5.3) we have (from Lemma 1.7)

$$\Pi_{1+} \frac{\left[\xi_1 - i(\tau^{4s} + |\xi'|^{4s} + a^{4s})^{\frac{1}{2(2s+m)}}\right]^{2s+m}}{\xi_1 + \lambda} ir(\tau, \xi') = \frac{\left[-\lambda - i(\tau^{4s} + |\xi'|^{4s} + a^{4s})^{\frac{1}{2(2s+m)}}\right]^{2s+m}}{\xi_1 + \lambda} ir(\tau, \xi').$$

Hence

$$\left|\Pi_{1+}\left\{\left[\xi_1 - i(\tau^{4s} + |\xi'|^{4s} + a^{4s})^{\frac{1}{2(2s+m)}}\right]^{2s+m} u_+(\tau, \xi)\right|^2_{L_2(R^{n+1})}$$

$$\leqslant C \int_{R^n} (|\lambda|^{2(2s+m)} + \tau^{4s} + |\xi'|^{4s} + a^{4s})|r(\tau, \xi')|^2 d\tau d\xi'$$

$$\leqslant C_1 \int_{R^n} (|\lambda|^{2m} + 1)(\tau^{4s} + |\xi'|^{4s} + a^{4s})|r(\tau, \xi')|^2 d\tau d\xi' \leqslant C_2 |r|^2_{Y_s^1}.$$

Here we have made use of the fact that

$$a^2 \leqslant |\lambda|^2 = |(\tau + ia)^2 - (\mathbf{A}'\xi', \xi')| \leqslant K(\tau^2 + |\xi'|^2 + a^2).$$

Thus $|u|_{R_s^+} \leqslant C_2 |r|^2_{Y_s^1}$.

Let us now verify the assertions concerning traces. We have

$$u_{x_1}^{(j)}(\tau, \xi') = r(\tau, \xi')(\exp ix_1\lambda)(i\lambda)^j, \quad j = 0,1,$$

and from the condition $r(t, x') \in Y_s^1$ it is clear that $u_{x_1}^{(j)}(t, x') \in Y_s^{1-j}$ for any $x_1 > 0$. By using exactly the same estimate as in the proof of assertions 2) and 3) of Lemma 5.2, we can now obtain the required limit relations. The proof of the lemma is complete.

REMARK 5.1. The weight function in the definitions of the norms of the spaces Y_s^i can be written in an equivalent form:

$$[\tau^2 + \xi_2^2 + |\xi''|^2 + a^2]^{2s} \approx [(\tau + k\xi_2)^2 + \xi_2^2 + |\xi''|^2 + a^2]^{2s}. \tag{5.8}$$

The symbol $\approx$ here is understood in the sense that the ratio of functions connected by it is bounded from above and below by positive constants.

Let us prove this remark. First, we note that

$$(\tau + k\xi_2)^2 + \xi_2^2 \leqslant 2\tau^2 + 2k^2\xi_2^2 + \xi_2^2 \leqslant (2k^2+1)(\tau^2+\xi_2^2).$$

Second, for $\epsilon = (k^2 + 1/2)(k^2 + 1)$ we have

$$(\tau + k\xi_2)^2 + \xi_2^2 \geqslant \tau^2 + (k^2+1)\xi_2^2 - |2k\tau\xi_2| \geqslant \tau^2 + (k^2+1)\xi_2^2 - \varepsilon\tau^2$$
$$- \frac{k^2}{\varepsilon}\xi_2^2 \geqslant \frac{1}{2(k^2+1)}(\tau^2+\xi_2^2),$$

which proves (5.8).

REMARK 5.2. The inclusion $R_s^+ \subset V_s^+$ holds, where V_s and V_s^+ are the spaces introduced in 4.1. The proof follows upon noting that

$$|A(\tau + ia, \xi)|^{2s} = |(\tau+ia)^2 - |\xi|^2|^2 \leqslant C(\xi_1^{4s} + \tau^{4s} + |\xi'|^{4s} + a^{4s}).$$

§6. Statement of the problem. Existence and uniqueness theorem ($|k| > \sqrt{a_{22}}$)

1. *Statement of the problem.*

0) Let $h^{(0)}(t, x')$ and $h^{(1)}(t, x')$ denote a pair of functions that are equal to zero for $t < 0$ and respectively defined for $x_2 > kt(R^n_{k+})$ and $x_2 < kt(R^n_{k-})$ and suppose they respectively have extensions $\hat{h}^{(0)}(t, x') \in Y_0^1$ and $\hat{h}^{(1)}(t, x') \in Y_0^0$ onto all of R^n. Find functions $u(t, x) \in \mathfrak{A}_0$ and $v^{(i)}(t, x') \in Y_0^{1-i}$, $i = 0, 1$, satisfying the following conditions:

1) $\operatorname{supp} v^{(0)}(t, x') \subset \overline{R^n_{k-}}$ and $\operatorname{supp} v^{(1)}(t, x') \subset \overline{R^n_{k+}}$ (the spaces $R^n_{k\pm}$ are as defined in §2.4) and $u(t, x) = 0$ for $t < x_1$.

2) $\lim_{x_1 \to +0} u^{(0)}_{x_1}(t, x') = \hat{h}^{(0)}(t, x') + v^{(0)}(t, x')$ in Y_0^1.

3) $\lim_{x_1 \to +0} u^{(1)}_{x_1}(t, x') = \hat{h}^{(1)}(t, x') + v^{(1)}(t, x')$ in Y_0^0.

As before, for the sake of brevity we omit a proof of the fact that the solution constructed below is equal to zero for $t < x_1$. It is carried out in the same way as in [11] for the case $|k| < \sqrt{a_{22}}$.

2. *Existence and uniqueness theorem.*

THEOREM 6.1. *Problem* 0)–3) *has a unique solution satisfying the estimate*

$$|u|_{R_0^+} \leqslant C(\inf|\hat{h}^{(0)}|_{Y_0^1} + \inf|\hat{h}^{(1)}|_{Y_0^0}), \tag{6.1}$$

where the infimums are respectively taken over all of the extensions of $h^{(0)}(t, x')$ *and* $h^{(1)}(t, x')$ *onto all of* R^n *in the classes* Y_0^1 *and* Y_0^0.

PROOF. According to Lemma 5.1, Corollary 5.1 and Lemma 5.3 it suffices to prove that the system

2′) $r(\tau, \xi') = \hat{h}^{(0)}(\tau, \xi') + v^{(0)}(\tau, \xi')$,

3′) $i\lambda r(\tau, \xi') = \hat{h}^{(1)}(\tau, \xi') + v^{(1)}(\tau, \xi')$

has a unique solution $\{r(t, x'), v^{(0)}(t, x'), v^{(1)}(t, x')\}$ in $Y_0^1 \times Y_0^1 \times Y_0^0$.

Suppose system 2′), 3′) has a solution. Making the substitution $\tau + k\xi_2 = \tau'$ and letting

$$r(\tau' - k\xi_2, \xi') = U^{(0)}(\tau', \xi'),$$
$$\hat{h}^{(i)}(\tau' - k\xi_2, \xi') = H^{(i)}(\tau', \xi'),$$
$$v^{(i)}(\tau' - k\xi_2, \xi') = V^{(i)}(\tau', \xi'), \quad i = 0,1,$$

we get

2″) $U^{(0)}(\tau', \xi') = H^{(0)}(\tau', \xi,) + V^{(0)}(\tau', \xi')$,

3″) $i\tilde{\lambda}(\tau' + ia, \xi')U^{(0)}(\tau', \xi') = H^{(1)}(\tau', \xi') + V^{(1)}(\tau', \xi')$.

We now consider the cases $k > \sqrt{a_{22}}$ and $k < -\sqrt{a_{22}}$ separately.

a) $k > \sqrt{a_{22}}$. As in the proof of Theorem 3.2, we find from 2″), 3″) that the following relations must be satisfied:

$$U^{(0)}(\tau', \xi') = \frac{1}{\tilde{\lambda}(\tau' + ia, \xi')}[\Pi_{2+}\tilde{\lambda}(\tau' + ia, \xi')H^{(0)}(\tau', \xi') - i\Pi_{2-}H^{(1)}(\tau', \xi')], \tag{6.2}$$

$$V^{(0)}(\tau', \xi') = \frac{-1}{\tilde{\lambda}(\tau' + ia, \xi')}\Pi_{2-}[\tilde{\lambda}(\tau' + ia, \xi') \times H^{(0)}(\tau', \xi') + iH^{(1)}(\tau', \xi')], \tag{6.3}$$

$$V^{(1)}(\tau', \xi') = \Pi_{2+}[i\tilde{\lambda}(\tau' + ia, \xi')H^{(0)}(\tau', \xi') - H^{(1)}(\tau', \xi')]. \tag{6.4}$$

Now let

$$r(\tau, \xi') = U^{(0)}(\tau + k\xi_2, \xi'), \tag{6.5}$$

$$v^{(i)}(\tau', \xi') = V^{(i)}(\tau + k\xi_2, \xi'), \quad i = 0, 1, \tag{6.6}$$

where $U^{(0)}$ and the $V^{(i)}$ are determined by (6.2)–(6.4). Then the formulas (5.4), (6.5), (6.6) and $u(t, x) = F^{-1}[u_+(\tau, \xi)]$ for $x_1 > 0$ determine a unique solution of problem 0)–3). The uniqueness of the solution follows from the necessity of 2″), 3″) and the uniqueness of the solution of this system. The proof that the indicated formulas actually determine a solution of problem 0)–3) proceeds in the same way as in the proof of Theorem 3.2. The independence of $u(t, x)$ from the choice of extensions of the functions $h^{(i)}(t, x')$ and the validity of estimate (6.1) are proved in exactly the same way as before, the latter by taking into account (5.7).

b) $k < -\sqrt{a_{22}}$ (we recall that in the definitions of the spaces it is necessary to take $m = 0$). From 2″), 3″) we now have

$$U^{(0)}(\tau', \xi') = \Pi_{2+}H^{(0)}(\tau', \xi') - i\Pi_{2-}\frac{H^{(1)}(\tau', \xi')}{\tilde{\lambda}(\tau' + ia, \xi')}, \tag{6.7}$$

$$V^{(0)}(\tau', \xi') = -\Pi_{2-}\left[H^{(0)}(\tau', \xi') + \frac{iH^{(1)}(\tau', \xi')}{\tilde{\lambda}(\tau' + ia, \xi')}\right], \tag{6.8}$$

$$V^{(1)}(\tau', \xi') = \tilde{\lambda}(\tau' + ia, \xi')\Pi_{2+}[iH^{(0)}(\tau', \xi') - (\tilde{\lambda}(\tau' + ia, \xi'))^{-1}H^{(1)}(\tau', \xi')]. \tag{6.9}$$

The formulas (5.4), (6.5) and (6.6), where $U^{(0)}$, $V^{(0)}$ and $V^{(1)}$ are now determined by (6.7) – (6.9), $u(t, x) = F^{-1}[u_+(\tau, \xi)]$ for $x_1 > 0$ and $v^{(i)}(t, x') = F^{-1}[v^{(i)}(\tau, \xi')]$ again determine a unique solution of problem 0)–3). The proof of this assertion as well as the facts that $u(t, x)$ does not depend on the choice of extensions of the $h^{(i)}$ and estimate (6.1) holds proceeds as in the proof of Theorem 3.2.

We note once more (see §6.1) that the proof of the fact that $u(t, x) = 0$ for $t < x_1$ is omitted for the sake of brevity. The proof of Theorem 6.1 is complete.

§7. Theorem on the smoothness of the solution ($|k| > \sqrt{a_{22}}$)

1. *Auxiliary facts.*

a) $k > \sqrt{a_{22}}$ (in the definitions of the spaces, $m = 1$).

1) If $\hat{h}^{(i)}(t, x') \in Y_s^{1-i}$, $i = 0, 1$, then $g_i(t, x') = F^{-1}[\lambda^{1-i}\hat{h}^{(i)}(\tau, \xi')] \in H_{2s}(R^n)$ (by the definition of Y_s^{1-i}). Therefore, if $\mathbf{1} = \{l_0, l_2, \cdots, l_n\} \in E^n$ is a unit vector, we have for any integer $p \in [0, 2s)$, as is well known,

$$\frac{\partial^p g_i}{\partial \mathbf{l}^p} = \left[l_0\frac{\partial}{\partial t} + \sum_{j=2}^{n} l_j\frac{\partial}{\partial x_j}\right]^p g_i \in H_{2s-p-1/2}(R^{n-1})$$

on each hyperplane perpendicular to $\mathbf{1}$.

2) If $\hat{h}_i^{(0)}(t, x') \in Y_s^1$, $i = 1, 2$, are different extensions of the function $h^{(0)}(t, x')$ defined for $x_2 > kt$, then

$$g(t, x') = F^{-1}[\lambda(\hat{h}_1^{(0)}(\tau, \xi') - \hat{h}_2^{(0)}(\tau, \xi'))] \in H_{2s}(R^n)$$

and $g(t, x') = 0$ for $x_2 > kt$. Let us prove this assertion. Let $h(t, x') = \hat{h}_1^{(0)}(t, x') - \hat{h}_2^{(0)}(t, x')$ (which implies $h(\tau, x') = 0$ for $x_2 > kt$). Then $g(\tau, \xi') = \lambda(\tau + ia, \xi')h(\tau, \xi')$. After making the substitution $\tau = \tau' - k\xi_2$ and letting $g(\tau' - k\xi_2, \xi') = G(\tau', \xi')$ and $h(\tau' - k\xi_2, \xi') = H(\tau', \xi')$, we get $G(\tau', \xi') = \tilde{\lambda}(\tau' + ia, \xi')H(\tau', \xi')$. By virtue of the properties of $h(t, x')$ the support of the function

$$H(t, x_2, x'') = F^{-1}[H(\tau', \xi')] = h(t, x_2 + kt, x'')$$

is contained in the halfspace $x_2 \leqslant 0$. The function $\tilde{\lambda}(\tau' + ia, z_2, \xi'')$, as we noted in §5.1, is analytic with respect to $z_2 = \xi_2 + i\eta_2$ for $\eta_2 \leqslant 0$.

Let $\varphi(t, x') \in C_0^\infty(R^n)$, $\operatorname{supp} \varphi(t, x') \subset R_+^n = \{(t, x') | x_2 > 0\}$. We then have

$$(2\pi)^n \langle G(t, x'), \varphi(t, x')\rangle = \langle G(\tau', \xi'), \varphi(\tau', \xi')\rangle$$
$$= \langle H(\tau', \xi'), \bar{\tilde{\lambda}}(\tau' + ia, \xi')\varphi(\tau', \xi')\rangle.$$

The function $\bar{\tilde{\lambda}}(\tau' + ia, \xi')$ is the complex conjugate of $\tilde{\lambda}(\tau' + ia, \xi')$, and can therefore be analytically continued into the halfplane $\eta_2 \geqslant 0$ of the complex z_2 $(= \xi_2 + i\eta_2)$ plane. Applying Corollary 1.3 to $\varphi(\tau', \xi_2, \xi'')$ and $\bar{\tilde{\lambda}}(\tau' + ia, z_2, \xi_2'')$, we now get that

$$\operatorname{supp} F^{-1}[\bar{\tilde{\lambda}}(\tau' + ia, \xi')\varphi(\tau', \xi')] \in \overline{R_+^n}.$$

But then

$$\langle H(\tau', \xi'), \bar{\tilde{\lambda}}(\tau' + ia, \xi')\varphi(\tau', \xi')\rangle = 0,$$

which implies that $G(t, x') = 0$ for $x_2 > 0$, i.e. $g(t, x') = 0$ for $x_2 > kt$.

From what has been proved it follows that

$$\left.\frac{\partial^p g}{\partial \mathbf{m}^p}\right|_{x_2 = kt} = 0, \quad 0 \leqslant p \leqslant 2s - 1, \qquad p \text{ an integer},$$

where

$$\mathbf{m} = \left\{\frac{k}{1+k^2}, -\frac{1}{1+k^2}, 0, \ldots, 0\right\} \in E^n \tag{7.1}$$

is the unit normal to the hyperplane $x_2 = kt$.

b) $k < -\sqrt{a_{22}}$ (in the definitions of the spaces, $m = 0$).

1′) If $\hat{h}^{(i)}(t, x') \in Y_{s'}^0$, $i = 0, 1$, then

$$g_i(t, x') = F^{-1}[\lambda^{-1}\hat{h}^{(i)}(\tau, \xi')] \in H_{2s}(R^n)$$

and

$$\frac{\partial^p g_i}{\partial \mathbf{l}^p} \in H_{2s-p-1/2}(R^{n-1})$$

on each hyperplane perpendicular to $\mathbf{1}$, where $\mathbf{1}$ is the same as in case a), 1).

2′) If $\hat{h}_i^{(1)}(t, x') \in Y_{s'}^0$, $i = 1, 2$, are different extensions of $h^{(1)}(t, x')$ defined for $x_2 < kt$, then

$$g(t, x') = F^{-1}[\lambda^{-1}(\tau + ia, \xi')(\hat{h}_1^{(1)}(\tau, \xi') - \hat{h}_2^{(1)}(\tau, \xi'))] \in H_{2s}(R^n)$$

and $g(t, x') = 0$ for $x_2 < kt$. The proof of the latter assertion is now based on the fact that $\tilde{\lambda}(\tau' + ia, z_2, \xi'')$ is analytic with respect to $z_2 = \xi_2 + i\eta_2$ for $\eta_2 \geqslant 0$ and is analogous to the proof of a), 2). It follows that if $\mathbf{m}$ is the same as in a), 2), then

$$\left.\frac{\partial^p g}{\partial \mathbf{m}^p}\right|_{x_2=kt}=0,\ 0\leqslant p\leqslant 2s-1.$$

2. *Smoothness theorem.*

THEOREM 7.1. *Suppose the functions $h^{(i)}(t, x')$, $i = 0, 1$, in problem 0)–3) respectively have extensions $\hat{h}^{(i)}(t, x') \in Y_s^{1-i}$, $s \geqslant 1$, onto all of R^n. Then the solution $u(t, x)$ of this problem will belong to R_s^+, $s \geqslant 1$, if the following condition is satisfied.*

In the case a) $k > \sqrt{a_{22}}$

$$\lim_{\varepsilon\to+0}\left\{\frac{\partial^p}{\partial \mathbf{m}^p}h^{(1)}(t, x')\right\}\Big|_{x_2=kt-\varepsilon} + \lim_{\varepsilon\to+0}\left\{\frac{\partial^p}{\partial \mathbf{m}^p}F^{-1}[-i\lambda\hat{h}^{(0)}(\tau, \xi')]\right\}\Big|_{x_2=kt+\varepsilon}=0,\ 0\leqslant p\leqslant 2s-1, \tag{7.2}$$

where the limits exist in $H_{2s-p-\frac{1}{2}}(R^{n-1})$, $\hat{h}^{(0)}(t, x')$ is an extension of $h^{(0)}(t, x')$ onto R^n in Y_s^1 and $\mathbf{m}$ *is defined by* (7.1).

In the case b) $k < -\sqrt{a_{22}}$

$$\lim_{\varepsilon\to+0}\left\{\frac{\partial^p}{\partial \mathbf{m}^p}h^{(0)}(t, x')\right\}\Big|_{x_2=kt+\varepsilon} + \lim_{\varepsilon\to+0}\left\{\frac{\partial^p}{\partial \mathbf{m}^p}F^{-1}\left[\frac{i\hat{h}^{(1)}(\tau, \xi')}{\lambda(\tau+ia, \xi')}\right]\right\}\Big|_{x_2=kt-\varepsilon}=0,\ 0\leqslant p\leqslant 2s-1, \tag{7.3}$$

where the limits exist in $H_{2s-p-\frac{1}{2}}(R^{n-1})$ and $\hat{h}^{(1)}(t, x')$ is an extension of $h^{(1)}(t, x')$ onto R^n in Y_s^0.

PROOF. We will examine case a) (case b) is considered analogously). By assumption we will regard the functions $\hat{h}^{(0)}(t, x')$ and $\hat{h}^{(1)}(t, x')$ in the formulas for the solution of problem 0)–3) as elements of Y_s^1 and Y_s^0 respectively. According to Lemma 5.3 the solution $u(t, x)$ will belong to R_s^+ if $r(t, x') \in Y_s^1$, i.e. taking into account (6.5),

$$\int_{R^n}|\lambda|^2(\tau^2+|\xi'|^2+a^2)^{2s}|U^{(0)}(\tau+k\xi_2, \xi')|^2\,d\tau d\xi'<\infty.$$

Using (5.8), we rewrite this condition in the form

$$\int_{R^n}|\lambda|^2((\tau+k\xi_2)^2+\xi_2^2+|\xi''|^2+a^2)^{2s}|U^{(0)}(\tau+k\xi_2, \xi')|^2\,d\tau d\xi'<\infty.$$

After making the substitution $\tau + k\xi_2 = \tau'$, we get

$$\int_{R^n}|\tilde{\lambda}(\tau'+ia, \xi')|^2|\xi_2^2+\tau'^2+|\xi''|^2+a^2|^{2s}|U^{(0)}(\tau', \xi')|^2\,d\tau' d\xi'<\infty, \tag{7.4}$$

where $U^{(0)}(\tau', \xi')$ is determined by (6.2). Taking into account the fact that $H^{(1)}(\tau', \xi') \in L_2(R^n)$ and hence

$$\Pi_{2-}H^{(1)}(\tau', \xi') = H^{(1)}(\tau', \xi') - \Pi_{2+}H^{(1)}(\tau', \xi'),$$

we rewrite (6.2) in the form

$$U^{(0)}(\tau', \xi') = -i\tilde{\lambda}^{-1}H^{(1)}(\tau', \xi') + i\tilde{\lambda}^{-1}\Pi_{2+}[-i\tilde{\lambda}H^{(0)}(\tau', \xi') + H^{(1)}(\tau', \xi')].$$

Here we have written $\tilde{\lambda} = \tilde{\lambda}(\tau' + ia, \xi')$ for brevity. The function standing under the sign of the operator Π_{2+} is denoted by $W(\tau', \xi')$. Condition (7.4) can now be rewritten in the form

$$\int_{R^n} (\xi_2^2 + \tau'^2 + |\xi''|^2 + a^2)^{2s} |\Pi_{2+}W(\tau', \xi') - H^{(1)}(\tau', \xi')|^2 d\tau' d\xi' < \infty,$$

i.e. $\Pi_{2+}W(\tau', \xi') - H^{(1)}(\tau', \xi') \in H_{2s}(R^n)$ (we recall that a space of Fourier transforms is denoted in the same way as the original space). By assumption, $\hat{h}^{(1)}(t, x') \in Y_s^0 = H_{2s}(R^n)$. Taking (5.8) into account, we therefore see that $H^{(1)}(\tau', \xi') \in H_{2s}(R^n)$. Hence $u(t, x)$ will belong to R_s^+ if

$$\Pi_{2+}W(\tau', \xi') = \Pi_{2+}[-i\tilde{\lambda}H^{(0)}(\tau', \xi') + H^{(1)}(\tau', \xi')] \in H_{2s}(R^n).$$

In the (t, x) representation this means that the function $W_+(t, x')$ equal to $W(t, x')$ for $x_2 > 0$ and vanishing for $x_2 \leqslant 0$ belongs to $H_{2s}(R^n)$. As we know, this is possible if and only if

$$\left\{\frac{\partial^p}{\partial x_2^p} W(t, x')\right\}\Big|_{x_2=0} = 0$$

as an element of $H_{2s-p-1/2}(R^{n-1})$ for integral $p \in [0, 2s)$. By virtue of the substitution $\tau' = \tau + k\xi_2$ we have

$$W(t, x_2, x'') = w(t, x_2 + kt, x''),$$

where

$$w(\tau, \xi') = -i\lambda\hat{h}^{(0)}(\tau, \xi') + \hat{h}^{(1)}(\tau, \xi').$$

Therefore

$$\frac{\partial^p W}{\partial x_2^p}(t, x_2, x'') = \frac{\partial^p}{\partial x_2} w(t, x_2 + kt, x'') = \frac{\partial^p}{\partial y_2^p} w(t, y_2, x'')\Big|_{y_2=x_2+kt},$$

and our condition takes the form

$$\frac{\partial^p W}{\partial x_2^p}(t, x')\Big|_{x_2=0} = \frac{\partial^p w}{\partial y_2^p}(t, y_2, x'')\Big|_{y_2=kt} = 0, \quad 0 \leqslant p \leqslant 2s - 1, \ p \text{ an integer}.$$

Inasmuch as the vector $\mathbf{1}_2 = \{0, 1, 0, \cdots, 0\}$ is not tangent to the hyperplane $x_2 = kt$, the latter condition means that the derivatives up to order $2s-1$ of $w(t, x')$ in any direction vanish on the hyperplane $x_2 = kt$. An equivalent form of this condition is

$$\left.\frac{\partial^p w(t, x')}{\partial \mathbf{m}^p}\right|_{x_2=kt} = 0, \qquad 0 \leqslant p \leqslant 2s-1, \tag{7.5}$$

where $\mathbf{m}$ is defined by (7.1); it implies the fulfilment of (7.4) and hence that $u(t, x) \in \mathfrak{A}_s \subset R_s^+$. Recalling the definition of $w(t, x')$, we find from (7.5) that

$$\left.\left\{\frac{\partial^p}{\partial \mathbf{m}^p}[F^{-1}[-i\lambda\hat{h}^{(0)}(\tau, \xi')]](t, x')\right\}\right|_{x_2=kt} + \left.\left\{\frac{\partial^p}{\partial \mathbf{m}^p}[\hat{h}^{(1)}(t, x')]\right\}\right|_{x_2=kt} = 0, \qquad 0 \leqslant p \leqslant 2s-1.$$

It remains to note that (i) the trace

$$\left.\left\{\frac{\partial^p}{\partial \mathbf{m}^p} g(t, x')\right\}\right|_{x_2=kt+\varepsilon} = g_\varepsilon^{(p)}$$

is a continuous function of ϵ with values in $H_{2s-p-1/2}(R^{n-1})$ for any $g(t, x') \in H_{2s}(R^n)$, and (ii)

$$\left.\left\{\frac{\partial^p}{\partial \mathbf{m}^p} \hat{h}^{(1)}(t, x')\right\}\right|_{x_2=kt-\varepsilon} = \left.\left\{\frac{\partial^p}{\partial \mathbf{m}^p} h^{(1)}(t, x')\right\}\right|_{x_2=kt-\varepsilon}$$

for $\epsilon > 0$. From this we obtain the required condition (7.2). The proof of Theorem 7.1 is complete.

REMARK 7.1. Conditions (7.2) and (7.3) depend only on the functions $h^{(0)}(t, x')$ and $h^{(1)}(t, x')$ themselves and not on the choice of their extensions. For the preliminary facts a), 2) and b), 2′) imply that (7.2) and (7.3) are satisfied for all extensions if they are satisfied for any.

REMARK 7.2. The compatibility conditions (7.2) and (7.3) have a local character for only one of the boundary functions and are nonlocal for the other. This can be seen, for example, by simply calculating the inverse Fourier transforms of the functions $-i\lambda\hat{h}^{(0)}(\tau, \xi')$ and $-i\lambda^{-1}\hat{h}^{(1)}(\tau, \xi')$. It turns out that the values of these functions at a point (t, x') depend on the values of $h^{(0)}(s, y')$ and $h^{(1)}(s, y')$ throughout (on the surface of) the cone of the past for $n = 2l$ ($n = 2l + 1$). And from this one sees once again that conditions (7.2) and (7.3) do not depend on the choice of extensions.

We note that if $|\lambda| \to \infty$, we obtain in the limit an ordinary mixed problem. In this case the nonlocal terms in (7.2) and (7.3) drop out and we obtain local conditions for the compatibility of the boundary functions with zero initial data.

REMARK 7.3. If either condition (7.2) or condition (7.3) is not satisfied for $p = 0$, the solution of problem 0)–3) will no longer belong to R_1^+, i.e. it will not be as smooth for $x_1 > 0$ as the functions $h^{(i)}(t, x')$, $i = 0, 1$, may be separately [11]. Qualitatively this means that here, in contrast to the case $|k| < \sqrt{a_{22}}$, the discontinuity of the boundary operator goes over into a discontinuity of the solution within the domain $x_1 > 0$.

BIBLIOGRAPHY

1. G. Fichera, *Sul problema della derivata obliqua e sul problema misto per l'equazione di Laplace,* Boll. Un. Mat. Ital. (3) 7 (1952), 367–377. MR 14, 750.

2. J. Peetre, *Mixed problems for higher order elliptic equations in two variables,* I, II, Ann. Scuola Norm. Sup. Pisa (3) 15 (1961), 337–353; 17 (1963), 1–12. MR 25 #4240; 28 #328.

3. M. I. Višik and G. I. Èskin, *General boundary value problems with discontinuous conditions at the boundary,* Dokl. Akad. Nauk SSSR 158 (1964), 25–28 = Soviet Math. Dokl. 5 (1964), 1154–1157. MR 30 #369.

4. ———, *Elliptic equations in convolution in a bounded domain and their applications,* Uspehi Mat. Nauk 22 (1967), no. 1 (133), 15–76 = Russian Math. Surveys 22 (1967), no. 1, 13–75. MR 35 #5757.

5. R. Courant and P. D. Lax, *The propagation of discontinuities in wave motion,* Proc. Nat. Acad. Sci. U.S.A. 42 (1956), 872–876. MR 18, 399.

6. V. P. Maslov, *Theory of perturbations and asymptotic methods,* Izdat. Moskov. Gos. Univ., Moscow, 1965; French transl., Dunod, Paris, 1972.

7. N. Dunford and J. T. Schwartz, *Linear operators. 1: General theory,* Pure and Appl. Math., vol. 7, Interscience, New York, 1958. MR 22 #8302.

8. H. J. Bremermann, *Distributions, complex variables and Fourier transforms,* Addison-Wesley, Reading, Mass., 1965; Russian transl., "Mir", Moscow, 1968. MR 34 #8174.

9. R.E.A.C. Paley and N. Wiener, *Fourier transforms in the complex domain,* Amer. Math. Soc. Colloq. Publ., vol. 19, Amer. Math. Soc., Providence, R.I., 1934.

10. V. S. Vladimirov, *Equations of mathematical physics,* "Nauka", Moscow, 1967; English transl., Pure and Appl. Math., vol. 3, Dekker, New York, 1971. MR 39 #599; 42 #3394.

11. V. I. Čehlov, *The mixed problem for a hyperbolic equation with discontinuous boundary operator,* Candidate's Dissertation, Steklov Institute of Math., Moscow, 1971. (Russian)